VIDEO NOTEBOOK WITH THE MATH COACH

BASIC COLLEGE MATHEMATICS

EIGHTH EDITION

John Tobey

North Shore Community College
Danvers, Massachusetts

Jeffrey Slater

North Shore Community College
Danvers, Massachusetts

Jamie Blair

Orange Coast College
Costa Mesa, California

Jennifer Crawford

Normandale Community College
Bloomington, Minnesota

Pearson

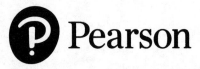

ISBN-13: 978-0-13-418762-4
ISBN-10: 0-13-418762-8

Contents

Contents

Preface

This *Video Notebook with the Math Coach* provides a convenient, ready-to-use format with ample work space to help your students to stay organized and take notes. The workbook provides an effective way to check to see if students understand the skills and concepts from each section of their textbook. The *Video Notebook with the Math Coach* is made up of two sections:

1. Worksheets with Math Coach and
2. Guided Learning Video Worksheets

The worksheets with the Math Coach are the Tobey series' popular extra-practice student supplement that correlates with the Math Coach feature in the textbook itself. The worksheets with the Math Coach include

- Key vocabulary terms and vocabulary practice problems
- Step-by-step worked examples
- Extra Student Practice problems (modeled after each guided example)
- Concept Check questions from each section of the book
- Additional exercises with space for students to show their work
- Math Coach problems for students to work along with the Math Coach videos

The second section makes up the latest addition to the Tobey series: the Guided Learning Video Worksheets. This video note-taking guide encourages students to follow along with the objective-level Guided Learning Videos in their MyMathLab course, prompting them to fill in the blanks as they go. Once completed, instructors have a way to verify that students have watched the video and understood the key concepts, and students have a study tool for future reference. The Guided Learning Video Worksheets include

- Understanding the Big Picture vocabulary terms and concept explanations
- Worked-out examples that follow along step-by-step with the Guided Learning Video
- Paused Student Practice with ample space for student work
- Two Active Video Lessons that prompt students to complete examples by themselves:

  o Active Video Lesson 1 supplies a worked-out solution
  o Active Video Lesson 2 supplies a static answer

The two sections of the *Video Notebook with the Math Coach* are flexible and can be used independently of each other or integrated with the Guided Learning Video Worksheets being completed at point of use. Both the Math Coach exercises and Guided Learning Videos are assignable in MyMathLab and can be integrated into your homework plan.

Chapter 1 Whole Numbers
1.1 Understanding Whole Numbers

Vocabulary
whole numbers   •   decimal system   •   digits   •   place-value system
period   •   expanded notation   •   standard notation

1. The whole numbers 0, 1, 2, 3, 4, 5, 6, 7, 8, 9 are called _____.

2. When we write very large numbers, we place a comma after every group of three digits, called a _____, moving from right to left.

3. A number system in which the position, or placement, of the digits in a number tells the value of the digits is called a _____.

4. A number system based on tens and ones is called the _____.

Example	**Student Practice**
1. Write 2378 in expanded notation. Sometimes it helps to say the number to yourself. two thousand, three hundred seventy-eight $2378 = 2000 + 300 + 70 + 8$	**2.** Write $2,032,410$ in expanded notation.
3. Write $500 + 30 + 8$ in standard notation. $\underline{5}00 + \underline{3}0 + \underline{8} = 538$	**4.** Write $100,000 + 4000 + 40 + 7$ in standard notation.

Example	Student Practice
5. Last year, the population of Central City was $1,509,637$. For the number $1,509,637$, answer parts **(a)** and **(b)**.	**6.** For the number $10,056,797$, answer parts **(a)** and **(b)**.

5. Last year, the population of Central City was $1,509,637$. For the number $1,509,637$, answer parts **(a)** and **(b)**.

(a) How many ten thousands are there?

Look at the digit in the ten thousands place. There are 0 ten thousands.

(b) What is the value of the digit 5?

The digit 5 is in the hundred thousands place. The value of the digit is $500,000$.

6. For the number $10,056,797$, answer parts **(a)** and **(b)**.

(a) How many ten millions are there?

(b) In what places is the digit 7?

---

7. Write a word name for $364,128,957$.

First write the number in a place-value chart.

Place-value Chart

Billions			Millions			Thousands			Ones		
			3	6	4	1	2	8	9	5	7
Hundreds	Tens	Ones	Hundreds	Tens	Ones	Hundreds	Tens	Ones	Hundreds	Tens	Ones

To write a word name, start from the left. Name the number in each period, followed by the name of the period, and a comma. The last period name, "ones," is not used.

three hundred sixty-four million, one hundred twenty-eight thousand, nine hundred fifty-seven.

8. Write a word name for $30,597,812$.

Place-value Chart

Billions			Millions			Thousands			Ones		
Hundreds	Tens	Ones	Hundreds	Tens	Ones	Hundreds	Tens	Ones	Hundreds	Tens	Ones

Example	Student Practice
9. Write a word name for each number. **(a)** 1695 To help us, we will put in the optional comma: $1,695$. one thousand, six hundred ninety-five **(b)** $200,470$ The word name is two hundred thousand, four hundred seventy. **(c)** $7,003,038$ The word name is seven million, three thousand, thirty-eight.	**10.** Write a word name for $10,897,908$.
11. Write each number in standard notation. **(a)** twenty-six thousand, eight hundred sixty-four. Write the number, separating each period with a comma. The number is $26,864$. **(b)** two billion, three hundred eighty-six million, five hundred forty-seven thousand, one hundred ninety Write the number, separating each period with a comma. The number is $2,386,547,190$.	**12.** Write the number in standard notation. eight trillion, nine hundred sixty-seven million, two hundred forty-one

	Example		Student Practice

13. Refer to the chart. What was the estimated population of Maine in 1780? Write the number in standard notation.

Estimated Population of the American Colonies from 1650 to 1780 (in thousands)

	Maine	New Hampshire	Vermont	Plymouth & Massachusetts	Rhode Island	Connecticut
1650	1	1	★	16	1	4
1670	★	2	★	35	2	13
1690	★	4	★	57	4	22
1700	★	5	★	56	6	26
1720	★	9	★	91	12	59
1740	★	23	★	152	25	90
1750	★	28	★	188	33	111
1770	31	62	10	235	58	184
1780	49	88	48	269	53	207

Look at the chart in the column marked "Maine" and move down to the row marked "1780."

The estimated population of Maine in 1780 was 49,000.

14. Refer to the chart. What was the estimated population of Vermont in 1770? Write the number in standard notation.

Estimated Population of the American Colonies from 1650 to 1780 (in thousands)

	Maine	New Hampshire	Vermont	Plymouth & Massachusetts	Rhode Island	Connecticut
1650	1	1	★	16	1	4
1670	★	2	★	35	2	13
1690	★	4	★	57	4	22
1700	★	5	★	56	6	26
1720	★	9	★	91	12	59
1740	★	23	★	152	25	90
1750	★	28	★	188	33	111
1770	31	62	10	235	58	184
1780	49	88	48	269	53	207

Extra Practice

1. Write 4965 in expanded notation.

2. Write $800 + 30 + 7$ in standard notation.

3. Write the word name for 4627.

4. What is the value of the digit 5 in 5829?

Concept Check

Explain why the zeros are needed when writing the following number in standard notation: three hundred sixty-eight million, five hundred twenty-two.

4

Name: _____   Date: _____

Instructor: _____   Section: _____

Chapter 1 Whole Numbers
1.2 Adding Whole Numbers

Vocabulary
addends • sum • identity property of zero • commutative property of addition
associative property of addition

1. The _____ says that when we add zero to another number, that number will be the sum.

2. The _____ states that the order in which we add numbers does not change the sum.

3. In an addition problem, the numbers being added are called _____.

4. The _____ states that the way we group numbers to be added does not change the sum.

Example	Student Practice
1. Add. $8+5$ $\begin{array}{r} 8 \\ +\ 5 \\ \hline 13 \end{array}$	**2.** Add. $9+7$
3. Add $3+4+8+2+5$. Rewrite the addition problem in a column format. Mentally add $3+4$. Then, add the result to 8. Then, repeat. $\begin{array}{r} 3 \\ 4 \\ 8 \\ 2 \\ +\ 5 \\ \hline 22 \end{array}$	**4.** Add $5+4+7+6+3$.

Example	Student Practice
5. Add.	**6.** Add.
$\begin{array}{r} 3 \\ 4 \\ 8 \\ 2 \\ +\ 6 \end{array}$	$\begin{array}{r} 2 \\ 7 \\ 1 \\ 9 \\ +\ 3 \end{array}$
Mentally group the numbers into tens. The sum of digits 8 and 2 is 10. The sum of digits 4 and 6 is 10. Thus, the sum is $10+10+3=23$.	
7. Add $4304+5163$.	**8.** Add $2417+6032$.
To add, add the corresponding digits in each place-value. Begin in the ones place, $4+3=7$. Repeat for the remaining place-values. $\begin{array}{r} 4304 \\ +\ 5163 \\ \hline 9467 \end{array}$	
9. Add $257+688+94$.	**10.** Add $58+6482+4251$.
$\begin{array}{r} {\scriptstyle 2\ 1} \\ 257 \\ 688 \\ +\ \ \ 94 \\ \hline 1039 \end{array}$ In the ones column add $7+8+4=19$. Because 19 is 1 ten and 9 ones, we place 9 in the ones column and carry 1 to the top of the tens column. Similarly, add the tens and the hundreds column, including any carried digits.	

Example	Student Practice

11. The bookkeeper for Smithville Trucking was examining the following data for the company checking account.

Monday: $23,416 was deposited and $17,389 was debited.
Tuesday: $44,823 was deposited and $34,089 was debited.
Wednesday: $16,213 was deposited and $20,057 was debited.

What was the total of all deposits during this period?

Begin by understanding the problem.

The word "total" implies that we will use addition. Since we don't need to know about the debits to answer this question, we use only the deposit amounts.

Now, calculate and state the answer.
Monday: $23,416 was deposited.
Tuesday: $44,823 was deposited.
Wednesday: $16,213 was deposited.

$$
\begin{array}{r}
{\scriptstyle 1\,1\ \ \ 1} \\
23,416 \\
44,823 \\
+\ 16,213 \\
\hline
84,452
\end{array}
$$

A total of $84,452 was deposited on those three days.

Finally, check your answer. You may add the numbers in reverse order to check.

12. The bookkeeper for a trucking company was examining the following data for the company checking account.

Wednesday: $15,624 was deposited and $14,938 was debited.
Thursday: $46,732 was deposited and $25,890 was debited.
Friday: $15,325 was deposited and $21,755 was debited.

What was the total of all deposits during this period?

7

Extra Practice

1. Add.

$$
\begin{array}{r}
11 \\
42 \\
92 \\
+\ 38 \\
\hline
\end{array}
$$

2. Add.

$$
\begin{array}{r}
6,327 \\
78,411 \\
+\ 29,073 \\
\hline
\end{array}
$$

3. Add.

$378+94+26+51+702$

4. Lake Huron, which is bordered by the United States and Canada, has an area of 23,000 square miles. Lake Tanganyika, which borders Tanzania and Zaire, measures 12,700 square miles. Lake Baikal in Russia has an area of 12,162 square miles. What is the total area of the three lakes?

Concept Check

Explain how you would use carrying when performing the following calculation: $4567+3189+895$.

Chapter 1 Whole Numbers
1.3 Subtracting Whole Numbers

Vocabulary

difference   •   subtrahend   •   minuend   •   borrowing   •   equation

1. The number being subtracted is called the _____.

2. The number being subtracted from is called the _____.

3. An _____ is a number sentence with an equals sign.

4. When we subtract one number from another, the answer is called the _____.

Example	Student Practice
1. Subtract.	**2.** Subtract.
(a) $8-2$	**(a)** $7-5$
$\begin{array}{r} 8 \\ -\ 2 \\ \hline 6 \end{array}$	
	(b) $11-6$
(b) $15-8$	
$\begin{array}{r} 15 \\ -\ 8 \\ \hline 7 \end{array}$	
3. Subtract $9867-3725$.	**4.** Subtract $7635-4512$.
7 ones $-$ 5 ones $=$ 2 ones 6 tens $-$ 2 tens $=$ 4 tens 8 hundreds $-$ 7 hundreds $=$ 1 hundred 9 thousands $-$ 3 thousands $=$ 6 thousands $\begin{array}{r} 9\ 8\ 6\ 7 \\ -\ 3\ 7\ 2\ 5 \\ \hline 6\ 1\ 4\ 2 \end{array}$	

Example	Student Practice
5. Subtract $864 - 548$.	**6.** Subtract $7053 - 3184$.

$$\begin{array}{ccc} & \overset{5}{\cancel{6}} & \overset{14}{\cancel{4}} \\ 8 & & \\ - \quad 5 & 4 & 8 \\ \hline 3 & 1 & 6 \end{array}$$

To subtract 8 ones from 4 ones, borrow 1 ten from the 6 tens and write 5 in the tens column to show what is left. Since 1 ten is borrowed, which is 10 ones, now we have 14 ones, and write 14 in the ones column.
Now subtract 8 ones from 14 ones to obtain 6 ones. Then continue to subtract from right to left.

7. The librarian knows that he has eight world atlases and that five of them are in full color. How many are not in full color?	**8.** The librarian knows that he has nine world atlases and that three of them are in full color. How many are not in full color?

Represent the unknown number as x and write an equation.

$8 = 5 + x$

Solve this equation by reasoning and by knowledge of the relationship between addition and subtraction.
$8 = 5 + x$ is equivalent to $8 - 5 = x$
We know that $8 - 5 = 3$. Then $x = 3$.
Check the answer by substituting 3 for x in the original equation.

$8 = 5 + x$

$8 = 5 + 3$ True

Thus, $x = 3$ checks, so our answer is correct. There are three atlases not in full color.

10
Copyright © 2017 Pearson Education, Inc.

Example	Student Practice

9. The number of real estate transfers in several towns during the years 2009 to 2011 is given in the following bar graph.

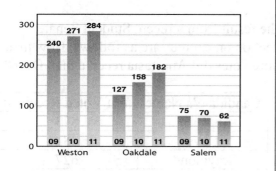

(a) What was the increase in homes sold in Weston from 2010 to 2011?

From the labels on the bar graph we see that 284 homes were sold in 2011 in Weston and 271 homes were sold in 2010. Thus the increase is $284 - 271 = 13$. There was an increase of 13 homes sold in Weston from 2010 to 2011.

(b) Between what two years did Oakdale have the greatest increase in sales?

Here make two calculations to decide where the greatest increase occurs.

    158   2010 sales
$-$ 127   2009 sales
     31   Sales increase
          from 2009 to 2010

    182   2011 sales
$-$ 158   2010 sales
     24   Sales increase
          from 2010 to 2011

The greatest increase in sales in Oakdale occurred from 2009 to 2010.

10. The number of real estate transfers in several towns during the years 2009 to 2011 is given in the following bar graph.

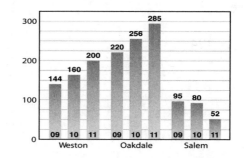

(a) What was the increase in homes sold in Weston from 2010 to 2011?

(b) Between what two years did Oakdale have the greatest increase in sales?

11

Extra Practice

1. Subtract.

$$20,342$$
$$-\ 19,297$$

2. Subtract.

$$167,002$$
$$-\ \ 89,370$$

3. Solve.

$$89 + x = 124$$

4. The result from a recent Student Senate election on campus are as follows. How many more votes did Anastasia receive than Dave?

Candidate's Name	# of Voted Received
Anastasia	873
Dave	482
Raymond	627

Concept Check

Explain how you would use borrowing when performing the calculation $12,345 - 11,976$.

Chapter 1 Whole Numbers
1.4 Multiplying Whole Numbers

Vocabulary
factors • product • multiplication property of zero • identity element for multiplication
commutative property of multiplication • associative property of multiplication
multiplication • partial products • power of 10 • distributive property

1. The _____ states that when we multiply three numbers, the multiplication can be grouped in any way.

2. The numbers that we multiply are called _____.

3. The _____ states the two numbers can be multiplied in either order with the same result.

4. The _____ states that the product of any number and zero yields zero as a result.

Example	**Student Practice**
1. Multiply. 5×7 5 $\times\ 7$ 35	**2.** Multiply. 8×4
3. Multiply. 4312×2 First multiply the ones column, then the tens column, and so on, moving right to left. 4312 $\times\ \ \ \ 2$ 8624 $2 \times 2 = 4$ in the ones column $2 \times 1 = 2$ in the tens column and so on.	**4.** Multiply. 3213×3

Example	Student Practice
5. Multiply 359×9.	**6.** Multiply 257×8.

5. Multiply 359×9.

$$
\begin{array}{r}
{}^{5\,8} \\
359 \\
\times \quad 9 \\
\hline
3231
\end{array}
$$

Since $9 \times 9 = 81$, leave the 1 in the ones column and carry the 8 to the top of the tens column.

Now, $9 \times 5 = 45$, so add 45 tens + 8 tens = 53 tens or 5 hundreds + 3 tens. Leave the 3 in the tens column and carry the 5 to the top of the hundreds column.

Now, $9 \times 3 = 27$, so add 27 hundreds + 5 hundreds to obtain 32 hundreds or 3 thousands + 2 hundreds.

---

7. Multiply.

(a) 12×3000

$$
\begin{aligned}
12 \times 3000 &= 12 \times 3 \times 1000 \\
&= 36 \times 1000 \\
&= 36,000
\end{aligned}
$$

(b) 25×600

$$
\begin{aligned}
25 \times 600 &= 25 \times 6 \times 100 \\
&= 150 \times 100 \\
&= 15,000
\end{aligned}
$$

(c) 430×260

$$
\begin{aligned}
430 \times 260 &= 43 \times 26 \times 10 \times 10 \\
&= 1118 \times 100 \\
&= 111,800
\end{aligned}
$$

8. Multiply.

(a) 11×5000

(b) 45×300

(c) 350×520

Example	Student Practice

9. Multiply 684×763.

$$
\begin{array}{r}
6\,8\,4 \\
\times 7\,6\,3 \\
\hline
2\,0\,5\,2 \\
4\,1\,0\,4 \\
4\,7\,8\,8 \\
\hline
5\,2\,1\,8\,9\,2
\end{array}
$$

Multiply $684 \times 3 = 2052$.

Multiply $684 \times 60 = 41,040$. Note that we omit the final zero.

Multiply $684 \times 700 = 478,800$. Note that we omit the final two zeros.

Finally, add the partial products.

Thus, $684 \times 763 = 521,892$.

---

10. Multiply 532×689.

---

11. Multiply $7 \times 20 \times 5 \times 6$.

Apply the Associative property.

$7 \times 20 \times 5 \times 6 = 7 \times (20 \times 5) \times 6$

Now, apply the Commutative property.

$7 \times (20 \times 5) \times 6 = 7 \times 6 \times (20 \times 5)$

Finally, simplify.

$$
\begin{aligned}
7 \times 6 \times (20 \times 5) &= 42 \times 100 \\
&= 4200
\end{aligned}
$$

Thus, $7 \times 20 \times 5 \times 6 = 4200$.

---

12. Multiply $9 \times 25 \times 3 \times 4$.

Example	Student Practice
13. The average annual salary of an employee at Software Associates is $42,132. There are 38 employees. What is the annual payroll?	**14.** The average annual salary of an employee at an automation firm is $54,312. There are 25 employees. What is the annual payroll?

$42,132
$\times$ 38
337 056
1 263 96
———————
1,601,016

The total annual payroll is $1,601,016.

Extra Practice

1. Multiply.

  20,609
$\times$ 1000

2. Multiply.

  1894
$\times$ 63

3. Multiply $13 \times 5 \times 41$.

4. Cameron worked 14 weeks during the summer as a house painter. He earned $635 per week. What is the total amount Cameron earned during the summer?

Concept Check
Explain what you do with the zeros when you multiply 3457×2008.

Chapter 1 Whole Numbers
1.5 Dividing Whole Numbers

Vocabulary
related sentences   •   divisor   •   dividend   •   quotient
remainder   •   division   •   undefined

1. Division by zero is _____.

2. The _____ is the number being divided in a division problem.

3. When two numbers do not divide exactly, a number called the _____ is left over.

4. The _____ is the answer to a division problem.

Example	**Student Practice**
1. Divide. $12 \div 4$ $$4\overline{)12}^{\,3}$$	**2.** Divide. $72 \div 6$
3. Divide, if possible. If it is not possible, state why. **(a)** $8 \div 8$ $$\frac{8}{8} = 1$$ Any number divided by itself is 1. **(b)** $9 \div 1$ $$\frac{9}{1} = 9$$ Any number divided by 1 remains unchanged.	**4.** Divide, if possible. If it is not possible, state why. **(a)** $0 \div 5$ **(b)** $14 \div 0$

Example	Student Practice
5. Divide. $3672 \div 7$	**6.** Divide. $2684 \div 5$

5. Divide. $3672 \div 7$

The divisor 7 can be divided into 36 5 times. Compute 5×7 and place the result under 36 and subtract.

$$
\begin{array}{r}
5 \\
7\overline{)3672} \\
\underline{35} \\
1
\end{array}
$$

Bring down the 7 and repeat the process with the result, 17.

$$
\begin{array}{r}
52 \\
7\overline{)3672} \\
\underline{35} \\
17 \\
\underline{14} \\
3
\end{array}
$$

Continue to repeat the process until 7 cannot be divided into the result.

$$
\begin{array}{r}
524 \\
7\overline{)3672} \\
\underline{35} \\
17 \\
\underline{14} \\
32 \\
\underline{28} \\
4
\end{array}
$$

The answer is $524 \text{ R } 4$.

6. Divide. $2684 \div 5$

Example	**Student Practice**
7. Divide. $293 \div 41$	**8.** Divide. $6798 \div 142$

The first digit of the divisor, 4, can be divided into the first two digits of the dividend, 29, 7 times. Try the answer 7 as the first number of the quotient and compute 7×41.

$$\begin{array}{r} 7 \\ 41\overline{)283} \\ 287 \end{array}$$

Notice that 287 is too large since it is larger than 283.

Since 7 is slightly too large, try 6 as the first number in the quotient and compute 6×41.

$$\begin{array}{r} 6 \\ 41\overline{)283} \\ 246 \end{array}$$

246 is not too large since it is less than 283.

Subtract 246 from 283.

$$\begin{array}{r} 6 \\ 41\overline{)283} \\ \underline{246} \\ 37 \end{array}$$

Since 41 does not divide into 36, the answer is $6 \text{ R } 37$.

Example	Student Practice
9. A car traveled from California to Texas, a distance of 1144 miles, in 22 hours. What was the average speed in miles per hour?	**10.** An airplane traveled 4745 miles in 13 hours. What was the average speed in miles per hour?

9. A car traveled from California to Texas, a distance of 1144 miles, in 22 hours. What was the average speed in miles per hour?

When doing a distance problem, it is helpful to remember that
distance ÷ time = rate . We need to divide 1144 miles by 22 hours to obtain the rate, or speed, in miles per hour.

$$
\begin{array}{r}
52 \\
22\overline{)1144} \\
\underline{110} \\
44 \\
\underline{44} \\
0
\end{array}
$$

The car traveled an average speed of 52 miles per hour.

10. An airplane traveled 4745 miles in 13 hours. What was the average speed in miles per hour?

Extra Practice

1. Divide. $2\overline{)0}$

2. Divide. $3\overline{)413}$

3. Divide. $21\overline{)2063}$

4. Divide. $41\overline{)8671}$

Concept Check

When performing the division problem $2956 \div 43$, you need to decide how many times 43 goes into 295. Explain how you would decide this.

Chapter 1 Whole Numbers
1.6 Exponents and the Order of Operations

Vocabulary
exponent   •   base   •   order of operations   •   evaluate

1. If numbers with exponents are added to other numbers, it is first necessary to _____, or find the value of, the number that is raised to a power.

2. In mathematics the _____ is a list of priorities for working with the numbers in computational problems.

3. An _____ is a "shorthand" number that saves writing multiplication of the same number.

4. In the expression 5^2, the number 5 is called the _____.

Example	Student Practice
1. Write the product in exponent form. $15 \times 15 \times 15$ $15 \times 15 \times 15 = 15^3$	**2.** Write the product in exponent form. $19 \times 19 \times 19 \times 19 \times 19$
3. Find the value of each expression. **(a)** 3^3 To find the value of 3^3, multiply the base 3 by itself 3 times. $3^3 = 3 \times 3 \times 3 = 27$ **(b)** 2^5 To find the value of 2^5 multiply the base 2 by itself 5 times. $2^5 = 2 \times 2 \times 2 \times 2 \times 2 = 32$	**4.** Find the value of each expression. **(a)** 5^4 **(b)** 1^{10}

Example	Student Practice
5. Find the value of each expression.	**6.** Find the value of each expression.

5. Find the value of each expression.

(a) $3^4 + 2^3$

$$3^4 + 2^3 = (3)(3)(3)(3) + (2)(2)(2)$$
$$= 81 + 8$$
$$= 89$$

(b) $5^3 + 7^0$

$$5^3 + 7^0 = (5)(5)(5) + 1$$
$$= 125 + 1$$
$$= 126$$

(c) $6^3 + 6$

$$6^3 + 6 = (6)(6)(6) + 6$$
$$= 216 + 6$$
$$= 222$$

6. Find the value of each expression.

(a) $5^4 + 6^2$

(b) $8^5 + 4^0$

(c) $4^5 + 4$

---

7. Evaluate $3^2 + 5 - 4 \times 2$.

First, evaluate the expression with exponents.

$$3^2 + 5 - 4 \times 2 = 9 + 5 - 4 \times 2$$

Now, multiply from left to right.

$$9 + 5 - 4 \times 2 = 9 + 5 - 8$$

Finally, add and subtract from left to right.

$$9 + 5 - 8 = 14 - 8$$
$$= 6$$

8. Evaluate $7 + 16 \div 4 - 8 + 5 \times 2$.

Example	Student Practice
9. Evaluate $5 + 12 \div 2 - 4 + 3 \times 6$.	**10.** Evaluate $6^4 + 4^2 - 5 \times 8$.

9. Evaluate $5 + 12 \div 2 - 4 + 3 \times 6$.

There are no numbers to raise to a power, we first do any multiplication or division in order from left to right.

Multiply or divide from left to right.

$$5 + 12 \div 2 - 4 + 3 \times 6 = 5 + 6 - 4 + 3 \times 6$$
$$= 5 + 6 - 4 + 18$$

Finally, add or subtract from left to right.

$$5 + 6 - 4 + 18 = 11 - 4 + 18$$
$$= 7 + 18$$
$$= 25$$

---

11. Evaluate $2 \times (7 + 5) \div 4 + 3 - 6$.

First, combine numbers inside the parentheses by adding the 7 to the 5.

$$2 \times (7 + 5) \div 4 + 3 - 6 = 2 \times 12 \div 4 + 3 - 6$$

Next, because multiplication and division have equal priority, work from left to right doing whichever of these operations comes first.

$$2 \times 12 \div 4 + 3 - 6 = 24 \div 4 + 3 - 6$$
$$= 6 + 3 - 6$$

Finally add or subtract from left to right.

$$6 + 3 - 6 = 9 - 6$$
$$= 3$$

12. Evaluate $4 \times (9 + 3) \div 8 + 7 - 2$.

Example	Student Practice
13. Evaluate $4^3 + 18 \div 3 - 2^4 - 3 \times (8-6)$.	**14.** Evaluate $5^3 + 24 \div 8 - 2^2 - 3 \times (10-4)$.

First, evaluate the expression in the parentheses.

$$4^3 + 18 \div 3 - 2^4 - 3 \times (8-6)$$
$$= 4^3 + 18 \div 3 - 2^4 - 3 \times 2$$

Simplify the expressions with exponents.

$$4^3 + 18 \div 3 - 2^4 - 3 \times 2$$
$$= 64 + 18 \div 3 - 16 - 3 \times 2$$

Multiply and divide from left to right.

$$64 + 18 \div 3 - 16 - 3 \times 2$$
$$= 64 + 6 - 16 - 6$$

Add and subtract from left to right.

$$64 + 6 - 16 - 6 = 48$$

Extra Practice

1. Write the expression in exponent form.

$$10 \times 10 \times 10 \times 10$$

2. Find the value of the given expression.

$$2^3 + 4^3 + 8^2$$

3. Evaluate.

$$50 \div 5 \times 2 + (11-8)^2$$

4. Evaluate.

$$9^2 + 4^2 \div 2^2 - 4^3 \div 2^2$$

Concept Check

Explain in what order you would do the steps to evaluate the expression $7 \times 6 \div 3 \times 4^2 - 2$.

Chapter 1 Whole Numbers
1.7 Rounding and Estimating

Vocabulary
round   •   number line   •   estimate   •   approximate equal to

1. A _____ is a picture where whole numbers are represented by points on a line.

2. In mathematics we often _____, or determine the approximate value of a calculation, if we need to do a quick check.

3. We use the symbol "≈" to mean _____.

4. To _____ a number, we first determine the place we are rounding to. Then we find which value is closest to the number that we are rounding.

Example	**Student Practice**
1. Round 2,445,360 to the nearest hundred thousand.	**2.** Round 5,364,289 to the nearest hundred thousand.

Example (continued)

1. Round 2,445,360 to the nearest hundred thousand.

Locate the hundred thousands round-off place.

2,4̲45,360

The first digit to the right is less than 5. So round down. Do not change the hundred thousands digit.

2,4|4|5,360

Replace all digits to the right with 0.

2,400,000

Thus, 2,445,360 rounded to the nearest hundred thousand is 2,400,000.

Example	Student Practice
3. Round as indicated.	**4.** Round as indicated.
(a) 561,328 to the nearest ten	**(a)** 452,319 to the nearest ten
First locate the digit in the tens place.	
561,3<u>2</u>8	
The digit to the right of the tens place is greater than 5. So round up.	
561,330	
(b) 51,362,523 to the nearest million	**(b)** 65,428,732 to the nearest million
First locate the digit in the millions place.	
5<u>1</u>,362,523	
The digit to the right of the millions place is less than 5. So round down.	
51,000,000	
5. Astronomers use the parsec as a measurement of distance. One parsec is approximately 30,900,000,000,000 kilometers. Round 1 parsec to the nearest trillion kilometers.	**6.** Astronomers use the astronomical unit as a measurement of distance. One astronomical unit is approximately 149,597,871 kilometers. Round to the nearest ten million kilometers.
First locate the digit in the trillions place.	
3<u>0</u>,900,000,000,000	
The digit to the right of the trillions place is greater than 5. So round up. 30,900,000,000,000 km is 31,000,000,000,000 km or 31 trillion km to the nearest trillion kilometers.	

Example	Student Practice

7. Phil and Melissa bought their first car last week. The selling price of this compact car was $8980. The dealer preparation charge was $289 and the sales tax was $449. Estimate the total cost that Phil and Melissa had to pay.

Round each number to have only one nonzero digit, and add the rounded numbers.

$$
\begin{array}{rr}
8980 & 9000 \\
289 & 300 \\
+\ 449 & +\ 400 \\
\hline
& 9700
\end{array}
$$

The total cost $\approx$ $9700. (The exact answer is $9718, so our answer is quite close.)

8. Mike and Julie bought a dishwasher last week. The selling price of this dishwasher was $3450. The dealer preparation charge was $150 and the sales tax was $245. Estimate the total cost that Mike and Julie had to pay.

---

9. Estimate the product. $56,789 \times 529$

Round each number so that there is one nonzero digit.

Then multiply the rounded numbers to obtain the estimate.

$$
\begin{array}{rr}
56,789 & 60,000 \\
\times\ \ \ 529 & \times\ \ \ \ \ 500 \\
\hline
& 30,000,000
\end{array}
$$

Therefore, the product is $\approx 30,000,000$. (This is reasonably close to the exact answer of 30,041,381.)

10. Estimate the product. $47,698 \times 475$

Example	Student Practice

11. John and Stephanie drove their Honda CR-V a distance of 778 miles. They used 25 gallons of gas. Estimate how many miles they can travel on 1 gallon of gas.

Divide 778 by 25 to obtain the number of miles John and Stephanie get with 1 gallon of gas. Round each number to a number with one nonzero digit and then perform the division.

$$\begin{array}{r} 26 \\ 25\overline{)778} \qquad 30\overline{)800} \\ \underline{60} \\ 200 \\ \underline{180} \\ 20 \quad \text{Remainder} \end{array}$$

The answer is $26 \text{ R } 20 \approx 27$ miles per gallon, which is close to the exact answer, 31 miles per gallon.

12. Jim and Maggie drove their car a distance of 685 miles. They used 37 gallons of gas. Estimate how many miles they can travel on 1 gallon of gas.

Extra Practice

1. Round 209,355 to the nearest thousand.

2. Estimate $86,982 - 79,984$.

3. Estimate $58,449 \times 1136$.

4. Allison determined that she spent $19 on cat food each week last year. Estimate how much she spent on cat food over the entire year. Remember that there are 52 weeks in a year.

Concept Check

Explain how to round $682,496,934$ to the nearest million.

Chapter 1 Whole Numbers
1.8 Solving Applied Problems Involving Whole Numbers

Vocabulary

mathematics blueprint   •   chart   •   table   •   bill of sale
solve and state the answer   •   check   •   understand the problem

1. The second step in solving an application problem is to _____, which involves performing the calculations.

2. It is helpful to create a(n) _____ so that you have the facts and a method of proceeding in a situation.

3. The first step in solving an application problem is to _____, which involves reading the problem carefully.

4. The third step in solving an application problem is to _____, which involves estimating the answer and comparing the exact answer with the estimate.

Example	Student Practice
1. Gerald made deposits of $317, $512, $84, and $161 into his checking account. He also made out checks for $100 and $125. What was the total of his deposits?	2. Rob made deposits of $225, $552, and $132 into his checking account. He also made out checks for $250 and $175. What was the total of his deposits?

Read over the problem carefully and create a Mathematics Blueprint. The goal is to find the total of Gerald's deposits, so we can ignore the checks. Add the four deposits to obtain the total.

$$
\begin{array}{r}
317 \\
512 \\
84 \\
+\ 161 \\
\hline
1074
\end{array}
$$

Use estimation to check. Our estimate is $1080, which is close to $1074. The total of the four deposits is $1074.

Example	Student Practice
3. One horsepower is the power needed to lift 550 pounds a distance of 1 foot in 1 second. How many pounds can be lifted 1 foot in 1 second by 7 horsepower?	**4.** One quart of liquid is about 946 milliliters. How many milliliters are there in 5 quarts?

3. One horsepower is the power needed to lift 550 pounds a distance of 1 foot in 1 second. How many pounds can be lifted 1 foot in 1 second by 7 horsepower?

Simplify the problem. If 1 horsepower can lift 550 pounds, how many pounds can be lifted by 7 horsepower?

Read over the problem carefully and create a Mathematics Blueprint.

To solve the problem, multiply the 7 horsepower by 550 pounds for each horsepower.

$$\begin{array}{r} 550 \\ \times\ \ \ 7 \\ \hline 3850 \end{array}$$

Thus, 7 horsepower moves 3850 pounds 1 foot in 1 second.

Check by estimating. Round 550 to 600 pounds.

$600 \times 7 = 4200$ pounds

Our estimate is 4200 pounds. 3850 is close to our estimate. So our answer is reasonable.

Student Practice

4. One quart of liquid is about 946 milliliters. How many milliliters are there in 5 quarts?

Example	Student Practice

5. When Lorenzo began his car trip, his gas tank was full and the odometer read 76,358 miles. He ended his trip at 76,668 miles and filled the gas tank with 10 gallons of gas. How many miles per gallon did he get with his car?

Read over the problem carefully and create a Mathematics Blueprint.

First subtract the odometer readings to obtain the miles traveled.

$$\begin{array}{r} 76,668 \\ -\ 76,358 \\ \hline 310 \end{array}$$

The trip was 310 miles. Next, divide the miles driven by the number of gallons.

$$\begin{array}{r} 31 \\ 10\overline{)310} \\ \underline{30} \\ 10 \\ \underline{10} \\ 0 \end{array}$$

Thus, Lorenzo obtained 31 miles per gallon on the trip.

Check by rounding to the nearest hundred for the values of mileage.

$76,668 \rightarrow 76,700$

$76,358 \rightarrow 76,400$

Now subtract the estimated values. Thus, we estimate the trip to be 300 miles. Then divide, $300 \div 10 = 30$. Our estimate is 30 miles per gallon. This is very close to our calculated value.

6. When Michael began his car trip, his gas tank was full and the odometer read 68,557 miles. He ended his trip at 68,809 miles and filled the gas tank with 12 gallons of gas. How many miles per gallon did he get with his car?

Extra Practice

1. Amy planned a two-day trip of 1137 miles to Florida. On the first day, Amy traveled 673 miles. How many miles must Amy travel on day two to complete her trip?

2. Four students rented a car for six days. The cost to rent the car was $22 per day. How much did each student have to pay, assuming each paid the same amount?

3. Mrs. Robichaud had a balance of $257 in her checking account when she went shopping. She bought 2 blouses for $29 each, a pair of jeans for $49, a pair of sandals for $24, and a jacket for $49. Assuming no sales tax, how much money did she have left in her checking account?

4. Sergio can bake 60 muffins in one hour. A large company has ordered 300 muffins for a company breakfast. How many hours will Sergio need to fill the order? How many minutes?

Concept Check

A company has purchased 38 new cars for the sales department for $836,543. Assuming that each car cost the same, explain how you would estimate the cost of each car.

MATH COACH

Mastering the skills you need to do well on the test.

Watch the MATH COACH videos in MyMathLab® or on You Tube™ while you work the problems below. These helpful hints will help you avoid making common errors on test problems.

Subtract Whole Numbers with Borrowing—Problem 8

$$501,760$$
$$-328,902$$

> **Helpful Hint:** It is wise to show the borrowing steps. This will help you avoid a borrowing error.

Look at your work for Problem 8. Examine your steps. Do your borrowing steps match the solution below?

Yes ____   No ____

Write out the borrowing steps:

$$
\begin{array}{r}
\overset{4}{\cancel{5}}\ \overset{9}{\cancel{0}}\ \overset{10}{1,}\ \overset{17}{\cancel{7}}\ \overset{5}{\cancel{6}}\ \overset{10}{\cancel{0}} \\
-\ 3\ \ 2\ \ 8,\ \ 9\ \ 0\ \ 2 \\
\end{array}
$$

If you answered No, be sure to write your borrowing steps carefully and then check each subtraction step for errors.

Multiply Whole Numbers with Several Digits—Problem 12

$$326$$
$$\times 592$$

> **Helpful Hint:** When multiplying, students sometimes make errors in alignment. Take the extra time to line up each column carefully. Your calculation accuracy will improve if you write your numbers about 50% larger than normal.

Did you line up each column accurately with the number shown below?

Yes ____   No ____

If you answered No, rework the problem using correct alignment. Then, be sure to check each step for multiplication and addition errors.

$$
\begin{array}{r}
326 \\
\times 592 \\
\hline
652 \\
2934 \\
1630 \\
\end{array}
$$

Order of Operations—Problem 19 $\quad 5+6^2-2\times(9-6)^2$

The first step is to combine numbers inside parentheses.

$$5+6^2-2\times(3)^2$$

The second step is to evaluate exponents in two places.

$$5+36-2\times9$$

Did you do the first step correctly?

Yes _____   No _____

Did you do the second step correctly?

Yes _____   No _____

Try to do problem 19 correctly now. Go slowly and write each step.

Estimate the Product—Problem 25 $\quad 4{,}867{,}010\times27{,}058$

Did you round each number so that it has only one nonzero digit?

Yes _____   No _____

Did you write down the correct number of zeros?

Yes _____   No _____

If you answered Problem 25 incorrectly, try to rework it now.

$$5{,}000{,}000\times30{,}000=150{,}000{,}000{,}000$$

Name: _____   Date: _____

Instructor: _____   Section: _____

Chapter 2 Fractions
2.1 Understanding Fractions

Vocabulary
fractions   •   undefined   •   numerator   •   denominator

1. In a fraction, the number on the bottom is called the _____.

2. One way to represent parts of a whole is with _____.

3. Division by zero is _____.

4. In a fraction, the number on top is called the _____.

Example	**Student Practice**
1. Use a fraction to represent the shaded or completed part of the whole shown.	2. Use a fraction to represent the shaded or completed part of the whole shown.

Example

1. Use a fraction to represent the shaded or completed part of the whole shown.

    (a)

    Three out of four circles are shaded.

    The fraction is $\dfrac{3}{4}$.

    (b)

    Five out of seven equal parts are shaded. The fraction is $\dfrac{5}{7}$.

    (c)

    One mile

    The mile is divided into five equal parts. The car, traveling from right to left, has traveled 1 part of 5 of the one-mile distance. The fraction is $\dfrac{1}{5}$.

Student Practice

2. Use a fraction to represent the shaded or completed part of the whole shown.

Example	Student Practice
3. Draw a sketch to illustrate.	**4.** Draw a sketch to illustrate.
(a) $\dfrac{7}{11}$ of an object	**(a)** $\dfrac{3}{14}$ of a group

Example

3. Draw a sketch to illustrate.

(a) $\dfrac{7}{11}$ of an object

The easiest figure to draw is a rectangular bar.

We divide the bar into 11 equal parts. We then shade in 7 parts to show $\dfrac{7}{11}$.

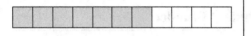

(b) $\dfrac{2}{9}$ of a group

We draw 9 circles of equal size to represent a group of 9.

We shade in 2 of the 9 circles to show $\dfrac{2}{9}$.

Student Practice

4. Draw a sketch to illustrate.

(a) $\dfrac{3}{14}$ of a group

(b) $\dfrac{5}{8}$ of a pizza

Example	Student Practice
5. Use a fraction to describe each situation.	**6.** Use a fraction to describe each situation.

5. Use a fraction to describe each situation.

(a) A baseball player gets a hit 5 out of 12 times at bat.

The denominator tells the total number of at-bats, and the numerator tells the number of at-bats in which the player got a hit.

The baseball player got a hit $\frac{5}{12}$ of his times at bat.

(b) There are 156 men and 185 women taking psychology this semester. Describe the part of the class that consists of women.

The total class is $156 + 185 = 341$. The fractional part that is women is 185 out of 341.

Thus, $\frac{185}{341}$ of the class is women.

(c) Robert Tobey found in the Alaska moose count that five-eighths of the moose observed were female.

Five-eighths of the moose observed were female.

The fraction is $\frac{5}{8}$.

---

6. Use a fraction to describe each situation.

(a) There were 27 rainy days and 61 sunny days during summer vacation. Use a fraction to describe the portion of days that were rainy during the vacation.

(b) In a big city, you can see the stars on two-sevenths of the evenings during the summer.

(c) A dog fetched 11 out of 18 balls.

Example	Student Practice
7. Wanda made 13 calls, out of which she made five sales. Albert made 17 calls, out of which he made six sales. Write a fraction that describes for both people together the number of calls in which a sale was made compared with the total number of calls. There are $5+6=11$ calls in which a sale was made. There were $13+17=30$ total calls. Thus, $\dfrac{11}{30}$ of the calls resulted in a sale.	**8.** Two students were gathering clothing to give to charity. Louise had 21 pairs of jeans that no longer fit. 10 were in good shape. Juan had 12 pairs of jeans that didn't fit him any more. 7 were still in good shape. Write a fraction that describes the number of good jeans to give away compared to the total number of jeans for the two students.

Extra Practice

1. Use a fraction to represent the shaded part of the object.

2. Use a fraction to represent the shaded portion of the set of objects.

3. Draw a sketch to illustrate the fractional part.

$\dfrac{0}{12}$ of an object.

4. In the parking lot, there were 15 black cars, 12 white cars, 14 blue cars, and 21 other colored cars. What fractional part of the cars were blue?

Concept Check

One hundred twenty new businesses have opened in Springfield in the last five years. Sixty-five of them were restaurants; the remaining ones were not. Thirty new restaurants went out of business; the other new restaurants did not. Of all the new businesses that were not restaurants, 25 of them went out of business; the others did not. Explain how you can find a fraction that represents the fractional part of the new businesses that did not go out of business.

Chapter 2 Fractions
2.2 Simplifying Fractions

Vocabulary
prime number • equivalent fractions • equality test for fractions • factor tree
common factor • simplified • reduced • in lowest terms • cross products
prime factorization • simplest form • composite number

1. A(n) _____ is a whole number greater than 1 that can be divided by whole numbers other than itself and 1.

2. Writing a number as a product of prime factors is called a(n) _____.

3. We say that two fractions are _____ when they are equal to the same value.

4. A(n) _____ is a whole number greater than 1 that cannot be evenly divided by whole numbers except by itself and 1.

Example	**Student Practice**
1. Write 60 as a product of prime factors. First write 60 as the product of any two factors. $60 = 6 \times 10$ Factor the two factors if they are not prime. $6 = 2 \times 3$ $10 = 2 \times 5$ All the factors are now prime. Rearrange the factors from least to greatest. $60 = 6 \times 10 = 2 \times 2 \times 3 \times 5$ or $60 = 2^2 \times 3 \times 5$	**2.** Write 84 as a product of prime factors.

Example	Student Practice
3. Simplify (write in lowest terms). $$\frac{15}{25}$$ The greatest common factor is 5. Divide the numerator and the denominator by 5. $$\frac{15}{25} = \frac{15 \div 5}{25 \div 5} = \frac{3}{5}$$	**4.** Simplify (write in lowest terms). $$\frac{14}{21}$$
5. Simplify the fraction $\frac{35}{42}$ by the method of prime factors. Factor 35 and 42 into prime factors. $$\frac{35}{42} = \frac{5 \times 7}{2 \times 3 \times 7}$$ The common prime factor is 7. Divide out 7. $$\frac{35}{42} = \frac{5 \times \overset{1}{\cancel{7}}}{2 \times 3 \times \underset{1}{\cancel{7}}}$$ Multiply the factors in the numerator and denominator to write the reduced or simplified form. $$\frac{35}{42} = \frac{5 \times 1}{2 \times 3 \times 1} = \frac{5}{6}$$ Thus, $\frac{35}{42} = \frac{5}{6}$, and $\frac{5}{6}$ is the simplest form.	**6.** Simplify the fraction $\frac{55}{88}$ by the method of prime factors.

Example	Student Practice
7. Are the fractions equal? Use the equality test.	**8.** Are the fractions equal? Use the equality test.

7. Are the fractions equal? Use the equality test.

(a) $\dfrac{2}{11} \overset{?}{=} \dfrac{18}{99}$

For two fractions, if $\dfrac{a}{b} = \dfrac{c}{d}$, then $a \times d = b \times c$. $a \times d$ and $b \times c$ are the cross products.

In this case, the two cross products are as follows.

$11 \times 18 = 198$
$2 \times 99 = 198$

Since $198 = 198$, we know that $\dfrac{2}{11} = \dfrac{18}{99}$.

(b) $\dfrac{3}{16} \overset{?}{=} \dfrac{12}{62}$

In this case, the two cross products are as follows.

$16 \times 12 = 192$
$3 \times 62 = 186$

Since $192 \neq 186$, we know that $\dfrac{3}{16} \neq \dfrac{12}{62}$.

8. Are the fractions equal? Use the equality test.

(a) $\dfrac{3}{14} \overset{?}{=} \dfrac{12}{56}$

(b) $\dfrac{3}{15} \overset{?}{=} \dfrac{16}{75}$

Extra Practice

1. Write the following number as a product of prime factors.

   28

2. Reduce the fraction $\frac{33}{55}$ by finding a common factor in the numerator and in the denominator and dividing by the common factor.

3. Reduce the fraction $\frac{36}{54}$ by the method of prime factors.

4. Are the following fractions equal?

   $$\frac{15}{21} \overset{?}{=} \frac{45}{84}$$

Concept Check

Explain how you would determine if the fraction $\frac{195}{231}$ can be reduced.

Chapter 2 Fractions
2.3 Converting Between Improper Fractions and Mixed Numbers

Vocabulary
proper fraction   •   improper fraction   •   prime   •   mixed numbers

1. If the numerator of a fraction is greater than or equal to the denominator, the fraction is a(n) _____.

2. A(n) _____ is the sum of a whole number greater than zero and a proper fraction.

3. If the numerator of a fraction is less than the denominator, the fraction is a(n) _____.

4. If the denominator of a fraction is _____, the fraction cannot be reduced unless the denominator is a factor of the numerator.

Example	Student Practice
1. Change each mixed number to an improper fraction.	**2.** Change each mixed number to an improper fraction.
(a) $3\frac{2}{5}$	**(a)** $7\frac{2}{9}$
Multiply the whole number by the denominator. Add the numerator to the product. Then, write the sum over the denominator.	
$3\frac{2}{5} = \frac{3\times5+2}{5} = \frac{15+2}{5} = \frac{17}{5}$	**(b)** $4\frac{2}{11}$
(b) $5\frac{4}{9}$	
Use the same procedure as before.	
$5\frac{4}{9} = \frac{5\times9+4}{9} = \frac{45+4}{9} = \frac{49}{9}$	

Example	Student Practice
3. Write the improper fraction as a mixed number.	**4.** Write the improper fraction as a mixed number.

$$\frac{13}{5}$$

$$\frac{31}{9}$$

Divide the denominator 5 into 13.

$$\begin{array}{r} 2 \quad \text{quotient} \\ 5\overline{)13} \\ \underline{10} \\ 3 \quad \text{remainder} \end{array}$$

The answer is of the following form.

$$\text{quotient}\,\frac{\text{remainder}}{\text{denominator}}$$

Thus, $\dfrac{13}{5} = 2\dfrac{3}{5}$.

5. Reduce the improper fraction.	**6.** Reduce the improper fraction.

$$\frac{22}{8}$$

$$\frac{42}{15}$$

Write the fraction in terms of prime factors. Then, divide the numerator and denominator by any common factors.

$$\frac{22}{8} = \frac{\overset{1}{\cancel{2}}\times 11}{\underset{1}{\cancel{2}}\times 2\times 2} = \frac{11}{4}$$

Example	Student Practice
7. Reduce the mixed number.	**8.** Reduce the mixed number.

7. Reduce the mixed number.

$$4\frac{21}{28}$$

We cannot reduce the whole number 4, only the fraction $\frac{21}{28}$.

$$\frac{21}{28} = \frac{3 \times \cancel{7}^{1}}{4 \times \cancel{7}_{1}} = \frac{3}{4}$$

Therefore, $4\frac{21}{28} = 4\frac{3}{4}$.

8. Reduce the mixed number.

$$7\frac{12}{18}$$

9. Reduce $\frac{945}{567}$ by first changing to a mixed number.

Divide the denominator 567 into 945.

$$567\overline{)945}^{\,1}$$
$$\underline{567}$$
$$378$$

So, $\frac{945}{567} = 1\frac{378}{567}$. Now reduce the fraction $\frac{378}{567}$.

$$\frac{378}{567} = \frac{2 \times 3 \times 3 \times 3 \times 7}{3 \times 3 \times 3 \times 3 \times 7} = \frac{2 \times \cancel{3}^{1} \times \cancel{3}^{1} \times \cancel{3}^{1} \times \cancel{7}^{1}}{3 \times \cancel{3}_{1} \times \cancel{3}_{1} \times \cancel{3}_{1} \times \cancel{7}_{1}}$$

So, $\frac{945}{567} = 1\frac{378}{567} = 1\frac{2}{3}$.

10. Reduce $\frac{294}{210}$ by first changing to a mixed number.

Extra Practice

1. Change the mixed number to an improper fraction.

   $3\dfrac{11}{14}$

2. Reduce the mixed number.

   $5\dfrac{12}{36}$

3. Reduce the improper fraction.

   $\dfrac{171}{9}$

4. Change to a mixed number and reduce.

   $\dfrac{630}{240}$

Concept Check

Explain how you change the mixed number $5\dfrac{6}{13}$ to an improper fraction.

Chapter 2 Fractions
2.4 Multiplying Fractions and Mixed Numbers

Vocabulary
proper fractions　•　improper fractions　•　numerators　•　denominator

1. To multiply two fractions, we multiply the _____ and multiply the denominators.

2. When you multiply two _____ together, you get a smaller fraction.

3. To multiply a fraction by a mixed number or to multiply two mixed numbers, first change each mixed number to a(n) _____ .

4. When multiplying a fraction by a whole number, it is more convenient to express the whole number as a fraction with a(n) _____ of 1.

Example	**Student Practice**
1. Multiply.	**2.** Multiply.
(a) $\dfrac{3}{8} \times \dfrac{5}{7}$	**(a)** $\dfrac{2}{3} \times \dfrac{7}{11}$
To multiply two fractions, we multiply the numerators and multiply the denominators.	
$\dfrac{3}{8} \times \dfrac{5}{7} = \dfrac{3 \times 5}{8 \times 7} = \dfrac{15}{56}$	
(b) $\dfrac{1}{11} \times \dfrac{2}{13}$	**(b)** $\dfrac{5}{6} \times \dfrac{13}{23}$
As before, multiply the numerators and multiply the denominators.	
$\dfrac{1}{11} \times \dfrac{2}{13} = \dfrac{1 \times 2}{11 \times 13} = \dfrac{2}{143}$	

Example	Student Practice
3. Simplify first and then multiply.	**4.** Simplify first and then multiply.

3. Simplify first and then multiply.

$$\frac{12}{35} \times \frac{25}{18}$$

First, find the prime factors.

$$\frac{12}{35} \times \frac{25}{18} = \frac{2 \cdot 2 \cdot 3}{5 \cdot 7} \times \frac{5 \cdot 5}{2 \cdot 3 \cdot 3}$$

Arrange the factors in order and write the product as one fraction.

$$\frac{12}{35} \times \frac{25}{18} = \frac{2 \cdot 2 \cdot 3 \cdot 5 \cdot 5}{2 \cdot 3 \cdot 3 \cdot 5 \cdot 7}$$

Divide the numerator and denominator by the common factors and multiply.

$$\frac{12}{35} \times \frac{25}{18} = \frac{\overset{1}{\cancel{2}} \cdot 2 \cdot \overset{1}{\cancel{3}} \cdot \overset{1}{\cancel{5}} \cdot 5}{\underset{1}{\cancel{2}} \cdot \underset{1}{\cancel{3}} \cdot 3 \cdot \underset{1}{\cancel{5}} \cdot 7} = \frac{10}{21}$$

4. Simplify first and then multiply.

$$\frac{16}{21} \times \frac{28}{96}$$

---

5. Multiply.

(a) $5 \times \dfrac{3}{8}$

To multiply a fraction by a whole number, express the whole number as a fraction with the denominator 1.

$$5 \times \frac{3}{8} = \frac{5}{1} \times \frac{3}{8} = \frac{15}{8} \text{ or } 1\frac{7}{8}$$

(b) $\dfrac{22}{7} \times 14$

$$\frac{22}{7} \times 14 = \frac{22}{\underset{1}{\cancel{7}}} \times \frac{\overset{2}{\cancel{14}}}{1} = \frac{44}{1} = 44$$

6. Multiply.

(a) $12 \times \dfrac{4}{5}$

(b) $\dfrac{7}{15} \times 3$

Example	Student Practice
7. Mr. and Mrs. Jones found that $\frac{2}{7}$ of their income went to pay federal income taxes. Last year they earned \$37,100. How much did they pay in taxes? Find $\frac{2}{7}$ of \$37,100 by multiplying $\frac{2}{7} \times 37,100$. $\frac{2}{7} \times 37,100 = \frac{2}{\cancel{7}_1} \times \cancel{37,100}^{5300} = \frac{2}{1} \times 5300$ $\qquad\qquad\qquad = 10,600$ They paid \$10,600 in taxes.	8. Juan sent his resume to 1200 companies during his job search. If $\frac{2}{5}$ of the companies replied to him, how many replies did he receive?
9. Multiply. (a) $\frac{5}{7} \times 3\frac{1}{4}$ To multiply a fraction by a mixed number first change the mixed number to an improper fraction, and then multiply. $\frac{5}{7} \times 3\frac{1}{4} = \frac{5}{7} \times \frac{13}{4} = \frac{65}{28}$ or $2\frac{9}{28}$ (b) $20\frac{2}{5} \times 6\frac{2}{3}$ To multiply two mixed numbers change the mixed numbers to improper fractions, and then multiply. $20\frac{2}{5} \times 6\frac{2}{3} = \frac{\cancel{102}^{34}}{\cancel{5}_1} \times \frac{\cancel{20}^{4}}{\cancel{3}_1} = \frac{136}{1} = 136$	10. Multiply. (a) $4\frac{2}{5} \times 7$ (b) $8\frac{1}{3} \times 4\frac{5}{9}$

Example	Student Practice
11. Find the value of x if $\dfrac{3}{7} \cdot x = \dfrac{15}{42}$.	**12.** Find the value of x if $x \cdot \dfrac{2}{5} = \dfrac{8}{35}$.

The variable x represents a fraction. We know that 3 times one number equals 15 and 7 times another equals 42.

Since $3 \cdot 5 = 15$ and $7 \cdot 6 = 42$, we know that $\dfrac{3}{7} \cdot \dfrac{5}{6} = \dfrac{15}{42}$.

Therefore, $x = \dfrac{5}{6}$.

Extra Practice

1. Multiply. Make sure all fractions are simplified in the final answer.

$\dfrac{8}{15} \times 5$

2. Multiply. Change any mixed number to an improper fraction before multiplying.

$2\dfrac{4}{5} \times 1\dfrac{3}{7}$

3. Multiply. Make sure all fractions are simplified in the final answer.

$\dfrac{16}{9} \times \dfrac{3}{8}$

4. Multiply. Make sure all fractions are simplified in the final answer.

$2\dfrac{4}{5} \times 3\dfrac{3}{4}$

Concept Check

Explain how you would multiply the whole number 6 times the mixed number $4\dfrac{3}{5}$.

Chapter 2 Fractions
2.5 Dividing Fractions and Mixed Numbers

Vocabulary

invert • fractions • mixed numbers • reciprocal

1. When a fraction has its denominator and numerator interchanged, the result is called a(n) _____.

2. When we _____ a fraction, we interchange the numerator and the denominator.

3. To divide two _____, we invert the second one and multiply.

4. If one or more _____ are involved in the division, they should be converted to improper fractions first.

Example	**Student Practice**
1. Divide.	**2.** Divide.
$\dfrac{5}{8} \div \dfrac{25}{16}$	$\dfrac{2}{3} \div \dfrac{5}{21}$
To divide two fractions, invert the second fraction and multiply.	
$\dfrac{5}{8} \div \dfrac{25}{16} = \dfrac{\cancel{5}^{1}}{\cancel{8}_{1}} \times \dfrac{\cancel{16}^{2}}{\cancel{25}_{5}} = \dfrac{2}{5}$	
3. Divide.	**4.** Divide.
$\dfrac{3}{7} \div 2$	$\dfrac{5}{9} \div 3$
Remember that for any whole number a, $a = \dfrac{a}{1}$.	
$\dfrac{3}{7} \div 2 = \dfrac{3}{7} \div \dfrac{2}{1} = \dfrac{3}{7} \times \dfrac{1}{2} = \dfrac{3}{14}$	

Example	Student Practice
5. Divide, if possible.	**6.** Divide, if possible.

5. Divide, if possible.

(a) $\dfrac{23}{25} \div 1$

$$\dfrac{23}{25} \div 1 = \dfrac{23}{25} \times \dfrac{1}{1} = \dfrac{23}{25}$$

(b) $\dfrac{3}{17} \div 0$

Division by zero is undefined.

6. Divide, if possible.

(a) $1 \div \dfrac{12}{7}$

(b) $0 \div \dfrac{5}{21}$

7. Divide.

$$3\dfrac{7}{15} \div 1\dfrac{1}{25}$$

If there are mixed numbers involved in a division, convert them to improper fractions first.

$$3\dfrac{7}{15} \div 1\dfrac{1}{25} = \dfrac{52}{15} \div \dfrac{26}{25} = \dfrac{\overset{2}{\cancel{52}}}{\underset{3}{\cancel{15}}} \times \dfrac{\overset{5}{\cancel{25}}}{\underset{1}{\cancel{26}}} = \dfrac{10}{3}$$

$$3\dfrac{7}{15} \div 1\dfrac{1}{25} = \dfrac{10}{3} \text{ or } 3\dfrac{1}{3}$$

8. Divide.

$$6\dfrac{2}{3} \div 3\dfrac{1}{2}$$

9. Divide.

$$\dfrac{10\dfrac{2}{9}}{2\dfrac{1}{3}}$$

The division of two mixed numbers may be indicated by a wide fraction bar.

$$\dfrac{10\dfrac{2}{9}}{2\dfrac{1}{3}} = 10\dfrac{2}{9} \div 2\dfrac{1}{3} = \dfrac{92}{9} \div \dfrac{7}{3} = \dfrac{92}{\underset{3}{\cancel{9}}} \times \dfrac{\overset{1}{\cancel{3}}}{7} = \dfrac{92}{21}$$

10. Divide.

$$\dfrac{2\dfrac{1}{8}}{5\dfrac{1}{2}}$$

Example	Student Practice
11. Find the value of x if $x \div \dfrac{8}{7} = \dfrac{21}{40}$.	**12.** Find the value of x if $x \div \dfrac{3}{5} = \dfrac{25}{27}$.

First change the division to an equivalent multiplication.

$$x \div \frac{8}{7} = \frac{21}{40} \quad \rightarrow \quad x \cdot \frac{7}{8} = \frac{21}{40}$$

The variable x represents a fraction. We know that some number times 7 equals 21 and another number times 8 equals 40.

Since $3 \cdot 7 = 21$ and $5 \cdot 8 = 40$, we know that $\dfrac{3}{5} \cdot \dfrac{7}{8} = \dfrac{21}{40}$.

Thus, $x = \dfrac{3}{5}$.

---

13. There are 117 milligrams of cholesterol in $4\dfrac{1}{3}$ cups of milk. How much cholesterol is in 1 cup of milk?

Divide 117 by $4\dfrac{1}{3}$ to find how much is in one cup.

$$117 \div 4\frac{1}{3} = 117 \div \frac{13}{3}$$

$$= \frac{\overset{9}{\cancel{117}}}{1} \times \frac{3}{\underset{1}{\cancel{13}}}$$

$$= \frac{27}{1}$$

Thus, there are 27 milligrams of cholesterol in 1 cup of milk.

14. Delores drove 321 miles in $8\dfrac{1}{4}$ hours. What was her average speed in miles per hour?

Extra Practice

1. Divide, if possible.

$$1 \div \frac{5}{8}$$

2. Divide, if possible.

$$3\frac{3}{8} \div 2\frac{1}{4}$$

3. Divide, if possible.

$$\frac{\dfrac{15}{6}}{\dfrac{5}{3}}$$

4. Divide, if possible.

$$\frac{\dfrac{4}{9}}{100}$$

Concept Check

Explain how you would divide the whole number 7 by the mixed number $3\frac{3}{5}$.

Chapter 2 Fractions
2.6 The Least Common Denominator and Creating Equivalent Fractions

Vocabulary

multiples • building fraction property • least common multiple (LCM)
least common denominator (LCD)

1. The _____ of two natural numbers is the smallest number that is a multiple of both.

2. The _____ of two or more fractions is the smallest number that can be divided evenly by each of the fractions' denominators.

3. The _____ of a number are the products of that number and the numbers 1, 2, 3, 4, 5, 6, 7, and so on.

4. The _____ states that for whole numbers a, b, and c where $b \neq 0$ and $c \neq 0$,
$$\frac{a}{b} = \frac{a}{b} \times 1 = \frac{a}{b} \times \frac{c}{c} = \frac{a \times c}{b \times c}.$$

Example	Student Practice
1. Find the least common multiple of 10 and 12.	**2.** Find the least common multiple of 9 and 12.
The multiples of 10 are 10, 20, 30, 40, 50, $\boxed{60}$, 70,…	
The multiples of 12 are 12, 24, 36, 48, $\boxed{60}$, 72, 84,…	
The first multiple that appears on both lists is the least common multiple (LCM). The LCM of 10 and 12 is 60.	
3. Find the least common multiple of 7 and 35.	**4.** Find the least common multiple of 44 and 11.
Because $7 \times 5 = 35$, 35 is a multiple of 7. So we can state immediately that the least common multiple of 7 and 35 is 35.	

Example	Student Practice
5. Determine the LCD for the pair of fractions. $\dfrac{7}{15}$ and $\dfrac{4}{5}$ Since 5 can be divided into 15, the LCD of $\dfrac{7}{15}$ and $\dfrac{4}{5}$ is 15. (Notice that the least common multiple of 5 and 15 is 15.)	**6.** Determine the LCD for the pair of fractions. $\dfrac{4}{7}$ and $\dfrac{17}{21}$
7. Find the LCD for $\dfrac{1}{4}$ and $\dfrac{4}{5}$. We see that $4 \times 5 = 20$. Also, 20 is the smallest number that can be divided without remainder by 4 and by 5. We know this because the least common multiple of 4 and 5 is 20. So the LCD = 20.	**8.** Find the LCD for $\dfrac{1}{3}$ and $\dfrac{6}{7}$.
9. Find the LCD of $\dfrac{5}{6}$ and $\dfrac{4}{15}$ by the three-step procedure. Write each denominator as a product of prime factors. $6 = 2 \times 3$ and $15 = 3 \times 5$ List all the prime factors that appear in either product. The prime factors are 2, 3, and 5. Form a product of the prime factors, using each the greatest number of times it appears in any one denominator. LCD $= 2 \times 3 \times 5 = 30$	**10.** Find the LCD of $\dfrac{7}{10}$ and $\dfrac{5}{14}$ by the three-step procedure.

Example	Student Practice
11. Find the LCD of $\frac{7}{12}$, $\frac{1}{15}$, and $\frac{11}{30}$.	**12.** Find the LCD of $\frac{7}{18}$, $\frac{5}{42}$, and $\frac{8}{21}$.

$$
\begin{array}{rcccccccc}
12 & = & 2 & \times & 2 & \times & 3 & & \\
15 & = & \downarrow & & \downarrow & & 3 & \times & 5 \\
30 & = & \downarrow & & 2 & \times & 3 & \times & 5 \\
 & & \downarrow & & \downarrow & & \downarrow & & \downarrow \\
\text{LCD} & = & 2 & \times & 2 & \times & 3 & \times & 5 \\
\text{LCD} & = & 60 & & & & & &
\end{array}
$$

---

13. Build each fraction to an equivalent fraction with the given LCD.

(a) $\frac{3}{4}$, LCD $= 28$

We know that $4 \times 7 = 28$, so the value c that we multiply the numerator and denominator is 7.

$$\frac{3}{4} \times \frac{c}{c} = \frac{?}{28}$$

$$\frac{3}{4} \times \frac{7}{7} = \frac{21}{28}$$

(b) $\frac{1}{3}$ and $\frac{4}{5}$, LCD $= 15$

$$\frac{1}{3} \times \frac{c}{c} = \frac{?}{15}$$

$$\frac{1}{3} \times \frac{5}{5} = \frac{5}{15}$$

$$\frac{4}{5} \times \frac{c}{c} = \frac{?}{15}$$

$$\frac{4}{5} \times \frac{3}{3} = \frac{12}{15}$$

14. Build the fraction to an equivalent fraction with the given LCD.

(a) $\frac{5}{7}$, LCD $= 35$

(b) $\frac{1}{2}$ and $\frac{8}{11}$, LCD $= 22$

Example	Student Practice
15. Answer parts **(a)** and **(b)**.	**16.** Answer parts **(a)** and **(b)**.

(a) Find the LCD of $\dfrac{1}{32}$ and $\dfrac{7}{48}$.

Find the prime factors of 32 and 48

$32 = 2 \times 2 \times 2 \times 2 \times 2$
$48 = 2 \times 2 \times 2 \times 2 \times 3$

Thus, the LCD will need a factor of 2 five times and a factor of 3 one time.

$LCD = 2 \times 2 \times 2 \times 2 \times 2 \times 3 = 96$

(b) Build the fractions to equivalent fractions with that LCD.

$32 \times 3 = 96 \rightarrow \dfrac{1}{32} \times \dfrac{c}{c} = \dfrac{1}{32} \times \dfrac{3}{3} = \dfrac{3}{96}$

$48 \times 2 = 96 \rightarrow \dfrac{7}{48} \times \dfrac{c}{c} = \dfrac{7}{48} \times \dfrac{2}{2} = \dfrac{14}{96}$

(a) Find the LCD of $\dfrac{5}{14}$ and $\dfrac{2}{35}$.

(b) Build the fractions to equivalent fractions with that LCD.

Extra Practice

1. Find the LCD for $\dfrac{11}{15}$ and $\dfrac{7}{18}$.

2. Find the LCD for $\dfrac{2}{21}$, $\dfrac{5}{6}$, and $\dfrac{3}{7}$.

3. Build the fraction to an equivalent fraction with the denominator specified.

$\dfrac{8}{9} = \dfrac{?}{72}$

4. Find the LCD of $\dfrac{8}{15}$ and $\dfrac{6}{25}$. Build to equivalent fractions with the LCD as the denominator.

Concept Check

Explain how you would find the least common denominator of the fractions $\dfrac{5}{6}$, $\dfrac{11}{14}$, and $\dfrac{2}{15}$.

Chapter 2 Fractions
2.7 Adding and Subtracting Fractions

Vocabulary
denominator   •   common denominators   •   reduce   •   least common denominator (LCD)

1. You must have _____ to add or subtract fractions, but you need not have them to multiply or divide fractions.

2. To add or subtract two fractions with the same _____, add or subtract the numerators and write the sum or difference over the common denominator.

3. If the two fractions do not have a common denominator, we build each fraction so that its denominator is the _____.

4. When adding or subtracting fractions, it is very important to remember to _____ the final answer.

Example	**Student Practice**
1. Add $\dfrac{5}{13}+\dfrac{7}{13}$.	**2.** Add $\dfrac{1}{15}+\dfrac{7}{15}$.
To add two fractions with the same denominator, add the numerators and write the sum over the common denominator.	
$\dfrac{5}{13}+\dfrac{7}{13}=\dfrac{12}{13}$	
3. Add $\dfrac{4}{9}+\dfrac{2}{9}$.	**4.** Add $\dfrac{1}{18}+\dfrac{11}{18}$.
$\dfrac{4}{9}+\dfrac{2}{9}=\dfrac{6}{9}$	
Reduce the final answer.	
$\dfrac{6}{9}=\dfrac{2}{3}$	

Example	Student Practice
5. Subtract. $$\frac{13}{16} - \frac{3}{16}$$ To subtract two fractions with the same denominator, subtract the numerators and write the sum over the common denominator. $$\frac{13}{16} - \frac{3}{16} = \frac{10}{16} = \frac{5}{8}$$	**6.** Subtract. $$\frac{8}{13} - \frac{6}{13}$$
7. Add $\dfrac{3}{8} + \dfrac{5}{6} + \dfrac{1}{4}$. The LCD of 8, 6, and 4 is 24. $$\frac{3}{8} \times \frac{3}{3} = \frac{9}{24}, \ \frac{5}{6} \times \frac{4}{4} = \frac{20}{24}, \text{ and } \frac{1}{4} \times \frac{6}{6} = \frac{6}{24}$$ $$\frac{3}{8} + \frac{5}{6} + \frac{1}{4} = \frac{9}{24} + \frac{20}{24} + \frac{6}{24}$$ $$= \frac{35}{24} \text{ or } 1\frac{11}{24}$$	**8.** Add $\dfrac{5}{18} + \dfrac{7}{24}$.
9. Subtract $\dfrac{17}{25} - \dfrac{3}{35}$. The LCD of 25 and 35 is 175. $$\frac{17}{25} \times \frac{7}{7} = \frac{119}{175} \text{ and } \frac{3}{35} \times \frac{5}{5} = \frac{15}{175}$$ $$\frac{17}{25} - \frac{3}{35} = \frac{119}{175} - \frac{15}{175} = \frac{104}{175}$$	**10.** Subtract $\dfrac{11}{14} - \dfrac{2}{21}$.

Example	Student Practice
11. John and Stephanie and the triplets have a house on $\dfrac{7}{8}$ acre of land. They have $\dfrac{1}{3}$ acre of land planted with grass. How much of the land is not planted with grass?	**12.** Deborah exercises on a route that is $\dfrac{7}{8}$ of a mile long. She walks $\dfrac{1}{5}$ of a mile to warm up and runs the rest. What fraction of a mile does she run?

$\frac{7}{8}$ acre of land

$\frac{1}{3}$ acre of grass

We need to subtract $\dfrac{1}{3}$ from $\dfrac{7}{8}$.

The LCD of 8 and 3 is 24.

$$\frac{7}{8}\times\frac{3}{3}=\frac{21}{24} \quad \text{and} \quad \frac{1}{3}\times\frac{8}{8}=\frac{8}{24}$$

$$\frac{7}{8}-\frac{1}{3}=\frac{21}{24}-\frac{8}{24}=\frac{13}{24}$$

13. Find the value of x in the equation $x+\dfrac{5}{6}=\dfrac{9}{10}$. Reduce your answer.	**14.** Find the value of x in the equation $\dfrac{3}{8}+x=\dfrac{7}{12}$. Reduce your answer.

The LCD for the two fractions is 30.

Rewrite $x+\dfrac{5}{6}=\dfrac{9}{10}$ as $x+\dfrac{25}{30}=\dfrac{27}{30}$.

Add 2 to 25 to get 27 in the numerator.

Thus, $\dfrac{2}{30}+\dfrac{25}{30}=\dfrac{27}{30}$. So, $x=\dfrac{2}{30}=\dfrac{1}{15}$.

Example	Student Practice
15. Add $\dfrac{11}{12}+\dfrac{13}{30}$ by using the product of the two denominators as a common denominator.	**16.** Add $\dfrac{7}{9}+\dfrac{1}{25}$ by using the product of the two denominators as a common denominator.

$$\frac{11}{12}\times\frac{30}{30}=\frac{330}{360} \text{ and } \frac{13}{30}\times\frac{12}{12}=\frac{156}{360}$$

$$\frac{11}{12}+\frac{13}{30}=\frac{330}{360}+\frac{156}{360}=\frac{486}{360}=\frac{27}{20} \text{ or } 1\frac{7}{20}$$

Extra Practice

1. Add $\dfrac{7}{25}+\dfrac{12}{45}$. Simplify the answer.

2. Subtract $\dfrac{8}{9}-\dfrac{5}{36}$. Simplify the answer.

3. Subtract $\dfrac{5}{12}-\dfrac{15}{42}$. Simplify the answer.

4. Add $\dfrac{23}{42}+\dfrac{1}{7}+\dfrac{1}{6}$. Simplify the answer.

Concept Check

Explain how you would subtract the fractions $\dfrac{8}{9}-\dfrac{3}{7}$.

Chapter 2 Fractions
2.8 Adding and Subtracting Mixed Numbers and the Order of Operations

Vocabulary
mixed numbers   •   borrow   •   order of operations   •   common denominator

1. If the fraction portions of the mixed numbers do not have a(n) _____, we must build the fraction parts to obtain one before adding.

2. Similar to simplifying expressions with whole numbers, follow the _____ when simplifying expressions involving fractions and mixed numbers.

3. When adding _____, it is best to add the fractions together and then add the whole numbers together.

4. Sometimes we must _____ before we can subtract.

Example	**Student Practice**
1. Add $3\dfrac{1}{8}+2\dfrac{5}{8}$.	**2.** Add $5\dfrac{7}{12}+6\dfrac{1}{12}$.

Add the fractions $\dfrac{1}{8}+\dfrac{5}{8}=\dfrac{6}{8}$.

Add the whole numbers $3+2=5$.

Perform the addition vertically.

$$3\dfrac{1}{8}$$
$$+2\dfrac{5}{8}$$
$$\overline{5\dfrac{6}{8}}$$

Reduce the answer.

$$5\dfrac{6}{8}=5\dfrac{3}{4}$$

Example	Student Practice

3. Add $1\dfrac{2}{7}+5\dfrac{1}{3}$.

The LCD of $\dfrac{2}{7}$ and $\dfrac{1}{3}$ is 21. Rewrite the fractions using the LCD.

$\dfrac{2}{7}\times\dfrac{3}{3}=\dfrac{6}{21}$ and $\dfrac{1}{3}\times\dfrac{7}{7}=\dfrac{7}{21}$

Rewrite the mixed numbers and perform the addition vertically.

$$\begin{array}{rcl} 1\dfrac{2}{7} & = & 1\dfrac{6}{21} \\[2mm] +5\dfrac{1}{3} & = & +5\dfrac{7}{21} \\ \hline & & 6\dfrac{13}{21} \end{array}$$

4. Add $3\dfrac{5}{6}+4\dfrac{1}{9}$.

---

5. Add $6\dfrac{5}{6}+4\dfrac{3}{8}$.

The LCD of $\dfrac{5}{6}$ and $\dfrac{3}{8}$ is 24.

$$\begin{array}{rcl} 6\;\boxed{\dfrac{5}{6}\times\dfrac{4}{4}} & = & 6\dfrac{20}{24} \\[3mm] +4\;\boxed{\dfrac{3}{8}\times\dfrac{3}{3}} & = & +4\dfrac{9}{24} \\ \hline & & 10\dfrac{29}{24} \end{array}$$

If the sum of the fractions is an improper fraction, we convert it to a mixed number and add the whole numbers.

$$10\dfrac{29}{24}=10+1\dfrac{5}{24}=11\dfrac{5}{24}$$

6. Add $7\dfrac{3}{4}+8\dfrac{7}{10}$.

Example	Student Practice

7. Subtract $8\dfrac{5}{7} - 5\dfrac{5}{14}$.

The LCD of $\dfrac{5}{7}$ and $\dfrac{5}{14}$ is 14. Rewrite the fractions using the LCD.

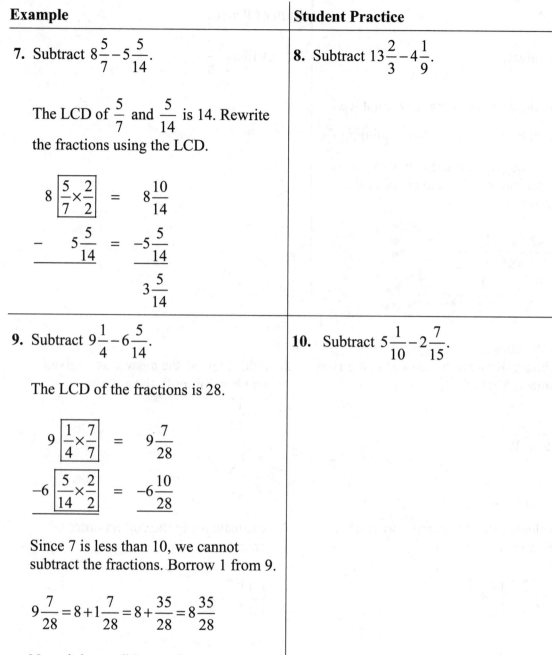

$$8\boxed{\dfrac{5}{7}\times\dfrac{2}{2}} \;=\; 8\dfrac{10}{14}$$

$$-\quad 5\dfrac{5}{14} \;=\; -5\dfrac{5}{14}$$

$$3\dfrac{5}{14}$$

8. Subtract $13\dfrac{2}{3} - 4\dfrac{1}{9}$.

---

9. Subtract $9\dfrac{1}{4} - 6\dfrac{5}{14}$.

The LCD of the fractions is 28.

$$9\boxed{\dfrac{1}{4}\times\dfrac{7}{7}} \;=\; 9\dfrac{7}{28}$$

$$-6\boxed{\dfrac{5}{14}\times\dfrac{2}{2}} \;=\; -6\dfrac{10}{28}$$

Since 7 is less than 10, we cannot subtract the fractions. Borrow 1 from 9.

$$9\dfrac{7}{28} = 8 + 1\dfrac{7}{28} = 8 + \dfrac{35}{28} = 8\dfrac{35}{28}$$

Now, it is possible to subtract.

$$9\dfrac{7}{28} \;=\; 8\dfrac{35}{28}$$

$$-6\dfrac{10}{28} \;=\; -6\dfrac{10}{28}$$

$$2\dfrac{25}{28}$$

10. Subtract $5\dfrac{1}{10} - 2\dfrac{7}{15}$.

Example	Student Practice
11. Evaluate $\dfrac{3}{4}-\dfrac{2}{3}\times\dfrac{1}{8}$.	**12.** Evaluate $\dfrac{7}{8}\div\dfrac{1}{4}+\dfrac{1}{6}$.

Use the order of operations. First we multiply. Then we subtract, building $\dfrac{3}{4}$ to an equivalent fraction with a denominator of 12, and simplify the answer.

$$\dfrac{3}{4}-\dfrac{2}{3}\times\dfrac{1}{8}=\dfrac{3}{4}-\dfrac{1}{12}$$

$$=\dfrac{9}{12}-\dfrac{1}{12}=\dfrac{8}{12}=\dfrac{2}{3}$$

Extra Practice

1. Subtract. Express the answer as a mixed number. Simplify the answer.

    $$7\dfrac{3}{5}-5\dfrac{7}{10}$$

2. Add. Express the answer as a mixed number. Simplify the answer.

    $$4\dfrac{11}{12}$$
    $$+\ 7\dfrac{3}{8}$$
    $$\overline{\phantom{+\ 7\dfrac{3}{8}}}$$

3. Evaluate using the correct order of operations.

    $$\dfrac{3}{2}\div\left(\dfrac{3}{4}-\dfrac{1}{6}\right)+\dfrac{3}{7}$$

4. Evaluate using the correct order of operations.

    $$\left(\dfrac{3}{8}\right)^{2}\div\dfrac{15}{8}$$

Concept Check

Explain how you would evaluate $\dfrac{4}{5}-\dfrac{1}{4}\times\dfrac{2}{3}$ using the correct order of operations.

Name: _____   Date: _____
Instructor: _____   Section: _____

Chapter 2 Fractions
2.9 Solving Applied Problems Involving Fractions

Vocabulary

diameter • problem solving • "do a similar, simpler problem" • estimation

1. One of the most important uses of _____ in mathematics is in the calculation of problems involving fractions.

2. A(n) _____ is a line segment that passes through the center of and intersects a circle twice. It can also mean the length of this segment.

Example	Student Practice
1. What is the inside diameter (distance across) of a concrete storm drain pipe that has an outside diameter of $4\frac{1}{8}$ feet and is $\frac{3}{8}$ foot thick?	**2.** A round picture frame has an outside diameter of $27\frac{2}{5}$ inches, and the frame itself is $3\frac{3}{5}$ inches in width. What is the diameter of the picture inside the frame?

Read the problem carefully and create a Mathematics Blueprint. Draw a picture.

The thickness of the pipe is used twice because it goes around the whole pipe.

Adding $\frac{3}{8} + \frac{3}{8} = \frac{6}{8}$ gives the total amount to subtract from the outer diameter.

Subtract $\frac{6}{8}$ from $4\frac{1}{8}$. Borrow 1 from 4.

$$4\frac{1}{8} - \frac{6}{8} = 3\frac{9}{8} - \frac{6}{8} = 3\frac{3}{8}$$

The inner diameter is $3\frac{3}{8}$ feet. Work backward to check the answer.

Example	Student Practice
3. A slipcover manufacturer uses $8\frac{1}{4}$ yards to make a slipcover for a chair. To make a slipcover for a recliner, $1\frac{1}{2}$ times that amount is needed. How many yards of fabric are needed to fill an order of 8 chair slipcovers and 10 recliner slipcovers?	**4.** A bakery uses $2\frac{3}{4}$ cups of white flour for each loaf of regular white bread that is baked. A French baguette requires $2\frac{1}{2}$ times as much flour. How much white flour does the bakery need to bake 20 loaves of regular white bread and 15 French baguettes?

Read the problem carefully and create a Mathematics Blueprint.

Find the amount of fabric needed for the chair slipcovers.

$$8 \times 8\frac{1}{4} = 8 \times \frac{33}{4} = 66 \text{ yards}$$

Find the amount of fabric needed for the recliner slipcovers.

For 1 recliner slipcover, the amount will be as follows.

$$1\frac{1}{2} \times 8\frac{1}{4} = \frac{3}{2} \times \frac{33}{4} = \frac{99}{8} \text{ yards}$$

For 10 slip covers, multiply by 10.

$$10 \times \frac{99}{8} = \frac{495}{4} = 123\frac{3}{4} \text{ yards}$$

Now add to find the total amount.

$$66 \text{ yards} + 123\frac{3}{4} \text{ yards} = 189\frac{3}{4} \text{ yards}$$

Thus, $189\frac{3}{4}$ yards of fabric are needed.

Check by estimation.

Example	Student Practice
5. A fishing boat traveled $69\frac{3}{8}$ nautical miles in $3\frac{3}{4}$ hours. How many knots (nautical miles per hour) did the fishing boat average?	**6.** A speed boat traveled $88\frac{3}{8}$ nautical miles in $1\frac{3}{4}$ hour. How many knots (nautical miles per hour) did the speed boat average?

5. A fishing boat traveled $69\frac{3}{8}$ nautical miles in $3\frac{3}{4}$ hours. How many knots (nautical miles per hour) did the fishing boat average?

Think of a simpler problem. If a boat traveled 70 nautical miles in 4 hours, how many knots did it average? We would divide distance by time.

$70 \div 4 =$ average speed

Do the same in the original problem.

$69\frac{3}{8} \div 3\frac{3}{4} =$ average speed

Read the problem carefully and create a Mathematics Blueprint.

Divide distance by time to get speed in knots.

$$69\frac{3}{8} \div 3\frac{3}{4} = \frac{555}{8} \div \frac{15}{4} = \frac{\overset{37}{\cancel{555}}}{\underset{2}{\cancel{8}}} \cdot \frac{\overset{1}{\cancel{4}}}{\underset{1}{\cancel{15}}} = \frac{37}{2}$$

Write the answer as a mixed number.

$$\frac{37}{2} = 18\frac{1}{2}$$

The speed of the boat was $18\frac{1}{2}$ knots.

Check by estimating.

Use $70 \div 4 = 17\frac{1}{2}$ knots.

The estimate is close to the calculated value. The answer is reasonable.

Extra Practice

You may benefit from using the Mathematics Blueprint for Problem Solving when solving the following exercises.

1. The local pizza parlor used $27\frac{1}{4}$ pounds of pizza dough on Monday, $23\frac{2}{3}$ pounds on Tuesday, and $25\frac{1}{2}$ pounds on Wednesday. How many pounds of pizza dough were used over the three days?

2. Renaldo bought $6\frac{3}{8}$ pounds of cheese for a party. The guests ate $4\frac{1}{2}$ pounds of cheese. How much did Renaldo have left over?

3. Lexi wonders if a tree in her front yard will grow to be 8 feet tall. When the tree was planted, it was $2\frac{1}{6}$ feet tall. It grew $1\frac{1}{4}$ feet the first year after being planted. How many more feet does it need to reach a height of 8 feet?

4. A triangle has three sides that measure $6\frac{1}{4}$ feet, $2\frac{5}{6}$ feet, and $7\frac{7}{12}$ feet. What is the perimeter (the total distance around) of the triangle?

Concept Check

A trail to a peak on Mount Washington is $3\frac{3}{5}$ miles long. Caleb started hiking on the trail and stopped after walking $1\frac{7}{8}$ miles to take a break. Explain how you would find how far he still has to go to get to the peak.

MATH COACH

Mastering the skills you need to do well on the test.

Watch the **MATH COACH** videos in MyMathLab® or on You Tube™ while you work the problems below. These helpful hints will help you avoid making common errors on test problems.

Simplifying Fractions—Problem 5 $\dfrac{225}{50}$

Helpful Hint: After you complete each step when simplifying a fraction, stop to see if your result can be simplified any further.

Did you divide the numerator and denominator by 25?

Yes _____ No _____

If you answered Yes, then you chose the best factor to divide by. But, if you answered No, then check your division.

If you answered Problem 5 incorrectly, rework it now using these suggestions.

Finding the Least Common Denominator—Problem 16 $\dfrac{3}{16}$ and $\dfrac{1}{24}$

Helpful Hint: Write each denominator as the product of prime factors. When forming the LCD, use each factor the greatest number of times that it appears in any one denominator.

Did you factor 16 into $2 \times 2 \times 2 \times 2$? Yes _____ No _____
If you answered No, stop and perform that step.

Did you factor 24 into $2 \times 2 \times 2 \times 3$? Yes _____ No _____
If you answered No, stop and perform that step.

If you answered Problem 16 incorrectly, rework it now using these suggestions. Remember to use the factor 2 four times.

Evaluate Using the Correct Order of Operations—Problem 25 $\left(\dfrac{1}{2}+\dfrac{1}{3}\right)\times\dfrac{7}{5}$

Helpful Hint: Write out each step separately. Be sure to combine operations in the parentheses first.

Did you first combine $\dfrac{1}{2}+\dfrac{1}{3}$? Yes _____ No _____

If you answered problem 25 incorrectly, rework it now using these suggestions.

Did you rewrite the fractions as $\dfrac{3}{6}+\dfrac{2}{6}$ before adding them together? Yes _____ No _____

If you answered No to either of these questions, stop and perform these operations first.

Solve Applied Problems Involving Fractions—Problem 27

A butcher has $18\dfrac{2}{3}$ pounds of steak that he wishes to place into packages that average $2\dfrac{1}{3}$ pounds each. How many packages can he make?

Helpful Hint: Remember to use the problem-solving steps and the Mathematics Blueprint for Problem Solving to help you set up the problem correctly. When dividing, remember to invert the second fraction and then multiply.

Did you translate the problem correctly to form the calculation $18\dfrac{2}{3}\div 2\dfrac{1}{3}$? Yes _____ No _____
If you answered No, stop and perform that calculation.

If you answered Problem 27 incorrectly, rework it now using these suggestions.

Did you remember to change the mixed numbers to fractions and write $\dfrac{56}{3}\div\dfrac{7}{3}$? Yes _____ No _____
If you answered No, stop and perform that calculation.

72

Chapter 3 Decimals
3.1 Using Decimal Notation

Vocabulary
decimal fractions   •   tropical year   •   whole-number part   •   decimal
part   •   decimal point

1. 1 _____ = 365.24122 days rounded to the nearest hundred-thousandth.

2. Fractions with 10, 100, 1000, and so on, in the denominator are called _____.

3. The dot in 1.683 is called the _____.

4. The digits 6, 8, and 3 in 1.683 are part of the _____ of the number.

Example	Student Practice
1. Write the word name for each decimal.	**2.** Write the word name for 435.607.
(a) 0.79	
Use a place value chart if necessary to determine the word name.	
0.79 = seventy-nine hundredths	
(b) 0.5308	
0.5308 = five thousand three hundred eight ten-thousandths	
(c) 1.6	
Note that the word "and" is used to indicate the decimal point.	
1.6 = one and six tenths	
(d) 23.765	
23.765 = twenty-three and seven hundred sixty-five thousandths	

Example	Student Practice
3. Write a word name for the amount on a check made out for $672.89 It is common to write the cents as a fraction over 100. Six hundred seventy-two and $\dfrac{89}{100}$ dollars	**4.** Write a word name for the amount on a check made out for $1643.99
5. Write as a decimal. **(a)** $\dfrac{8}{10}$ Since there is one zero in the denominator, we need one decimal place in the decimal number. $\dfrac{8}{10} = 0.8$ **(b)** $\dfrac{74}{100}$ $\dfrac{74}{100} = 0.74$ **(c)** $1\dfrac{3}{10}$ $1\dfrac{3}{10} = 1.3$ **(d)** $2\dfrac{56}{1000}$ $2\dfrac{56}{1000} = 2.056$	**6.** Write $12\dfrac{61}{1000}$ as a decimal.

Example	Student Practice
7. Write in fraction notation.	**8.** Write 165.033 in fraction notation.

(a) 0.51

$$0.51 = \frac{51}{100}$$

(b) 18.1

$$18.1 = 18\frac{1}{10}$$

(c) 0.7611

$$0.7611 = \frac{7611}{10,000}$$

(d) 1.363

$$1.363 = 1\frac{363}{1000}$$

9. Write in fraction notation. Reduce whenever possible.	**10.** Write 45.096 in fraction notation.

(a) 2.6

$$2.6 = 2\frac{6}{10} = 2\frac{3}{5}$$

(b) 361.007

$$361.007 = 361\frac{7}{1000}$$
(cannot be reduced)

(c) 0.525

$$0.525 = \frac{525}{1000} = \frac{105}{200} = \frac{21}{40}$$

Example	Student Practice
11. A chemist found that the concentration of lead in a water sample was 5 parts per million. What fraction would represent the concentration of lead?	**12.** A chemist found that the concentration of arsenic in a water sample was 16 parts per ten-million. What fraction would represent the concentration of arsenic?

As a fraction, this is $\dfrac{5}{1,000,000}$. We can reduce this by dividing the numerator and denominator by 5. Thus,

$$\frac{5}{1,000,000} = \frac{1}{200,000}.$$

The concentration of lead in the water sample is $\dfrac{1}{200,000}$.

Extra Practice

1. Write the word name for 24.0049.

2. Write $1238.09 as you would on a check.

3. Write $18\dfrac{45}{1000}$ as a decimal.

4. Write 0.256 in fractional notation. Reduce if possible.

Concept Check

Explain how you know how many zeros to put in your answer if you need to write $\dfrac{953}{100,000}$ as a decimal.

Name: _____   Date: _____
Instructor: _____   Section: _____

Chapter 3 Decimals
3.2 Comparing, Ordering, and Rounding Decimals

Vocabulary
number line   •   inequality symbols   •   round   •   order

1. The symbols "<" and ">" are called _____.

2. When asked to _____ a set of decimals, you compare the values and list them from smallest to largest, or largest to smallest.

3. To make a number like $234.123 useful, we _____ to the nearest cent.

4. To study the order of all numbers, we can place the numbers on a(n) _____.

Example	**Student Practice**
1. Write an inequality statement with 0.167 and 0.166.	**2.** Write an inequality statement with 8.217 and 8.219.
The numbers in the tenths place are the same. They are both 1.	
0.1̲66 0.1̲67	
The numbers in the hundredths place are the same. They are both 6.	
0.16̲6 0.16̲7	
The numbers in the thousandths place differ.	
0.16̲6 0.16̲7	
Since $7 > 6$, $0.167 > 0.166$.	

Example	Student Practice
3. Fill in the blank with one of the symbols $<, =, >$.	**4.** Fill in the blank with one of the symbols $<, =, >$.

3. (continued)

0.77 _____ 0.777

We begin by adding a zero to the first decimal.

0.77<u>0</u>   0.77<u>7</u>

We see that the tenths and hundredths digits are equal, but the thousandths digit differ. Since $0 < 7$, we have $0.770 < 0.777$.

4. (continued)

2.0031 _____ 2.003

---

5. Place the following 5 decimal numbers in order from smallest to largest.

1.834, 1.83, 1.381, 1.38, 1.8

First we add zeroes to make the comparison easier.

1.834, 1.830, 1.381, 1.380, 1.800

Now we rearrange with smallest first.

1.380, 1.381, 1.800, 1.830, 1.834

6. Place the following 5 decimal numbers in order from smallest to largest.

3.21, 3.312, 3.213, 3.31, 3.2

---

7. Round 156.37 to the nearest tenth.

We locate the tenths place.

156.<u>3</u>7

Note that 7, the next place to the right, is greater than 5. We round up to 156.4 and drop the digits to the right.

The answer is 156.4.

8. Round 6.76 to the nearest tenth.

Example	Student Practice
9. Round to the nearest thousandth. **(a)** 0.06358 Note that the digit to the right of the thousandths place is 5. We round up to 0.064 and drop all the digits to the right. **(b)** 128.37448 Note that the digit to the right of the thousandths place is less than 5. We round to 128.374 and drop all the digits to the right.	**10.** Round 45.236832 to the nearest thousandth.
11. Round to the nearest hundredth. Fred and Linda used 203.9964 kilowatt-hours of electricity in their house in May. Since the digit to the right of the hundredths place is greater than 5, we round up. This affects the next two positions. The result is 204.00 kilowatt-hours. Notice that we have the two zeroes to the right of the decimal place to show we have rounded to the nearest hundredth.	**12.** Round to the nearest hundredth. Jarrett and Mandy used 759.9984 kilowatt-hours of electricity in their house in August.
13. To complete her income tax return, Marge needs to round these figures to the nearest dollar. Medical bills $779.86 Taxes $563.49 Retirement contributions $674.38 Contributions to charity $534.77 $779.86 rounds to $780 $563.49 rounds to $563 $674.38 rounds to $674 $534.77 rounds to $535	**14.** Round these figures to the nearest dollar. Monthly paycheck $2953.61 Sound system $431.23 Electric bill $152.50

Extra Practice

1. Fill in the blank with one of the symbols $<, =, >$. 23.0097 _____ 23.009

2. Round 13.57430 to the nearest thousandth.

3. Place the following 5 decimal numbers in order from smallest to largest.

   12.99, 12.987, 12.887, 12.909, 12.88889

4. Round these figures to the nearest dollar.
   Monthly paycheck $2485.45
   Video game system $389.89
   Phone bill $104.53

Concept Check

Explain how you would round 34.958365 to the nearest ten-thousandth.

Chapter 3 Decimals
3.3 Adding and Subtracting Decimals

Vocabulary

zeros   •   decimals   •   place value   •   decimal point

1. When adding _____, write the numbers to be added vertically and line up the decimal points.

2. When subtracting decimals vertically, subtract all digits with the same _____, starting with the right column and moving left. Borrow when necessary.

3. When adding or subtracting decimals vertically, place the _____ of the sum or difference in line with the ones in the addends or subtrahends.

4. Additional _____ may be placed to the right of the decimal point if not all of the numbers in a decimal calculation have the same number of decimal places.

Example	**Student Practice**
1. Add. $5.3 + 26.182 + 0.0007 + 624$ Write the numbers to be added vertically and line up the decimal points. Note that the decimal point is understood to be to the right of the digit 4. Place extra zeros to the right of the decimal point. Then, add as normal, carrying when necessary. 5.3000 $\overset{1}{2}6.1820$ 0.0007 $\underline{+624.0000}$ 655.4827	**2.** Add. $345 + 46.234 + 12.2 + 0.004$

Example	Student Practice
3. Barbara checked her odometer before the summer began. It read 49,645.8 miles. She traveled 3852.6 miles that summer in her car. What was the odometer reading at the end of the summer?	**4.** A car odometer read 761,683.2 miles before a trip of 1726.9 miles. What was the final odometer reading?

$$\begin{array}{r} \overset{1\ 1\qquad 1}{49,645.8} \\ +\ \ \ 3,852.6 \\ \hline 53,498.4 \end{array}$$

The odometer read 53,498.4 miles.

Example	Student Practice
5. During his first semester at Tarrant County Community College, Kelvey deposited checks into his checking account in the amounts of $98.64, $157.32, $204.81, $36.07, and $229.89. What was the sum of his five deposits?	**6.** During the semester, Marty deposited the following checks into his account: $32.94, $184.96, $218.28, $93.01, $393.52. What was the sum of his five deposits?

$$\begin{array}{r} \overset{2\ 3\ 2\ 2}{\$\ 98.64} \\ 157.32 \\ 204.81 \\ 36.07 \\ +\ \ 229.89 \\ \hline \$726.73 \end{array}$$

Example	Student Practice
7. Subtract.	**8.** Subtract.

$$\begin{array}{r} 1076.320 \\ -\ \ 983.518 \end{array}$$

$$\begin{array}{r} 7234.613 \\ -\ \ 955.967 \end{array}$$

$$\begin{array}{r} \cancel{1}\ \overset{9}{\cancel{\overset{16}{\cancel{0}}}}\ \overset{17}{\cancel{7}}\ \overset{5}{\cancel{6}}.\ \overset{13}{\cancel{3}}\ \overset{1}{\cancel{2}}\ \overset{10}{\cancel{0}} \\ -\qquad 9\ \ 8\ \ 3.\ \ 5\ \ 1\ \ 8 \\ \hline 9\ \ 2.\ \ 8\ \ 0\ \ 2 \end{array}$$

Example	Student Practice
9. Subtract.	**10.** Subtract $10 - 9.395$

(a) $12 - 8.362$

$$
\begin{array}{r}
\overset{11}{\cancel{1}}\ \overset{9}{\cancel{2}}.\ \overset{9}{\cancel{0}}\ \overset{9}{\cancel{0}}\ \overset{9}{\cancel{0}} \\
-\quad\ \ 8.\ 3\ 6\ 2 \\
\hline
3.\ 6\ 3\ 8
\end{array}
$$

(b) $156.381 - 99.82$

$$
\begin{array}{r}
\overset{14}{\cancel{1}}\ \overset{15}{\cancel{5}}\ \cancel{6}.\ \overset{13}{\cancel{3}}\ 8\ 1 \\
-\quad\ 9\ 9.\ 8\ 2\ 0 \\
\hline
5\ 6.\ 5\ 6\ 1
\end{array}
$$

---

11. On Tuesday, Don Ling filled the gas tank in his car. The odometer read 56,098.5. He drove for four days. The next time he filled the tank, the odometer read 56,420.2. How many miles had he driven?

Subtract 56,098.5 from 56,420.2 to determine how many miles Don drove.

$$
\begin{array}{r}
5\ 6,\ \overset{3}{\cancel{4}}\ \overset{11}{\cancel{2}}\ \overset{9}{\cancel{0}}.\ \overset{12}{\cancel{2}} \\
-\ 5\ 6,\ 0\ 9\ 8.\ 5 \\
\hline
3\ 2\ 1.\ 7
\end{array}
$$

He had driven 321.7 miles.

12. Johnny went on vacation. Before the trip, his odometer read 63,779.3 miles. After his vacation, his odometer read 65,146.2 miles. How many miles did he drive on vacation?

Example	Student Practice
13. Find the value of x if $x + 3.9 = 14.6$.	**14.** Find the value of x if $x - 8.2 = 4.9$.

Recall that the letter x is a variable. It represents a number that is added to 3.9 to obtain 14.6. We can find the number x if we calculate $14.6 - 3.9$.

$$
\begin{array}{r}
\overset{3}{\cancel{1}} \overset{16}{\cancel{4}}. \overset{}{\cancel{6}} \\
-\quad 3.\ 9 \\
\hline
1\ \ 0.\ 7
\end{array}
$$

Thus $x = 10.7$. To check, replace x with 10.7 and simplify.

Extra Practice

1. Subtract $42.367 - 23.01$.

2. Add $105.36 + 12.7 + 89.49 + 41$.

3. At the local restaurant, Alicia bought a soda for $0.99, a medium French fry for $1.29, and a chicken sandwich for $2.89. How much did Alicia pay for her meal?

4. A car odometer read 58,762.3 miles before a trip of 659.3 miles. What was the final odometer reading?

Concept Check
Explain how you perform the correct borrowing and correct use of the decimal point if you subtract $567.45 - 345.9872$.

Chapter 3 Decimals
3.4 Multiplying Decimals

Vocabulary

decimal places   •   product   •   factor   •   power of 10

1. The number of decimal places in the _____ of two decimals is the sum of the number of decimal places in the two factors.

2. When multiplying a decimal by a whole number, you need to remember that a whole number has no _____ .

3. You use multiplying by a _____ when you convert a larger unit of measure to a smaller unit of measure in the metric system.

4. When multiplying decimals, it is usually easier to place the _____ with the smaller number of nonzero digits underneath the other.

Example	**Student Practice**
1. Multiply 0.07×0.3.	**2.** Multiply 0.4×0.08.
Multiply the numbers just as you would multiply whole numbers $7 \times 3 = 21$.	
Find the sum of the number of decimal places in the two factors, 0.07 has 2 decimal places and 0.3 has 1 decimal place. The sum is $1 + 2 = 3$ decimal places.	
Place the decimal point in the product so that the product has the same number of decimal places as the sum. You may need to write zeros to the left of the product.	
$\begin{array}{r} 0.07 \\ \times\ 0.3 \\ \hline 0.021 \end{array}$	

Example	Student Practice
3. Multiply.	**4.** Multiply 34.72×0.826.

(a) 0.38×0.26

$$\begin{array}{r} 0.38 \quad \text{2 decimal places} \\ \times \quad 0.26 \quad \text{2 decimal places} \\ \hline 228 \qquad\quad \\ 76 \qquad\quad \\ \hline 0.0988 \quad \text{4 decimal places} \\ (2+2=4) \end{array}$$

(b) 12.64×0.572

$$\begin{array}{r} 12.64 \quad \text{2 decimal places} \\ \times \; 0.572 \quad \text{3 decimal places} \\ \hline 2528 \qquad\quad \\ 8848 \qquad\quad \\ 6\,320 \qquad\quad \\ \hline 7.23008 \quad \text{5 decimal places} \\ (2+3=5) \end{array}$$

---

5. Multiply 5.261×45.

When multiplying a decimal by a whole number, you need to remember that a whole number has no decimal places.

$$\begin{array}{r} 5.261 \quad \text{3 decimal places} \\ \times \qquad 45 \quad \text{0 decimal places} \\ \hline 26\,305 \qquad\quad \\ 210\,44 \qquad\quad \\ \hline 236.745 \quad \text{3 decimal places } (3+0=3) \end{array}$$

6. Multiply 0.3577×43.

Example	Student Practice

7. Uncle Roger's rectangular front lawn measures 50.6 yards wide and 71.4 yards long. What is the area of the lawn in square yards?
 	
We multiply the length by the width.
 	
    71.4   1 decimal place
$\times$ 50.6   1 decimal place
  42 84
3570 0
3612.84 2 decimal places
 	
The area of the lawn is 3612.84 square yards.
	8. Walt's rectangular driveway measures 9.1 yards by 18.4 yards. What is the area of the driveway in square yards?

9. Multiply. **(a)** 2.671×10 Note that there is one zero in 10, so move the decimal point one place to the right. $2.671 \times 10 = 26.71$ **(b)** 37.85×100 $37.85 \times 100 = 3785$	**10.** Multiply 0.3042×100.
11. Multiply. $0.076 \times 10,000$ Note that there are four zeros in 10,000, so move the decimal point four places to the right. One extra zero will be needed. $0.076 \times 10,000 = 760$	**12.** Multiply. $7.0846 \times 10,000,000$

Example	Student Practice
13. Multiply 3.68×10^3. Note there is an exponent of 3, so the decimal point will move three places to the right. $3.68 \times 10^3 = 3680$	**14.** Multiply 5.981×10^4.
15. Change 2.96 kilometers to meters. Since we are going from a larger unit of measure to a smaller one, we multiply. There are 1000 meters in 1 kilometer. Multiply 2.96 by 1000. $2.96 \times 1000 = 2960$ 2.96 kilometers is equal to 2960 meters.	**16.** Change 92.04 kilometers to meters.

Extra Practice

1. Multiply.

$$\begin{array}{r} 8.239 \\ \times \quad 4.9 \\ \hline \end{array}$$

2. Multiply.

$$\begin{array}{r} 7890.456 \\ \times \quad 10,000 \\ \hline \end{array}$$

3. Multiply.

$$\begin{array}{r} 5787.256 \\ \times \quad 125 \\ \hline \end{array}$$

4. How many inches are in 50 meters? Remember, one meter is about 39.37 inches.

Concept Check

Explain how you know where to put the decimal point in the answer when you multiply 3.45×9.236.

Chapter 3 Decimals
3.5 Dividing Decimals

Vocabulary

divisor • dividend • quotient • round

1. The first step when dividing by a decimal is to make the _____ a whole number by moving the decimal point to the right.

2. The number 26.8 in $4\overline{)26.8}$ (quotient 6.7) is called the _____.

3. When dividing decimals, to _____ to the nearest thousandth, we carry out the division to the ten-thousandths place.

4. The number 6.7 in $4\overline{)26.8}$ (quotient 6.7) is called the _____.

Example	Student Practice
1. Divide.	**2.** Divide.
$9\overline{)0.3204}$	$6\overline{)5.082}$
Place the decimal above the decimal point in the dividend. Divide as with whole numbers.	

$$\begin{array}{r} 0.0356 \\ 9\overline{)0.3204} \\ \underline{27} \\ 50 \\ \underline{45} \\ 54 \\ \underline{54} \\ 0 \end{array}$$

Example	Student Practice
3. Divide and round the quotient to the nearest thousandth. $12.67 \div 39$	**4.** Divide and round the quotient to the nearest thousandth. $2.2 \div 7$

$$\begin{array}{r} 0.3248 \\ 39{\overline{\smash{\big)}\,12.6700}} \end{array}$$

$$\underline{11\,7}$$

$$97$$

$$\underline{78}$$

$$190$$

$$\underline{156}$$

$$340$$

$$\underline{312}$$

$$28$$

Now we round 0.3248 to 0.325. The answer is rounded to the nearest thousandth.

5. Divide $0.08{\overline{\smash{\big)}\,1.632}}$.	**6.** Divide $3.567 \div 0.041$.

Move each decimal point two places to the right. Mark the new position by a caret $(\wedge)$ and place the decimal point of the answer directly above the caret.

$$0.08_{\wedge}{\overline{\smash{\big)}\,1.63_{\wedge}2}}$$

Then, perform the division.

$$\begin{array}{r} 20.4 \\ 0.08_{\wedge}{\overline{\smash{\big)}\,1.63_{\wedge}2}} \end{array}$$

$$\underline{1\,6}$$

$$3\,2$$

$$\underline{3\,2}$$

$$0$$

Example	Student Practice
7. Divide $0.0032\overline{)7.68}$.	**8.** Divide $2.1\overline{)0.0525}$.

7. (continued)

$$
\begin{array}{r}
2400. \\
0.0032_\wedge\overline{)7.6800_\wedge} \\
\underline{6\ 4} \\
1\ 28 \\
\underline{1\ 28} \\
000
\end{array}
$$

Note that two extra zeroes are needed in the dividend as we move the decimal point four places to the right.

---

9. Find $2.9\overline{)431.2}$ rounded to the nearest tenth.

Calculate to the hundredths place and round the answer to the nearest tenth.

$$
\begin{array}{r}
14\ 8\ .68 \\
2.9_\wedge\overline{)431.2_\wedge00} \\
\underline{29} \\
141 \\
\underline{116} \\
25\ 2 \\
\underline{23\ 2} \\
2\ 0\ 0 \\
\underline{17\ 4} \\
2\ 60 \\
\underline{2\ 32} \\
28
\end{array}
$$

The answer rounded to the nearest tenth is 148.7

---

10. Find $5.42\overline{)0.81}$ rounded to the nearest thousandth.

Example	Student Practice
11. Find the value of n if $0.8 \times n = 2.68$.	**12.** Find the value of n if $0.13 \times n = 0.351$.

If we divide 2.68 by 0.8, we will find the value of n.

$$
\begin{array}{r}
3\,.35 \\
0.8_\wedge \overline{)2.6_\wedge 80} \\
\underline{24} \\
2\ 8 \\
\underline{2\ 4} \\
40 \\
\underline{40} \\
0
\end{array}
$$

Thus, the value of n is 3.35. We can check by substituting the value of $n = 3.35$ into the equation to see if it makes the statement true.

$$0.8 \times n = 2.68$$

$$0.8 \times 3.35 \overset{?}{=} 2.68$$

$$2.68 = 2.68$$

Extra Practice

1. Divide $5\overline{)58.5}$.

2. Divide $36.935 \div 0.83$.

3. Divide $1.32\overline{)234.5}$ and round your answer to the nearest hundredth.

4. Find the value of n. $1.3 \times n = 93.6$

Concept Check

Explain how you would know where to place the decimal point in the answer if you divide $0.173 \div 0.578$.

Chapter 3 Decimals
3.6 Converting Fractions to Decimals and the Order of Operations

Vocabulary
terminating decimals   •   repeating decimals   •   decimal form   •   equivalent

1. To convert a fraction to a(n) _____ decimal, divide the denominator into the numerator.

2. Decimals that have a digit or a group of digits that repeats are called _____.

3. If we are required to place a fraction and a decimal in order, it is usually easiest to convert the fraction to _____ and then compare.

4. When converting a fraction to a decimal and the division operation eventually yields a remainder of zero, that decimal is called a _____.

Example	**Student Practice**
1. Write as an equivalent decimal. $\dfrac{3}{8}$	**2.** Write as an equivalent decimal. $\dfrac{17}{40}$

Divide the denominator into the numerator until the remainder becomes zero.

$$\begin{array}{r} 0.375 \\ 8\overline{)3.000} \\ \underline{2\ 4} \\ 60 \\ \underline{56} \\ 40 \\ \underline{40} \\ 0 \end{array}$$

Therefore, $\dfrac{3}{8} = 0.375$.

Example	Student Practice

3. Write as an equivalent decimal. $\dfrac{13}{22}$

$$\begin{array}{r} 0.59090 \\ 22\overline{)13.00000} \\ 11\ 0 \\ \hline 2\ 00 \\ 1\ 98 \\ \hline 200 \\ 198 \\ \hline 20 \end{array}$$

Thus, $\dfrac{13}{22} = 0.5909090\ldots = 0.5\overline{90}$.

4. Write $\dfrac{31}{30}$ as an equivalent decimal.

---

5. Write as an equivalent decimal. $3\dfrac{7}{15}$

Recall that $3\dfrac{7}{15}$ means $3 + \dfrac{7}{15}$.

$$\begin{array}{r} 0.466 \\ 15\overline{)7.000} \\ 60 \\ \hline 100 \\ 90 \\ \hline 100 \\ 90 \\ \hline 10 \end{array}$$

Thus, $\dfrac{7}{15} = 0.4\overline{6}$ and $3\dfrac{7}{15} = 3.4\overline{6}$.

6. Write $5\dfrac{7}{11}$ as an equivalent decimal.

Example	Student Practice
7. Express $\dfrac{5}{7}$ as a decimal rounded to the nearest thousandth.	**8.** Express $\dfrac{29}{28}$ as a decimal rounded to the nearest thousandth.

$$\begin{array}{r} 0.7142 \\ 7\overline{)5.0000} \\ \underline{4\;9} \\ 10 \\ \underline{7} \\ 30 \\ \underline{28} \\ 20 \\ \underline{14} \\ 6 \end{array}$$

We round 0.7142 to 0.714.

9. Fill in the blank with one of the symbols $<$, $=$, or $>$. $\dfrac{7}{16}$ _____ 0.43	**10.** Fill in the blank with one of the symbols $<$, $=$, or $>$. $\dfrac{7}{8}$ _____ 0.876

Write the fraction as an equivalent decimal.

$$\begin{array}{r} 0.4375 \\ 16\overline{)7.0000} \\ \underline{6\;4} \\ 60 \\ \underline{48} \\ 120 \\ \underline{112} \\ 80 \\ \underline{80} \\ 0 \end{array}$$

Since $0.43\boxed{7}5 > 0.43\boxed{0}0$, $\dfrac{7}{16} > 0.43$.

Example	Student Practice
11. Evaluate. $(8-0.12) \div 2^3 + 5.68 \times 0.1$	**12.** Evaluate. $6.3 \div (1+0.05) + (7.3-6.8)^2$

First do subtraction inside the parentheses. Then simplify the expression with an exponent. Next from left to right do division and multiplication. Finish by adding the final two numbers.

$$(8-0.12) \div 2^3 + 5.68 \times 0.1$$
$$= 7.88 \div 2^3 + 5.68 \times 0.1$$
$$= 7.88 \div 8 + 5.68 \times 0.1$$
$$= 0.985 + 0.568$$
$$= 1.553$$

Extra Practice

1. Write $4\frac{39}{40}$ as an equivalent decimal or a decimal approximation. Round your answer to the nearest thousandth if needed.

2. Write $\frac{11}{7}$ as an equivalent decimal or a decimal approximation. Round your answer to the nearest thousandth if needed.

3. Write as an equivalent decimal. If a repeating decimal is obtained, use notation such as $0.\overline{7}$, $0.1\overline{6}$, or $0.2\overline{45}$. $\frac{4}{11}$

4. Evaluate.
$$(2.3)^3 + (14-7.62) \div (0.8-0.6)$$

Concept Check
Explain how you would perform the operations in the calculation
$45.78 - (3.42-2.09)^2 \times 0.4$.

Chapter 3 Decimals
3.7 Estimating and Solving Applied Problems Involving Decimals

Vocabulary

estimate   •   reasonable   •   solve and state the answer   •   check

1. The final step of analyzing applied problem situations is to _____ the answer.

2. To avoid making an error in solving applied problems, it is wise to make a(n) _____ .

3. When we encounter real-life applied problems, it is important to know an answer is _____ .

4. The second step of analyzing applied problem situations is to _____ .

Example	**Student Practice**
1. Estimate.	**2.** Estimate.
(a) $184,987.09 + 676,393.95$	**(a)** $0.0782 - 0.0477$
Round to one nonzero digit to estimate.	
$184,987.09 + 676,393.95$ $\approx 200,000 + 700,000 = 900,000$	
(b) $138.85 \div 5.887$ $\approx 100 \div 6$	**(b)** 8924×0.023
$\begin{array}{r} 16 \\ 6\overline{)100} \\ \underline{6} \\ 40 \\ \underline{36} \\ 4 \end{array}$	
Thus, $138.85 \div 5.887 \approx 16\frac{4}{6} \approx 17$.	

Example	Student Practice
3. A laborer is paid $10.38 per hour for a 40-hour week and 1.5 times that wage for any hours worked beyond the standard 40. If he works 47 hours in a week, what will he earn?	**4.** A laborer is paid $11.78 per hour for a 40-hour week and 1.5 times that wage for any hours worked beyond the standard 40. If she works 49 hours in a week, what will she earn?

3. (continued)

Read the problem carefully and create a Mathematics Blueprint.

Compute his regular and overtime pay and add the results.

Regular pay + Overtime pay = Total pay

Compute his regular pay.

$$\begin{array}{r} 10.38 \\ \times \quad 40 \\ \hline 415.20 \end{array}$$

Compute his overtime pay rate.

$$\begin{array}{r} 10.38 \\ \times \quad 1.5 \\ \hline 5\ 190 \\ \underline{10\ 38} \\ 15.570 \end{array}$$

Calculate how much he earned doing 7 hours of overtime work.

$15.57 \times 7 = 108.99$

Add the two amounts.

$\$415.20 + \$108.99 = \$524.19$

Check by estimating his regular and overtime pay and adding the results.

Example	Student Practice
5. A chemist is testing 36.85 liters of cleaning fluid. She wishes to pour it into several smaller containers that each hold 0.67 liter of fluid.	**6.** A butcher divides 21.45 pounds of prime steak into small, equal-sized packages. Each package contains 1.65 pounds of prime steak.

(a) How many containers will the chemist need?

She has 36.85 liters of cleaning fluid and she wants to put it into equal-sized containers each holding 0.67, liter. Divide the total number of liters 36.85 by the amount each container holds, 0.67.

$$
\begin{array}{r}
55. \\
0.67_\wedge \overline{)36.85_\wedge} \\
\underline{33\ 5} \\
3\ 35 \\
\underline{3\ 35}
\end{array}
$$

The chemist will need 55 containers.

(b) How much does the cleaning fluid in each container cost? (Round your answer to the nearest cent.)

Each container holds 0.67 liter. If one liter costs \$3.50, then to find the cost of one container, multiply $0.67 \times \$3.50$.

$$
\begin{array}{r}
3.50 \\
\times\ 0.67 \\
\hline
2450 \\
\underline{2100\ \ } \\
2.3450
\end{array}
$$

Rounding to the nearest cent, each container would cost \$2.35.

(a) How many packages of steak will he have?

(b) Prime steak sells for \$8.51 per pound. How much will each package of prime steak cost? (Round your answer to the nearest cent.)

Extra Practice

1. First round each number to one nonzero digit. Then perform the calculation using the rounded numbers to obtain an estimate. $636,700 \div 77,411.88$

2. Andrea opened a savings account with $300. During the month, she went shopping and spent $123.49 on clothing and $89.23 on groceries. At this point, Andrea really wanted to purchase a new CD player for $99.99. Does she have enough money?

3. Lina measures her backyard and finds that it is 75.4 feet long and 45.2 feet wide. What is the area of her backyard?

4. A water tower contains 5750 gallons of water. It empties at a rate of 27.8 gallons per minute. How long will it take to completely empty the water tower?

Concept Check

Explain how you would solve the following problem. The Classic Chocolate Company has 24.7 pounds of chocolate. They wish to place it in individual boxes that each hold 1.3 pounds of chocolate. How many boxes will they need?

MATH COACH

Mastering the skills you need to do well on the test.

Watch the **MATH COACH** videos in MyMathLab® or on YouTube™ while you work the problems below. These helpful hints will help you avoid making common errors on test problems.

Adding Decimals—Problem 9 $17 + 2.1 + 16.8 + 0.04 + 1.59$

Helpful Hint: Place additional zeros at the end of each number so that each decimal has two digits to the right of each decimal point. Include the decimal point before adding zeros to any whole numbers.

Did you change the problem to
$17.00 + 2.10 + 16.80 + 0.04 + 1.59$?
Yes _____ No _____

If you answered No, stop and make those changes now.

If you answered No, stop and write the numbers carefully in a column with the decimal points exactly aligned.

Did you line up the decimal points carefully when you wrote the numbers in a column? Yes _____ No _____

$$
\begin{array}{r}
17.00 \\
2.10 \\
16.80 \\
0.04 \\
+1.59 \\
\hline
\end{array}
$$

Subtracting Decimals—Problem 11 $72.3 - 1.145$

Helpful Hint: Place additional zeros at the end of each number so that each decimal has three digits to the right of the decimal point. Be sure to align the decimal points when you write the subtraction in column form.

Did you change the problem to $72.300 - 1.145$?
Yes _____ No _____

If you answered No, stop and make those changes now.

If you answered No, stop and write the numbers carefully in a column with the decimal points exactly aligned.

Did you line up the decimal points carefully when you wrote the numbers one beneath the other? Yes _____ No _____

$$
\begin{array}{r}
72.300 \\
-1.145 \\
\hline
\end{array}
$$

Writing Fractions in Decimal Form—Problem 17 $\dfrac{7}{8}$

Helpful Hint: Divide the denominator into the numerator until the remainder becomes zero. Be careful with each division step.

In your first step of division, was the result a quotient of 8?

If you answered No, try to go back and complete Problem 17 correctly now.

Yes _____ No _____

When you multiplied $8 \cdot 8$, was the result a product of 64?

Yes _____ No _____

When you subtracted 64 from 70, was the result a difference of 6?

Yes _____ No _____

If you answered No to any of these questions, stop and do the first part of the division problem again.

Did you write the problem as $8\overline{)7.000}$ and complete a total of three division steps?

Yes _____ No _____

Order of Operations with Decimals—Problem 19 $19.36 \div (0.24 + 0.26) \times (0.4)^2$

Helpful Hint: Be sure to write out each step. Combine operations in parentheses first. Then raise numbers to a power.

After the first step, your result should be $19.36 \div 0.5 \times (0.4)^2$. Did you get that result?

Now try to answer the entire Problem 19 correctly using these suggestions.

Yes _____ No _____

If you answered No, stop and perform that step correctly.

After the second step, your result should be $19.36 \div 0.5 \times 0.16$.

Did you get that result? Yes _____ No _____

If you answered No, stop and perform that step correctly.

Chapter 4 Ratio and Proportion
4.1 Ratios and Rates

Vocabulary
ratio   •   simplest form   •   rate   •   unit rate

1. A ratio is in _____ when the two numbers do not have a common factor and both numbers are whole numbers.

2. A _____ is a comparison of two quantities with different units.

3. A _____ is a rate in which the denominator is the number 1.

4. A _____ is the comparison of two quantities that have the same units.

Example	**Student Practice**
1. Write in simplest form. Express your answer as a fraction.	2. Write in simplest form. Express your answer as a fraction.
(a) The ratio of 15 hours to 20 hours	**(a)** The ratio of 24 hours to 48 hours
Write the ratio as a fraction.	
$$\frac{15}{20}$$	
Reduce the fraction.	
$$\frac{15}{20} = \frac{3}{4}$$	
(b) The ratio of 36 hours to 30 hours	**(b)** The ratio of 72 hours to 96 hours
$$\frac{36}{30} = \frac{6}{5}$$	
(c) 125:150	**(c)** 84:108
$$\frac{125}{150} = \frac{5}{6}$$	

Example	Student Practice
3. Martin earns $350 weekly. However, he takes home only $250 per week in his paycheck.	**4.** Jerry earns $400 weekly. However, he takes home only $275 per week in his paycheck.

3. Martin earns $350 weekly. However, he takes home only $250 per week in his paycheck.

$350.00   gross pay
45.00   withheld for federal tax
20.00   witheld for state tax
<u>35.00</u>   withheld for retirement
$250.00   take-home pay

(a) What is the ratio of the amount withheld for federal tax to gross pay?

The ratio of the amount withheld for federal tax to gross pay is $\dfrac{45}{350} = \dfrac{9}{70}$.

(b) What is the ratio of the amount withheld for state tax to the amount withheld for federal tax?

The ratio of the amount withheld for state tax to the amount withheld for federal tax is $\dfrac{20}{45} = \dfrac{4}{9}$.

4. Jerry earns $400 weekly. However, he takes home only $275 per week in his paycheck.

$400.00   gross pay
55.00   withheld for federal tax
37.00   witheld for state tax
<u>33.00</u>   withheld for retirement
$275.00   take-home pay

(a) What is the ratio of the amount withheld for federal tax to gross pay?

(b) What is the ratio of the amount withheld for state tax to the amount withheld for federal tax?

5. Recently an automobile manufacturer spent $946,000 for a 48-second television commercial shown on a national network. What is the rate of dollars spent to seconds of commercial time?

The rate is given below.

$$\frac{946,000 \text{ dollars}}{48 \text{ seconds}} = \frac{59,125 \text{ dollars}}{3 \text{ seconds}}$$

6. Recently a dish washer manufacturer spent $476,000 for a 24-second television commercial shown on a national network. What is the rate of dollars spent to seconds of commercial time?

Example	Student Practice
7. A car traveled 301 miles in seven hours. Find the unit rate. The rate $\dfrac{301 \text{ miles}}{7 \text{ hour}}$ can be simplified. We find $301 \div 7 = 43$. Thus, $\dfrac{301 \text{ miles}}{7 \text{ hours}} = \dfrac{43 \text{ miles}}{1 \text{ hour}}$. The denominator is 1. We write our answer as 43 miles/hour.	**8.** A van traveled 246 miles in six hours. Find the unit rate.
9. A grocer purchased 200 pounds of apples for \$68. He sold the 200 pounds of apples for \$86. How much profit did he make per pound of apples? Compute the total profit. $\begin{array}{r} \$86 \text{ selling price} \\ -\ 68 \text{ cost} \\ \hline \$18 \text{ profit} \end{array}$ The rate that compares profit to pounds of apples sold is $\dfrac{18 \text{ dollars}}{200 \text{ pounds}}$. Find $18 \div 200$. $\begin{array}{r} 0.09 \\ 200\overline{)18.00} \\ \underline{18\ 00} \\ 0 \end{array}$ The unit rate of profit is \$0.09 per pound.	**10.** A lady purchased 300 pounds of oranges for \$73. She sold the 300 pounds of oranges for \$88. How much profit did she make per pound of oranges?

Example	Student Practice
11. Hamburger at a local butcher shop is packaged in large and extra-large packages. A large package costs $7.86 for 6 pounds and an extra-large package is $10.08 for 8 pounds. What is the unit rate in dollars per pound for each size package?	**12.** Hot dogs at a stall are packaged in large and extra-large packages. A large package costs $6.58 for 2 pounds and an extra-large package is $15.06 for 6 pounds. What is the unit rate in dollars per pound for each size package?

Compute the unit rate for the large package.

$$\frac{7.86 \text{ dollars}}{6 \text{ pounds}} = \$1.31 / \text{pound}$$

Compute the unit rate for the extra-large package.

$$\frac{10.08 \text{ dollars}}{8 \text{ pounds}} = \$1.26 / \text{pound}$$

Extra Practice

1. Write 228 yards to 12 yards in simplest form. Express your answer as a fraction.

2. Write 13 pizzas for 104 people as a rate in simplest form.

3. A retailer purchased 15 washing machines for $3525. He sold them for $5700. How much profit did he make per washing machine?

4. Jonathon had 450 at-bats, and had 126 hits. What is his ratio of hits to at-bats?

Concept Check

At a company picnic, there were 663 cans of soda for 231 people. Explain how you would write that as a rate in simplest form.

Chapter 4 Ratio and Proportion
4.2 The Concept of Proportions

Vocabulary

proportion • equality test for fractions • rates • cross products

1. To determine whether a statement is a proportion, we use the _____.

2. A _____ states that two ratios or two rates are equal.

3. For the proportion $\dfrac{4}{14} = \dfrac{6}{21}$, $14 \times 6 = 84$ and $4 \times 21 = 84$ are called _____.

4. When you write a proportion, be sure that the similar units for the _____ are in the same position in the fractions.

Example	Student Practice
1. Write a proportion to express the following: If four rolls of wallpaper measure 300 feet, then eight rolls of wallpaper will measure 600 feet. When you write a proportion, order is important. The similar units for the rates must be in the same position in the fractions. $\dfrac{4 \text{ rolls}}{300 \text{ feet}} = \dfrac{8 \text{ rolls}}{600 \text{ feet}}$	**2.** Write a proportion to express the following: If 3 gallons of paint can paint 200 square feet of wall, then 6 gallons of paint can paint 400 square feet of wall.
3. Determine whether or not the equation $\dfrac{14}{18} \overset{?}{=} \dfrac{35}{45}$ is a proportion. Compute the cross products. $18 \times 35 = 630$ $14 \times 45 = 630$ The cross products are equal, so this is a proportion.	**4.** Determine whether or not the equation $\dfrac{16}{20} \overset{?}{=} \dfrac{25}{55}$ is a proportion.

Example	Student Practice
5. Determine which equations are proportions.	**6.** Determine which equations are proportions.

(a) $\dfrac{5.5}{7} \overset{?}{=} \dfrac{33}{42}$

Compute the cross products.

$7 \times 33 = 231$
$5.5 \times 42 = 231$

The cross products are equal.

Thus, $\dfrac{5.5}{7} = \dfrac{33}{42}$.

This is a proportion.

(b) $\dfrac{5}{8\frac{3}{4}} \overset{?}{=} \dfrac{40}{72}$

Compute the cross products.

$8\dfrac{3}{4} \times 40 = \dfrac{35}{\cancel{4}_1} \times \cancel{40}^{10} = 35 \times 10 = 350$

$5 \times 72 = 360$

The cross products are not equal.

Thus, $\dfrac{5}{8\frac{3}{4}} \neq \dfrac{40}{72}$.

This is not a proportion.

(a) $\dfrac{7.2}{9} \overset{?}{=} \dfrac{17.6}{22}$

(b) $\dfrac{3}{5\frac{7}{8}} \overset{?}{=} \dfrac{48}{64}$

Example	Student Practice
7. Complete parts **(a)** and **(b)**.	**8.** Complete parts **(a)** and **(b)**.

7. Complete parts **(a)** and **(b)**.

(a) Is the rate $\dfrac{\$86}{13 \text{ tons}}$ equal to the rate $\dfrac{\$79}{12 \text{ tons}}$?

We want to know whether $\dfrac{86}{13} = \dfrac{79}{12}$.

Compute the cross products.

$13 \times 79 = 1027$
$86 \times 12 = 1032$

The cross products are not equal.

This is not a proportion.

Thus, the two rates are not equal.

(b) Is the rate $\dfrac{3 \text{ American dollars}}{2 \text{ British pounds}}$ equal to the rate $\dfrac{27 \text{ American dollars}}{18 \text{ British pounds}}$?

We want to know whether $\dfrac{3}{2} = \dfrac{27}{18}$.

Compute the cross products.

$2 \times 27 = 54$

$3 \times 18 = 54$

The cross products are equal.

This is a proportion.

Thus, the two rates are equal.

8. Complete parts **(a)** and **(b)**.

(a) Is the rate $\dfrac{52 \text{ kilometers}}{15 \text{ hours}}$ equal to the rate $\dfrac{85 \text{ kilometers}}{11 \text{ hours}}$?

(b) Is the rate $\dfrac{\$5}{7 \text{ kilograms}}$ equal to the rate $\dfrac{\$25}{35 \text{ kilograms}}$?

Extra Practice

1. Write a proportion.

   60 is to 12 as 95 is to 19.

2. Determine whether the equation is a proportion.

   $$\frac{2\frac{1}{3}}{3} \overset{?}{=} \frac{11\frac{2}{3}}{15}$$

3. Determine whether the equation is a proportion.

   $$\frac{3}{8} \overset{?}{=} \frac{18.6}{48.8}$$

4. In the 8 A.M. section of mathematics, there are 12 female students and 16 male students enrolled. In the 1 P.M. section of mathematics, there are 12 male students and 9 female students enrolled. Is the ratio of male students to female students the same in both mathematics classes?

Concept Check

Explain how to determine if $\dfrac{33 \text{ chairs}}{45 \text{ employees}} \overset{?}{=} \dfrac{165 \text{ chairs}}{225 \text{ employees}}$ is a proportion.

Name: _____   Date: _____

Instructor: _____   Section: _____

Chapter 4 Ratio and Proportion
4.3 Solving Proportions

Vocabulary

variable   •   equation   •   units of measure   •   proportion

1. A(n) _____ has an equals sign. This indicates that the values on each side of it are equivalent.

2. We represent a number that we do not know by a(n) _____.

3. In a(n) _____, the same units should be in the same position in the fractions.

4. In real-life situations it is helpful to write the _____ in the proportion.

Example	**Student Practice**
1. Solve for *n*.	**2.** Solve for *n*.
$24 \times n = 240$	$12 \times n = 360$
Divide each side by 24.	
$\dfrac{24 \times n}{24} = \dfrac{240}{24}$	
Since $24 \div 24 = 1$ and $240 \div 24 = 10$, $n = 10$.	
3. Solve for *n*.	**4.** Solve for *n*.
$143 = 13 \times n$	$159 = 3 \times n$
Divide each side by 13.	
$\dfrac{143}{13} = \dfrac{13 \times n}{13}$	
$11 = n$	

111

Copyright © 2017 Pearson Education, Inc.

Example	Student Practice
5. Solve for n. $18.2 = 2.6 \times n$ Divide each side by 2.6. $\dfrac{18.2}{2.6} = \dfrac{2.6 \times n}{2.6}$ Perform the division. $\begin{array}{r} 7. \\ 2.6_\wedge \overline{)18.2_\wedge} \\ \underline{18\ 2} \\ 0 \end{array}$ Since $18.2 \div 2.6 = 7$, $7 = n$.	**6.** Solve for n. $37.8 = 4.2 \times n$
7. Find the value of n in $\dfrac{125}{2} \overset{?}{=} \dfrac{150}{n}$. Find the cross products. $125 \times n = 2 \times 150$ $125 \times n = 300$ Divide each side by 125. $\dfrac{125 \times n}{125} = \dfrac{300}{125}$ $n = 2.4$ Check. $\dfrac{125}{2} \overset{?}{=} \dfrac{150}{2.4}$ $125 \times 2.4 \overset{?}{=} 2 \times 150$ $300 = 300$ It is a proportion, $n = 2.4$.	**8.** Find the value of n in $\dfrac{130}{5} \overset{?}{=} \dfrac{195}{n}$.

Example	Student Practice
9. Find the value of n in $\dfrac{n}{20} = \dfrac{\frac{3}{4}}{5}$.	**10.** Find the value of n in $\dfrac{n}{42} = \dfrac{\frac{5}{6}}{7}$.

9. (continued)

Find the cross products.

$$5 \times n = 20 \times \dfrac{3}{4}$$

Simplify.

$$5 \times n = 15$$

Divide each side by 5.

$$\dfrac{5 \times n}{5} = \dfrac{15}{5}$$
$$n = 3$$

Check by verifying that this is a proportion.

---

11. If 5 grams of non-icing additive are placed in 8 liters of diesel fuel, how many grams n should be added to 12 liters of diesel fuel?

We need to find the value of n in

$$\dfrac{n \text{ grams}}{12 \text{ liters}} = \dfrac{5 \text{ grams}}{8 \text{ liters}}.$$

$$8 \times n = 12 \times 5$$
$$8 \times n = 60$$
$$\dfrac{8 \times n}{8} = \dfrac{60}{8}$$
$$n = 7.5$$

The answer is 7.5 grams. 7.5 grams of the additive should be added to 12 liters of the diesel fuel. Check by verifying that this is a proportion.

12. If 3 grams of non-icing additive are placed in 4 liters of diesel fuel, how many grams n should be added to 14 liters of diesel fuel?

Example	Student Practice

13. Find the value of n in
$$\frac{141 \text{ miles}}{4.5 \text{ hours}} = \frac{67 \text{ miles}}{n \text{ hours}}.$$ Round to the nearest tenth.

$$141 \times n = 67 \times 4.5$$
$$141 \times n = 301.5$$
$$\frac{141 \times n}{141} = \frac{301.5}{141}$$

If we calculate to four decimal places, we have $n = 2.1382$. Rounding to the nearest tenth, $n \approx 2.1$ hours.

14. Find the value of n in
$$\frac{163 \text{ pounds}}{\$5.30} = \frac{73 \text{ pounds}}{\$n}.$$ Round to the nearest tenth.

Extra Practice

1. Solve for n.

$$n \times 9 = 20.7$$

2. Find the value of n. Check your answer.

$$\frac{14}{56} = \frac{n}{12}$$

3. Find the value of n. Round your answer to the nearest tenth if necessary.

$$\frac{12\frac{3}{5}}{100} = \frac{n}{5}$$

4. A photograph is 4 inches wide and 6 inches tall. If you want to make an enlargement that is 18 inches wide, how tall will the enlargement be?

Concept Check

Explain how you would solve the proportion $\dfrac{2\frac{1}{2}}{3\frac{3}{4}} = \dfrac{16\frac{1}{2}}{n}$.

114

Chapter 4 Ratio and Proportion
4.4 Solving Applied Problems Involving Proportions

Vocabulary

estimate   •   units   •   proportion   •   cross products

1. You can set up an applied problem in several different ways as long as the _____ are in correctly corresponding positions.

2. Usually, a mistake will be obvious when compared with a(n) _____.

Example	Student Practice
1. Ted's car can go 245 miles on 7 gallons of gas. Ted wants to take a trip of 455 miles. Approximately how many gallons of gas will this take? Read the problem carefully and create a Mathematics Blueprint. Let $n=$ the unknown number of gallons. $$\frac{245 \text{ miles}}{7 \text{ gallons}} = \frac{455 \text{ miles}}{n \text{ gallons}}$$ Find the cross products. $245 \times n = 7 \times 455$ Simplify. $245 \times n = 3185$ Divide both sides by 245. $$\frac{245 \times n}{245} = \frac{3185}{245}$$ $$n = 13$$ Ted will need approximately 13 gallons of gas for the trip.	**2.** Peter's van can go 125 miles on 5 gallons of gas. Peter wants to take a trip of 375 miles. Approximately how many gallons of gas will this take?

Example	Student Practice
3. In a certain gear, Alice's 18-speed bicycle has a gear ratio of three revolutions of the pedal for every two revolutions of the bicycle wheel. If her bicycle wheel is turning at 65 revolutions per minute, how many times must she pedal per minute?	**4.** In a certain gear, Isha's 18-speed bike has a gear ratio of seven revolutions of the pedal for every 5 revolutions of the bike wheel. If her bike wheel is turning 97 revolutions per minute, how many times must she pedal per minute?

Read the problem carefully and create a Mathematics Blueprint.

Let $n =$ the number of revolutions of the pedal.

$$\frac{3 \text{ revolutions of the pedal}}{2 \text{ revolutions of the wheel}} = \frac{n \text{ revolutions of the pedal}}{65 \text{ revolutions of the wheel}}$$

Find the cross products.

$$3 \times 65 = 2 \times n$$

Simplify.

$$195 = 2 \times n$$

Divide both sides by 2.

$$\frac{195}{2} = \frac{2 \times n}{2}$$
$$97.5 = n$$

Alice will pedal at the rate of 97.5 revolutions per 1 minute.

Example	Student Practice

5. A biologist catches 42 fish in a lake and tags them. She then quickly returns them to the lake. In a few days she catches a new sample of 50 fish. Of those 50 fish, 7 have her tag. Approximately how many fish are in the lake?

Read the problem carefully and create a Mathematics Blueprint.

Let $n =$ the number of fish in the lake.

$$\frac{42 \text{ fish tagged in 1st sample}}{n \text{ fish in lake}} = \frac{7 \text{ fish tagged in 2nd sample}}{50 \text{ fish caught in 2nd sample}}$$

Find the cross products

$$42 \times 50 = 7 \times n$$

Simplify.

$$2100 = 7 \times n$$

Divide both sides by 7.

$$\frac{2100}{7} = \frac{7 \times n}{7}$$
$$300 = n$$

Assuming that no tagged fish died and that the tagged fish mixed throughout the population of fish in the lake, we estimate that there are 300 fish in the lake.

6. Biologists captured, tagged, and released 32 fish in a pond. A month later, they captured 30 fish. Six of these 30 fish had tags from the earlier capture. Approximately how many fish live in the pond?

Extra Practice

1. On a map, the distance from one city to another is 7 inches. If the actual distance is 315 miles, what is the actual distance between two other cities that are 11 inches apart on the map?

2. The Pizza Express sold 78 pizzas on Friday night when 110 customers were served. Next Friday, they expect 275 customers. How many pizzas should they be prepared to serve?

3. Leon counts 45 yellow flowers in one rack of 300 flowers at a garden center. Assuming that the other racks contain the same proportion of yellow flowers, how many of the 2500 flowers at the garden center are yellow?

4. Sandra went to the bank to get some Canadian money for a trip across the border. At the time, one U.S. dollar was worth 1.027 Canadian dollars. If she exchanged US$520, how much in Canadian dollars did she receive?

Concept Check

When Fred went to France he discovered that 70 euros were worth 104 American dollars. He brought 400 American dollars on his trip. Explain how he would find what that is worth in euros.

MATH COACH

Mastering the skills you need to do well on the test.

Watch the **MATH COACH** videos in MyMathLab® or on YouTube™ while you work the problems below. These helpful hints will help you avoid making common errors on test problems.

Write as a Unit Rate—Problem 7

5400 feet per 22 telephone poles

> **Helpful Hint:** Write the word phrase as a ratio. Then divide the denominator into the numerator to find the unit rate. Round your answer to the nearest hundredth.

Did you divide the denominator (22) into the numerator (5400)?

Yes _____ No _____

If you answered No, please perform the correct division.

Did you stop your division steps with 245 as your answer?

Yes _____ No _____

If you answered Yes, you need to continue your calculation further since the problem requires an answer rounded to the nearest hundredth. Remember to include the units with your answer.

If you did not solve Problem 7 correctly, please rework the problem using these suggestions.

Write as a Proportion—Problem 12   3 hours is to 180 miles as 5 hours is to 300 miles

> **Helpful Hint:** Write the appropriate units in the top and bottom of each fraction. Compare the two fractions to make sure the same unit appears in both numerators and the same unit appears in both denominators.

Did you write 3 hours on the top of one fraction and 5 hours on the top of the other fraction?

Yes _____ No _____

If you answered No, then stop and make this correction.

Did you write 180 miles in the denominator of the fraction that has 3 hours in the numerator?

If you answered No, then stop and make this correction.

If you answered Problem 12 incorrectly, then please rework the problem using these suggestions.

119

Determine if an Equation is a Proportion—Problem 16 $\dfrac{\$0.74}{16 \text{ ounces}} \overset{?}{=} \dfrac{\$1.84}{40 \text{ ounces}}$

Helpful Hint: Find the cross products and see if they are equal.

Did you multiply 0.74×40 to obtain 29.60?

Yes _____ No _____

If you answered No, perform this calculation.

Did you multiply 16×1.84 to obtain 29.44?

Yes _____ No _____

If you answered No, perform this calculation.

If you did not solve Problem 16 correctly, please rework the problem using these suggestions.

Now go back and state whether or not the equation is a proportion.

Solving Applied Problems using Proportions—Problem 27 If 9 inches on a map represent 57 miles, what distance does 3 inches represent?

Helpful Hint: Take the time to follow all the rules of calculations with fractions.

Did you translate the problem into the proportion
$\dfrac{9 \text{ inches}}{57 \text{ miles}} = \dfrac{3 \text{ inches}}{n \text{ miles}}$?

Yes _____ No _____

If you answered No, reread the problem and try to translate the situation again.

Did you find the cross products to obtain the equation $9 \times n = 57 \times 3$?

Yes _____ No _____

Did you multiply 57 and 3 to obtain 171 on the right side of the equation?

Yes _____ No _____

If you answered No, perform the multiplication again.

Did you divide both sides of the equation by 9 to find the value for n?

Yes _____ No _____

If you answered No, perform this step to solve for n.

If you did not solve Problem 27 correctly, please rework the problem using these suggestions.

(Remember to include the proper units with your final answer.)

Chapter 5 Percent
5.1 Understanding Percent

Vocabulary

percent   •   dividing by 100   •   multiplying by 100   •   parts per 100

1. To change a decimal number to a percent, we are _____.

2. We use the symbol % for percent. It means _____.

3. When you remove a percent sign (%), you are _____.

4. The word _____ means per 100.

Example	**Student Practice**
1. Recently 100 college students were surveyed about their intentions for voting in the next presidential election. 39 students intended to vote for the Republican candidate, 28 students intended to vote for the Democratic candidate, and 22 students were undecided about which candidate to vote for. The remaining 11 students admitted that they were not planning to vote.	**2.** Recently 100 college students were surveyed about their intentions for voting in the next class elections. 42 students intended to vote in the next class elections, 32 students did not intend to vote for the next class elections, and 26 students were undecided as to vote or not.
(a) What percent of the students intended to vote for the Democratic candidate? $\dfrac{28}{100} = 28\%$	**(a)** What percent of the students intended to vote for the class elections?
(b) What percent of the students intended to vote for the Republican candidate? $\dfrac{39}{100} = 39\%$	**(b)** What percent of the students were undecided as to vote or not?

Example	Student Practice
3.	**4.**
(a) Write $\dfrac{386}{100}$ as a percent.	**(a)** Write $\dfrac{465}{100}$ as a percent.
$\dfrac{386}{100} = 386\%$	
(b) Twenty years ago, four car tires for a full-size car cost $100. Now the average price for four car tires for a full-size car is $270. Write the present cost as a percent of the cost 20 years ago.	**(b)** Fifteen years ago, two truck tires cost $100. Now the average price for two truck tires is $355. Write the present cost as a percent of the cost 15 years ago.
The ratio is $\dfrac{\$270 \text{ for four tires now}}{\$100 \text{ for four tires then}}$.	
$\dfrac{270}{100} = 270\%$	
The present cost of four car tires for a full-size car is 270% of the cost 20 years ago.	
5. Write as a percent.	**6.** Write as a percent.
(a) $\dfrac{0.9}{100}$	**(a)** $\dfrac{0.12}{100}$
$\dfrac{0.9}{100} = 0.9\%$	
(b) $\dfrac{0.002}{100}$	**(b)** $\dfrac{0.00005}{100}$
$\dfrac{0.002}{100} = 0.002\%$	

Example	Student Practice
7. Write as a decimal.	**8.** Write as a decimal.
(a) 38%	**(a)** 52%
$38\% = \dfrac{38}{100} = 0.38$	
(b) 6%	**(b)** 7%
$6\% = \dfrac{6}{100} = 0.06$	

9. Write as a decimal.	**10.** Write as a decimal.
(a) 26.9%	**(a)** 92.5%
Drop the percent symbol and move the decimal point two places to the left.	
$26.9\% = 0.269$	
(b) 7.2%	**(b)** 8.5%
$7.2\% = 0.072$	
Note that we need to add an extra zero to the left of the seven.	
(c) 0.13%	**(c)** 0.47%
$0.13\% = 0.0013$	
Here we added zeros to the left of the one.	
(d) 158%	**(d)** 551%
$158\% = 1.58$	

Example	Student Practice
11. Write as a percent.	**12.** Write as a percent.
(a) 0.08	**(a)** 0.005
Move the decimal point two places to the right and write the percent symbol at the end of the number.	
$0.08 = 8\%$	
(b) 6.31	**(b)** 5.82
$6.31 = 631\%$	
(c) 0.055	**(c)** 0.0072
$0.055 = 5.5\%$	
(d) 0.001	**(d)** 0.0008
$0.001 = 0.1\%$	

Extra Practice

1. Write as a percent.

$$\frac{41}{100}$$

2. Write as a percent.

$$\frac{0.013}{100}$$

3. Write as a decimal.

0.83%

4. Write as a percent.

0.00076

Concept Check

Explain how you would change 0.00072% to a decimal.

Chapter 5 Percent
5.2 Changing Between Percents, Decimals, and Fractions

Vocabulary

decimal   •   fraction   •   mixed number   •   percent

1. It may be helpful to write a percent as a _____ before you write it as a fraction in simplest form.

2. If the percent is greater than 100%, the simplified fraction is usually changed to a _____.

3. A convenient way to change a fraction to a _____ is to write the fraction in decimal form first.

4. By using the definition of percent, we can write any percent as a _____ whose denominator is 100.

Example	Student Practice
1. Write 75% as a fraction in simplest form. $75\% = \dfrac{75}{100} = \dfrac{3}{4}$	**2.** Write 92% as a fraction in simplest form.
3. Write 43.5% as a fraction in simplest form. First, change the percent to a decimal. $43.5\% = 0.435$ Next, change the decimal to a fraction. $0.435 = \dfrac{435}{1000}$ Finally, reduce the fraction. $\dfrac{435}{1000} = \dfrac{87}{200}$	**4.** Write 52.85% as a fraction in simplest form.

Example	Student Practice
5. Write 138% as a mixed number. $$138\% = 1.38 = 1\frac{38}{100} = 1\frac{19}{50}$$	**6.** Write 448% as a mixed number.
7. In the fiscal 2011 budget of the United States, approximately $19\frac{1}{6}\%$ of the budget was designated for social security. Write this percent as a fraction. First, change the percent to a fraction. $$19\frac{1}{6}\% = \frac{19\frac{1}{6}}{100} = 19\frac{1}{6} \div 100$$ Now, divide. $$19\frac{1}{6} \div \frac{100}{1} = \frac{115}{6} \div \frac{100}{1} = \frac{115}{6} \cdot \frac{1}{100} = \frac{23}{120}$$ Thus, $\frac{23}{120}$ of the budget was designated for social security.	**8.** Approximately $16\frac{2}{13}\%$ of the 2011 budget for an automobile company was designated for advertising. Write the percent as a fraction.
9. Write $\frac{3}{8}$ as a percent. A convenient way to change a fraction to a percent is to write the fraction in decimal form and then convert the decimal to a percent. Calculate $3 \div 8$. $$3 \div 8 = 0.375$$ Thus, $\frac{3}{8} = 0.375 = 37.5\%$.	**10.** Write $\frac{3}{16}$ as a percent.

Example	Student Practice
11. Write $\dfrac{1}{6}$ as a percent. Round to the nearest hundredth of a percent.	**12.** Write $\dfrac{1}{3}$ as a percent. Round to the nearest hundredth of a percent.

We find that $\dfrac{1}{6} = 0.1666\ldots$ by calculating $1 \div 6$.

We will need a four-place decimal so that we will obtain a percent to the nearest hundredth.

If we round the decimal to the nearest ten-thousandth, we have $\dfrac{1}{6} \approx 0.1667$.

If we change this to a percent, we have $\dfrac{1}{6} \approx 16.67\%$. This is correct to the nearest hundredth of a percent.

13. Express $\dfrac{11}{12}$ as a percent containing a fraction.	**14.** Express $\dfrac{5}{14}$ as a percent containing a fraction.

We will stop the division after two steps and write the remainder in fraction form.

$$
\begin{array}{r}
0.91 \\
12\overline{)11.00} \\
\underline{108} \\
20 \\
\underline{12} \\
8
\end{array}
$$

Thus, $\dfrac{11}{12} = 0.91\dfrac{8}{12} = 0.91\dfrac{2}{3}$.

Expressed as a percent, $0.91\dfrac{2}{3} = 91\dfrac{2}{3}\%$.

15. Complete the following table of equivalent notations. Round decimals to the nearest ten-thousandth.

Fraction	Decimal	Percent
		$17\frac{1}{5}\%$

16. Complete the following table of equivalent notations. Round decimals to the nearest ten-thousandth.

Fraction	Decimal	Percent
		$11\frac{3}{8}\%$

The number is written as a percent. Write the number as a decimal and then as a fraction.

$$\frac{17\frac{1}{5}}{100} = \frac{86}{5} \times \frac{1}{100} = \frac{86}{500}$$

Divide to obtain the decimal, $86 \div 500 = 0.172$.

Simplify to obtain the fraction, $\frac{86}{500} = \frac{43}{250}$.

The completed table is shown below.

Fraction	Decimal	Percent
$\frac{43}{250}$	0.172	$17\frac{1}{5}\%$

Extra Practice

1. Write 78.2% as a fraction or as a mixed number and simplify.

2. Write $4\frac{1}{2}\%$ as a fraction or as a mixed number and simplify.

3. Write $\frac{8}{5}$ as a percent. Round to the nearest hundredth of a percent if necessary.

4. Write $3\frac{5}{6}$ as a percent. Round to the nearest hundredth of a percent if necessary.

Concept Check

Explain how you would change $8\frac{3}{8}\%$ to a decimal.

Chapter 5 Percent
5.3A Solving Percent Problems Using Equations

Vocabulary
amount = percent × base   •   equation   •   of   •   is

1. The word _____ represents any multiplication symbol: "×," "( )," or "·."

2. To solve a percent problem, express it as a(n) _____ with an unknown quantity.

3. The word _____ represents the mathematical symbol "=".

4. Percent problems are usually of the form _____, where either the percent, the base, or the amount are unknown.

Example	Student Practice
1. Translate into an equation. Find 0.6% of 400. "Find" translates to "$n =$." Find 0.6% of 400. ↓ ↓ ↓ ↓ $n =$ 0.6% × 400	**2.** Translate into an equation. Find 0.7% of 800.

3. Translate into an equation.
 	
(a) 35% of what is 60?
 	
35%   of   what   is   60?
  ↓   ↓   ↓   ↓   ↓
35%   × n =   60
 	
(b) 7.2 is 120% of what?
 	
7.2   is   120%   of   what?
  ↓   ↓   ↓   ↓   ↓
7.2   =   120%   × n
	4. Translate into an equation.
	
	(a) 52% of what is 650?
	
	(b) 5.4 is 145% of what?

Example	Student Practice
5. When Rick bought a new Toyota Yaris, he had to pay a sales tax of 5% on the cost of the car, which was $12,000. What was the sales tax?	**6.** When Jim bought a new Toyota Camry, he had to pay sales tax of 7% on the cost of the car, which was $20,000. What was the sales tax?

5. (continued)

Translate into an equation.

What is 5% of $12,000?

$\downarrow \quad \downarrow \quad \downarrow \quad \downarrow \quad \quad \downarrow$

$n \quad = \quad 5\% \quad \times \quad \$12,000$

Change the percent to decimal form and multiply to find n.

$n = 5\% \times \$12,000$

$\quad = 0.05 \times \$12,000$

$\quad = \$600$

The sales tax was $600.

---

7. Dave and Elsie went out to dinner. They gave the waiter a tip that was 15% of the total bill. The tip the waiter received was $6. What was the total bill (not including the tip)?	**8.** The coach of the university baseball team said that 40% of the players of his team are left-handed. Eight people on the team are left-handed. How many people are on the team?

7. (continued)

15% of what is $6

$\downarrow \quad \downarrow \quad \downarrow \quad \downarrow \quad \downarrow$

$15\% \quad \times \quad n \quad = \quad \$6$

Find n.

$15\% \times n = 6$

$0.15 \times n = 6$

$\dfrac{0.15n}{0.15} = \dfrac{6}{0.15}, n = 40$

The total bill for the meal was $40.

Example	Student Practice
9. What percent of 5000 is 3.8?	**10.** What percent of 8000 is 4.2?

9. What percent of 5000 is 3.8?

Translate into an equation.

$$n \times 5000 = 3.8$$

Now, solve the equation.

$$n \times 5000 = 3.8$$
$$5000n = 3.8$$
$$\frac{5000n}{5000} = \frac{3.8}{5000}$$
$$n = \frac{3.8}{5000} = 0.00076$$

Finally, express the decimal as a percent.

$$n = 0.076\%$$

11. In a basketball game for the Atlanta Hawks, Jamal Crawford made 10 of his 24 shots. What percent of his shots did he make? (Round to the nearest tenth of a percent.)

This is equivalent to "10 is what percent of 24?" Translate this into an equation.

$$10 = n \times 24$$

Now, solve the equation.

$$10 = n \times 24$$
$$10 = 24n$$
$$\frac{10}{24} = \frac{24n}{24}$$
$$0.41666... = n$$

Jamal Crawford made about 41.7% of his shots in this game.

12. In her most recent softball practice, Megan caught 13 out of the 18 fly balls that came her way. What percent of fly balls did she catch, to the nearest hundredth of a percent? (Round to the nearest tenth of a percent.)

Extra Practice

1. Translate into an equation.

   30% of what is 27?

2. Solve.

   28 is 0.4% of what?

3. Kaitlin spelled 34 words correctly on her last spelling test, receiving a grade of 85%. How many words did the test include?

4. Jasmine invited 34 friends to her Labor Day barbecue, but only 26 were able to attend. To the nearest tenth of a percent, what percent of those invited came to Jasmine's barbecue?

Concept Check

Explain how to solve the following problem using an equation. Jason found that 85% of all people who purchased Mustangs at Danvers Ford were previous Mustang owners. Last year 120 people purchased Mustangs at Danvers Ford. How many of them were previous Mustang owners?

Chapter 5 Percent
5.3B Solving Percent Problems Using Proportions

Vocabulary

percent proportion   •   percent number   •   base   •   amount

1. We use the letter p (a variable) to represent the _____.

2. The _____ is the entire quantity or the total involved.

3. The _____, which we represent by the letter a, is the part being compared to the whole.

4. The _____ is defined as $\dfrac{\text{amount}}{\text{base}} = \dfrac{\text{percent number}}{100}$.

Example	**Student Practice**
1. Identify the percent number p. Find 16% of 370. The value of p is 16.	**2.** Identify the percent number p. What percent of 72 is 8?
3. Identify the base b and the amount a. **(a)** 20% of 320 is 64. The base is the entire quantity. It follows the word "of." Here $b = 320$. The amount is the part compared to the whole. Here $a = 64$. **(b)** 12 is 60% of what? The amount 12 is the part of the base. Here $a = 12$. The base is unknown. Represent the base by the variable b.	**4.** Identify the base b and the amount a. **(a)** 40% of 480 is 192. **(b)** 18 is 90% of what?

Example	Student Practice
5. Find p, b, and a. What percent of 30 is 18? The value of p is not known. Let p represent the unknown percent. The base usually follows the word "of." Here $b = 30$. The amount is 18. Thus, $a = 18$.	**6.** Find p, b, and a. What is 75% of 550?

7. Find 260% of 40.

The percent $p = 260$. The number that is the base usually appears after the word "of." The base $b = 40$. Note that the amount is unknown. We use the variable a.

Thus, $\dfrac{a}{b} = \dfrac{p}{100}$ becomes $\dfrac{a}{40} = \dfrac{260}{100}$.

If we reduce the fraction on the right-hand side, we have $\dfrac{a}{40} = \dfrac{13}{5}$.

Now, cross-multiply and simplify.

$5a = (40)(13)$
$5a = 520$

Finally, divide each side of the equation by 5.

$\dfrac{5a}{5} = \dfrac{520}{5}$
$a = 104$

Thus, 260% of 40 is 104.

8. Find 360% of 60.

Example	Student Practice
9. George and Barbara purchased some no-load mutual funds. The account manager charged a service fee of 0.2% of the value of the mutual funds. George and Barbara paid this fee, which amounted to $53. When they got home they could not find the receipt that showed the exact value of the mutual funds that they purchased. Can you find the value of the mutual funds that they purchased?	**10.** Chuck and Sally purchased some no-load mutual funds. The account manager charged a service fee of 0.7% of the value of the mutual funds. Chuck and Sally paid this fee, which amounted to $140. When they got home they could not find the receipt that showed the exact value of the mutual funds that they purchased. Can you find the value of the mutual funds that they purchased?

The basic situation here is that 0.2% of some number is $53. This is equivalent to saying $53 is 0.2% of what? If we want to answer this question, we need to identify a, b, and p.

The percent $p = 0.2$. Note that the base is unknown. We use the variable b. The amount $a = 53$.

Thus, $\dfrac{a}{b} = \dfrac{p}{100}$ becomes $\dfrac{53}{b} = \dfrac{0.2}{100}$.

Now cross-multiply and simplify.

$$(53)(100) = 0.2b$$
$$5300 = 0.2b$$

Finally, divide each side by 0.2.

$$\frac{5300}{0.2} = \frac{0.2b}{0.2}$$
$$26,500 = b$$

Thus $53 is 0.2% of $26,500. Therefore, the value of the mutual funds was $26,500.

Example	Student Practice
11. What percent of 4000 is 160?	**12.** What percent of 7500 is 675?

11. What percent of 4000 is 160?

Note that the percent is unknown. Use the variable p. The base $b = 4000$. The amount $a = 160$.

Thus, $\dfrac{a}{b} = \dfrac{p}{100}$ becomes $\dfrac{160}{4000} = \dfrac{p}{100}$.

If we reduce the fraction on the left-hand side, we have $\dfrac{1}{25} = \dfrac{p}{100}$.

Now cross-multiply and solve to find p.

$$100 = 25p$$
$$\frac{100}{25} = \frac{25p}{25}$$
$$4 = p$$

Thus 4% of 4000 is 160.

Extra Practice

1. Solve for the amount a.

25% of 275 is what?

2. Solve for the base b.

30 is 40% of what?

3. Solve for the percent p.

What percent of 560 is 28?

4. Myra's family set out on a 1200-mile car trip and made their first stop after traveling only 111 miles. What percent of the total distance had they traveled when they made this stop?

Concept Check

Explain how to solve the following problem using a proportion: Alice purchased some stock and was charged a service fee of 0.7% of the value of the stock. The fee she paid was $140. What was the value of the stock that she purchased?

Chapter 5 Percent
5.4 Solving Applied Percent Problems

Vocabulary
markup problems   •   discount   •   percent   •   base

1. The amount of a _____ is the product of the discount rate and the list price.

2. Percents can be added if the base (whole) is the same. Such problems are called _____ .

3. Some percent problems give you an amount and a percent and ask you to find the _____ .

Example	Student Practice
1. Of all the computers manufactured last month, an inspector found 18 that were defective. This is 2.5% of all the computers manufactured last month. How many computers were manufactured last month?	**2.** Of all the monitors manufactured last month, an inspector found 40 that were defective. This is 5% of all the monitors manufactured last month. How many monitors were manufactured last month?

There are two ways to solve this problem. **Method A** is to translate to an equation. **Method B** is to use the percent proportion. In this problem, we will use **Method A**. The problem is equivalent to "2.5% of the number of computers is 18." Translate this to an equation and solve.

$$2.5\% \times n = 18$$
$$0.025n = 18$$
$$\frac{0.025n}{0.025} = \frac{18}{0.025}$$
$$n = 720$$

Thus, 720 computers were manufactured last month.

Example	Student Practice
3. How much sales tax will you pay on a plasma HDTV television priced at $499 if the sales tax is 5%?	**4.** How much sales tax will you pay on an OLED HDTV television priced at $800 if the sales tax is 6.5%?

3. How much sales tax will you pay on a plasma HDTV television priced at $499 if the sales tax is 5%?

Method A Translate to an equation.

What is 5% of $499?

$$\downarrow \quad \downarrow \quad \downarrow \quad \downarrow \quad \downarrow$$

$$n \quad = \quad 5\% \quad \times \quad 499$$

Solve for n to find the tax.

$$n = (0.05)(499)$$
$$n = 24.95$$

The tax is $24.95.

Method B Use the percent proportion.

The percent $p = 5$. The base $b = 499$. The amount a is unknown.

Thus, $\dfrac{a}{b} = \dfrac{p}{100}$ becomes $\dfrac{a}{499} = \dfrac{5}{100}$.

Reduce the fraction on the right side, cross-multiply and solve to find a.

$$\frac{a}{499} = \frac{1}{20}$$
$$20a = 499$$
$$\frac{20a}{20} = \frac{499}{20}$$
$$a = 24.95$$

The tax is $24.95. Thus, by either method, the amount of the sales tax is $24.95. Check by estimating.

138

Example	Student Practice
5. Walter and Mary Ann are going out to a restaurant. They have a limit of $63.25 to spend for the evening. They want to tip the waitress 15% of the cost of the meal. How much money can they afford to spend on the meal itself? (Assume there is no tax.)	**6.** Peter and Julie have $76.05 to spend at a restaurant, including a 17% tip. How much can they spend on the meal itself? (Assume there is no tax.)

Read over the problem carefully and create a Mathematics Blueprint.

Let n = the cost of the meal. 15% of the cost = the amount of the tip. We want to add the percents of the meal.

$$\boxed{\begin{array}{c}\text{Cost of}\\\text{meal } n\end{array}} + \boxed{\begin{array}{c}\text{tip of 15\%}\\\text{of the cost}\end{array}} = \boxed{\$63.25}$$

$$(100\% \text{ of } n) + (15\% \text{ of } n) = \$63.25$$

Note that 100% of n added to 15% of n is 115% of n.

$$115\% \text{ of } n = \$63.25$$
$$1.15 \times n = 63.25$$

Now, divide both sides by 1.15.

$$\frac{1.15 \times n}{1.15} = \frac{63.25}{1.15}$$
$$n = 55$$

They can spend up to $55.00 on the meal itself.

Verify by estimating whether or not the answer is reasonable.

Example	Student Practice
7. Jeff purchased a flat-panel LCD TV on sale at a 35% discount. The list price was $430.00. What was the amount of the discount?	8. John bought a car that lists for $14,600 at an 8% discount. What was the amount of the discount?

$$\text{Discount} = \text{discount rate} \times \text{list price}$$
$$= 35\% \times 430$$
$$= 0.35 \times 430$$
$$= 150.5$$

The discount was $150.50.

Extra Practice

1. Allison bought a new CD player. The sales tax in her state is 5%, and she paid $4.50 in tax. What was the price of the CD player before tax?

2. Alicia earns $908 per month. Her cell phone bill last month was $137. What percent of her total income went toward paying her cell phone bill? (Round to the nearest hundredth of a percent.)

3. Arline rented a storage space that usually costs $75.00, but received a 15% discount. What was the amount of the discount, and how much did she pay for the space?

4. When Sondra bought a used car priced at $5125.00, she negotiated a 13% discount from the seller. How much did she save on the car and how much did she pay for it?

Concept Check

Explain how to solve the following problem. Sam works in sales for a pharmaceutical company. He can spend 23% of his budget for travel expenses. He can spend 14% for entertainment of clients. He can spend 17% of his budget for advertising. Last year he had a total budget of $80,000. Last year he spent a total of $48,000 for travel expenses, entertainment, and advertising. Did he stay within his budget allowance for those items?

Chapter 5 Percent
5.5 Solving Commission, Percent of Increase or Decrease, and Interest Problems

Vocabulary

commission • commission rate • percent of increase • principal
percent of decrease • interest rate • interest • compound interest

1. Money paid for the use of money is called _____.

2. The amount of money you get that is a percentage of the value of your sales is called your _____.

3. The _____ is the amount deposited or borrowed.

4. The _____ is equal to $\dfrac{\text{amount of decrease}}{\text{original amount}}$.

Example	Student Practice
1. A salesperson has a commission rate of 17%. She sells $32,500 worth of goods in a department store in two months. What is her commission? Commission is calculated by multiplying the percentage (called a commission rate) by the value of the sales. Commission = commission rate× value of sales $\text{Commission} = 17\% \times \$32,500$ $= 0.17 \times 32,500$ $= 5525$ Her commission is $5525.00. Verify by estimating whether or not the answer is reasonable.	**2.** A real estate salesperson earns a commission rate of 7% when he sells a $125,000 home. What is his commission?

Example	Student Practice
3. The population of Center City increased from 50,000 to 59,500. What was the percent of increase?	**4.** A new car is sold for $25,000. A year later its price had decreased to $20,500. What is the percent of decrease?

3. The population of Center City increased from 50,000 to 59,500. What was the percent of increase?

Read over the problem carefully and create a Mathematics Blueprint.

Calculate the amount of increase.

$$59,500 - 50,000 = 9500$$

Now, calculate the percent of increase.

$$\text{Percent of increase} = \frac{\text{amount of increase}}{\text{original amount}}$$

$$= \frac{9500}{50,000}$$

$$= 0.19$$

$$= 19\%$$

The percent of increase is 19%.

4. A new car is sold for $25,000. A year later its price had decreased to $20,500. What is the percent of decrease?

---

5. Find the simple interest on a loan of $7500 borrowed at 13% for one year.

Use the formula for simple interest, $I = P \times R \times T$.

Identify the values of P, R, and T.

$P = \text{principal} = \$7500$, $R = \text{rate} = 13\%$, and $T = \text{time} = 1 \text{ year}$.

$$I = 7500 \times 13\% \times 1$$

$$= 7500 \times 0.13$$

$$= 975$$

The interest is $975.

6. Find the simple interest on a loan of $9200 borrowed at 15% for one year.

Example	Student Practice
7. Find the simple interest on a loan of $2500 that is borrowed at 9% for each of the following.	**8.** Find the simple interest on a loan of $3200 that is borrowed at 13% for each of the following.
(a) Three years	**(a)** Five years

(a) Three years

Use the formula for simple interest, $I = P \times R \times T$.

Identify the values of P, R, and T.

$P = \$2500$, $R = 9\%$, and $T = 3$ years

$I = 2500 \times 0.09 \times 3$

$\quad = 225 \times 3$

$\quad = 675$

The interest for three years is $675.

(b) Three months

(b) Six months

Use the formula for simple interest, $I = P \times R \times T$. The values $P = \$2500$ and $R = 9\%$ stay the same.

Note that three months $= \dfrac{1}{4}$ year.

The time period must be in years to use the formula.

$I = 2500 \times 0.09 \times \dfrac{1}{4}$

$\quad = 225 \times \dfrac{1}{4}$

$\quad = \dfrac{225}{4}$

$\quad = 56.25$

The interest for three months is $56.25.

Extra Practice

1. Amy works part time for a telemarketing firm selling magazine subscriptions. She is paid $400 per month plus 3% of her total sales. Last month, she sold $4800 in magazine subscriptions. What was her total income for the month?

2. The enrollment in the Evening Division of a local university increased from 4200 to 5544 students per semester. What was the percent of increase?

3. Chuck received a notice in the mail that his rent was being increased from $850 to $867 per month. What percent increase does this change represent? Round to the nearest hundredth.

4. Maria opened a savings account and made a deposit of $2800. Her account earns 3.6% interest annually. She did not add any more money within the year. At the end of one year, how much interest did she earn? How much money did she have total in the bank? Round to the nearest hundredth.

Concept Check

Explain how to find simple interest on a loan of $5800 borrowed at an annual rate of 16% for a period of three months.

MATH COACH

Mastering the skills you need to do well on the test.

Watch the MATH COACH videos in MyMathLab°or on You Tube™ while you work the problems below. These helpful hints will help you avoid making common errors on test problems.

Write a Decimal as a Percent—Problem 14   3.024

> **Helpful Hint:** Remember to move the decimal point two places to the right and then write the % symbol at the end of the number.

Did you move the decimal point to the right and add the % symbol?

That is why the decimal point moves to the right two places.

Yes _____   No _____

If you answered No, please stop and complete this process correctly.

Note that when writing a decimal as a percent, we are multiplying the decimal by 100.

Solving Percent Problems when the Base is Unknown—Problem 21   16% of what number is 800?

> **Helpful Hint:** Translate the problem into an equation first. This will show you that division is a necessary step in solving the problem.

Were you able to translate the problem into the equation $16\% \times n = 800$?

If you answered No, please go back and complete the division carefully to solve for n.

Yes _____   No _____

If you answered No, please stop and complete that step correctly first.

Were you able to go to the second step and write $0.16 \times n = 800$?

Yes _____   No _____

If you answered No, go back and complete that step.

Were you able to divide both sides of the equation by 0.16?

Yes _____   No _____

Solving Percent Problems when the Percent is Unknown—Problem 24

What percent is 15 of 75?

> **Helpful Hint:** Translate the problem into an equation first. Remember that you must add the % symbol to your answer.

Were you able to translate the problem into the equation $15 = n \times 75$?

Yes _____ No _____

If you answered No, stop and reason through the steps to see if you can obtain that equation.

Were you able to divide both sides of the equation by 75?

Yes _____ No _____

If you answered No, go back and complete the division carefully to solve for n.

Do you see why it is necessary to change the answer from a decimal to a percent? Yes _____ No _____

If you answered Problem 24 incorrectly, please go back and rework the problem using these suggestions.

Solving Applied Percent Problems—Problem 29

A total of 5160 people voted in the city election. This was 43% of the registered voters. How many registered voters are in the city?

> **Helpful Hint:** Write a percent statement that describes the situation in the applied problem first. Then consider whether you want to use percent proportions or percent equations to solve the problem.

Were you able to write a percent statement to describe the problem situation, such as 43% of the registered voters equals 5160? Yes _____ No _____

If you answered No, please read the problem again and try to write a similar statement.

Were you able to translate the statement into the percent equation $43\% \times n = 5160$ or the percent proportion

$$\frac{5160}{b} = \frac{43}{100}?$$ Yes _____ No _____

If you answered No, please reread your statement and see if you can rewrite it to describe this equation or proportion.

If you answered Problem 29 incorrectly, please go back and rework the problem using these suggestions.

Chapter 6 Measurements
6.1 American Units

Vocabulary
unit fraction   •   metric system   •   proportion   •   American units

1. The idea of multiplying by a unit fraction is the same as solving a(n) _____.

2. In the United States, except in science, most measurements are made in _____.

3. The two main systems of measurement are the _____ and the American System.

4. A ratio of measurements for which the measurement in the numerator is equivalent to the measurement in the denominator is called a(n) _____.

Example	**Student Practice**
1. Answer rapidly the following questions.	**2.** Answer rapidly the following questions.
(a) How many inches in a foot? 12	**(a)** How many days in a week?
(b) How many yards in a mile? 1760	**(b)** How many ounces in a pound?
(c) How many seconds in a minute? 60	**(c)** How many feet in a yard?
(d) How many hours in a day? 24	**(d)** How many minutes in an hour?
(e) How many pounds in a ton? 2000	**(e)** How many quarts in a gallon?
(f) How many cups in a pint? 2	**(f)** How many feet in a mile?

Example	Student Practice
3. Convert 8800 yards to miles.	**4.** Convert 840 minutes to hours.

3. Multiply by a unit fraction which relates miles to yards. There are 1760 yards in 1 mile. The unit fraction is $\dfrac{1 \text{ mile}}{1760 \text{ yards}}$.

$$8800 \text{ yards} \times \frac{1 \text{ mile}}{1760 \text{ yards}} = \frac{8800}{1760} \text{ miles}$$
$$= 5 \text{ miles}$$

Example	Student Practice
5. Convert.	**6.** Convert.
(a) 26.48 miles to yards	**(a)** 17.37 yards to inches

5. (a) Use a unit fraction.

$$26.48 \text{ miles} \times \frac{1760 \text{ yards}}{1 \text{ mile}}$$
$$= 46{,}604.8 \text{ yards}$$

6. (b) $9\dfrac{3}{5}$ quarts to gallons

5. (b) $3\dfrac{2}{3}$ feet to yards

Use a unit fraction. Be sure to convert the mixed number to an improper fraction before performing the calculations.

$$3\frac{2}{3} \text{ feet} \times \frac{1 \text{ yard}}{3 \text{ feet}} = \frac{11}{3} \times \frac{1}{3} \text{ yard}$$
$$= 1\frac{2}{9} \text{ yard}$$

Example	Student Practice
7. Lynda's new car weighs 2.43 tons. How many pounds is that?	**8.** Barry's new truck weighs 4.39 tons. How many pounds is that?

7.
$$2.43 \text{ tons} \times \frac{2000 \text{ pounds}}{1 \text{ ton}} = 4860 \text{ pounds}$$

Example	Student Practice
9. The chemistry lab has 34 quarts of weak hydrochloric acid. How many gallons of this acid are in the lab? (Express your answer as a decimal.) Multiply by a unit fraction which relates quarts to gallons, with gallons in the numerator and quarts in the denominator. $$34 \text{ quarts} \times \frac{1 \text{ gallon}}{4 \text{ quarts}} = \frac{34}{4} \text{ gallons}$$ Divide to write as a decimal. $34 \div 4 = 8.5$ There are 8.5 gallons of hydrochloric acid in the lab.	10. The length of a skateboard ramp is 165 inches. What is the length of the ramp in feet? (Express your answer as a decimal.)
11. A window is 4 feet 5 inches wide. How many inches is that? 4 feet 5 inches means 4 feet and 5 inches. Change the 4 feet to inches and add the 5 inches. This time use a unit fraction with inches in the numerator and feet in the denominator to convert the 4 feet to inches. $$4 \text{ feet} \times \frac{12 \text{ inches}}{1 \text{ foot}} = 48 \text{ inches}$$ Add the result to 5 inches to find the total width of the window. 48 inches + 5 inches = 53 inches The window is 53 inches wide.	12. A refrigerator has a volume of 305 gallons and 3 quarts. How many quarts is that? Change the 305 gallons to quarts and add the 3 quarts.

Example	Student Practice
13. The Charlotte all-night garage charges $1.50 per hour for parking both day and night. A businessman left his car there for $2\frac{1}{4}$ days. How much was he charged?	**14.** Zeno is doing a plumbing job in his home. The hardware store charges 20 cents per inch for the copper pipe that he wants to use. He needs $25\frac{3}{4}$ feet of the copper pipe for his job. How much will the pipe cost him?

First convert days to hours. Then, multiply by $1.50 per hour. To make the calculations easier, write the number of days as 2.25.

$$2.25 \, \cancel{days} \times \frac{24 \, \cancel{hours}}{1 \, \cancel{day}} \times \frac{\$1.50}{1 \, \cancel{hour}} = \$81$$

The business man was charged $81. The check is left to the student.

Extra Practice

1. Convert. When necessary, express your answer as a decimal.

$15,840$ yards $=$ _____ miles

2. Convert. When necessary, express your answer as a decimal.

7.25 tons $=$ _____ pounds

3. Convert. When necessary, express your answer as a decimal.

$8\frac{1}{2}$ hours $=$ _____ minutes

4. Convert. When necessary, express your answer as a decimal.

200 cups $=$ _____ gallons

Concept Check
Explain how you would convert 250 pints to quarts.

Chapter 6 Measurements
6.2 Metric Measurements: Length

Vocabulary

meter   •   metric system   •   millimeter   •   kilometer

1. The _____ of measurement is used in most industrialized nations of the world.

2. A _____ is one thousand meters.

3. In the metric system, the basic unit of length is the _____.

4. A _____ is one thousandth of a meter.

Example	**Student Practice**
1. Write the prefixes that mean **(a)** thousand and **(b)** tenth. Use the prefix chart.	**2.** Write the prefixes that mean **(a)** hundredth and **(b)** ten.

Prefix	**Meaning**
kilo-	thousand
hecto-	hundred
deka-	ten
deci-	tenth
centi-	hundredth
milli-	thousandth

(a) The prefix *kilo-* is used for thousand.

(b) The prefix *deci-* is used for tenth.

Example	**Student Practice**
3. Change 5 kilometers to meters. To go from kilometer to meter, we move three places to the right. 5 kilometer = 5.000. meters = 5000 meters	**4.** Change 3 kilometers to centimeters.

Example	Student Practice
5. Change 56 millimeters to kilometers. To go from millimeters to kilometers, we move six places to the left. 56 millimeters = 0.000056. kilometers = 0.000056 kilometers	**6.** Change 247 meters to kilometers.
7. Bob measured the width of a doorway in his house. He wrote down "73." What unit of measurement did he use? Select the most reasonable choice. **(a)** 73 kilometers **(b)** 73 meters **(c)** 73 centimeters The most reasonable choice is **(c)**, 73 centimeters. The other two units would be much too long. A meter is close to a yard, and a doorway would not be 73 yards wide. A kilometer is even larger than a meter.	**8.** Cassandra had a vegetable garden, and she measured the length of the row of watermelon that she had planted. She wrote down "21." What unit of measurement did she use? Select the most reasonable choice. **(a)** 21 kilometers **(b)** 21 meters **(c)** 21 centimeters
9. Convert. km hm dam m dm cm mm **(a)** 982 cm to m Move the decimal point two places to the left because we are going from a smaller unit to a larger unit. 982 cm = 9.82. m = 9.82 m **(b)** 5.2 m to mm Move the decimal point three places to the right because we are going from a larger unit to a smaller unit. 5.2 m = 5.200. mm = 5200 mm	**10.** Convert. km hm dam m dm cm mm **(a)** 3.8 mm to dm **(b)** 526 dam to cm

Example	Student Practice
11. Convert.	**12.** Convert.

km hm dam m dm cm mm   km hm dam m dm cm mm

(a) 426 decimeters to kilometers

We are converting from a smaller unit, dm, to a larger one, km. Therefore, there will be fewer kilometers than decimeters. We move the decimal point four places to the left.

$$426 \text{ dm} = 0.\underline{0426}. \text{ km}$$
$$= 0.0426 \text{ km}$$

(a) 9.65 kilometers to millimeters

(b) 9.47 hectometers to meters

We are converting a larger unit, hm, to a smaller one, m. We move the decimal point two places to the right.

$$9.47 \text{ hm} = 9.\underline{47}. \text{ m}$$
$$= 947 \text{ m}$$

(b) 467 centimeters to meters

---

13. Add. $125 \text{ m} + 1.8 \text{ km} + 793 \text{ m}$

First we change the kilometer measurement to a measurement in meters.

$$1.8 \text{ km} = 1800 \text{ m}$$

Then we add.

$$
\begin{array}{r}
125 \text{ m} \\
1800 \text{ m} \\
+ \quad 793 \text{ m} \\
\hline
2718 \text{ m}
\end{array}
$$

14. Add. $617 \text{ cm} + 3.77 \text{ m} + 243 \text{ cm}$

Extra Practice

1. Perform the conversion.

    11.5 kilometers = _____ meters

2. Perform the conversion.

    137 centimeters = _____ meters

3. Perform the conversion.

    71 meters = _____ millimeters

4. Change to a convenient unit of measure and add.

    $4.75 \text{ km} + 60 \text{ m} + 832 \text{ cm} =$ _____ meters

Concept Check

Explain how you would convert 5643 centimeters to kilometers.

Chapter 6 Measurements
6.3 Metric Measurements: Volume and Weight

Vocabulary
liter   •   mass   •   gram   •   megagram

1. The measure for the amount of material in an object is called _____.

2. The metric ton is also called a _____.

3. A _____ is the weight of water in a box that is 1 centimeter on each side.

4. A _____ is defined as the volume of a box $10 \text{ cm} \times 10 \text{ cm} \times 10 \text{ cm}$, or 1000 cm^3.

Example	**Student Practice**
1. Convert.	**2.** Convert.
(a) $3 \text{ L} = \_\_\_\_ \text{ mL}$	**(a)** $0.59 \text{ L} = \_\_\_\_ \text{ mL}$

The prefix "milli-" is three places to the right. Move the decimal point three places to the right.

$3 \text{ L} = 3.000. = 3000 \text{ mL}$

(b) $24 \text{ kL} = \_\_\_\_ \text{ L}$	**(b)** $5.3 \text{ kL} = \_\_\_\_ \text{ L}$

$24 \text{ kL} = 24.000. = 24,000 \text{ L}$

(c) $0.084 \text{ L} = \_\_\_\_ \text{ mL}$	**(c)** $27 \text{ L} = \_\_\_\_ \text{ mL}$

$0.084 \text{ L} = 0.084. = 84 \text{ mL}$

Example	Student Practice
3. Convert.	**4.** Convert.
(a) 26.4 mL = ____ L	**(a)** 6.79 cL = ____ L

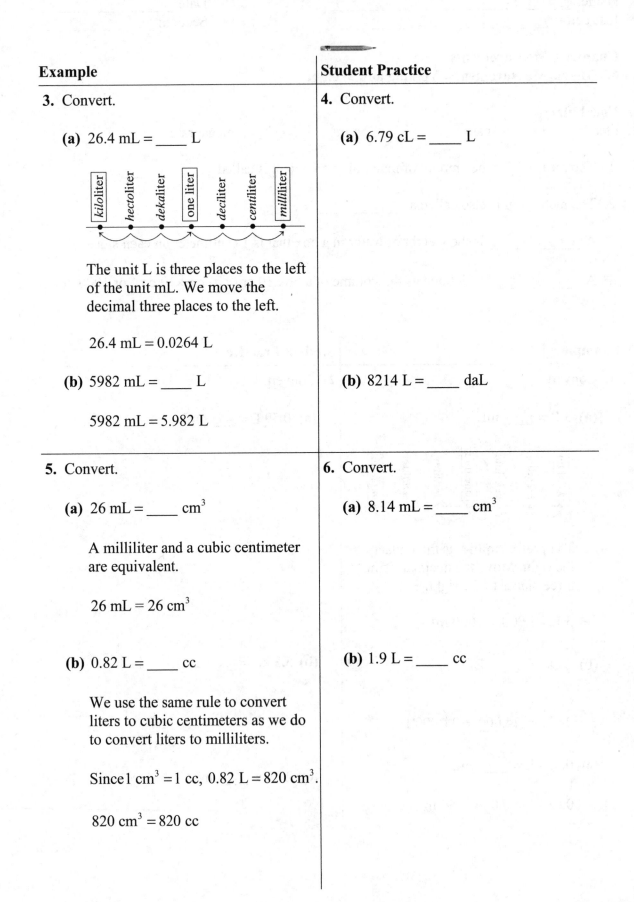

The unit L is three places to the left of the unit mL. We move the decimal three places to the left.

26.4 mL = 0.0264 L

(b) 5982 mL = ____ L	**(b)** 8214 L = ____ daL

5982 mL = 5.982 L

---

5. Convert.	**6.** Convert.
(a) 26 mL = ____ cm^3	**(a)** 8.14 mL = ____ cm^3

A milliliter and a cubic centimeter are equivalent.

26 mL = 26 cm^3

(b) 0.82 L = ____ cc	**(b)** 1.9 L = ____ cc

We use the same rule to convert liters to cubic centimeters as we do to convert liters to milliliters.

Since 1 cm^3 = 1 cc, 0.82 L = 820 cm^3.

820 cm^3 = 820 cc

Example	Student Practice
7. Convert.	**8.** Convert.
(a) 2 t = _____ kg	**(a)** 1.73 kg = _____ g

$$2\ t = 2000\ kg$$

(b) 0.42 kg = _____ g

$$0.42\ kg = 420\ g$$

(b) 0.39 t = _____ kg

---

9. Convert.

(a) 283 kg = _____ t

$$283\ kg = 0.283\ t$$

(b) 7.98 mg = _____ g

$$7.98\ mg = 0.00798\ g$$

10. Convert.

(a) 81.4 mg = _____ g

(b) 61.7 kg = _____ t

---

11. If a chemical costs $0.03 per gram, what will it cost per kilogram?

Since there are 1000 grams in a kilogram, a chemical that costs $0.03 per gram would cost 1000 times as much per kilogram.

$$1000 \times \$0.03 = \$30.00$$

The chemical would cost $30.00 per kilogram.

12. If an alternative heating fuel costs $4500 per metric ton, how much will it cost per kilogram?

Example	Student Practice
13. Select the most reasonable weight for Tammy's Toyota. (a) 820 t (b) 820 g (c) 820 kg (d) 820 mg The most reasonable answer is **(c)**, 820 kg. The other weight values are much too large or much too small. Since a kilogram is slightly more than 2 pounds, we see that this weight, 820 kg, most closely approximates the weight of a car.	**14.** Select the most reasonable volume for Ryan's fish tank. (a) 8.2 kL (b) 8.2 L (c) 8.2 mL (d) 8.2 hL

Extra Practice

1. Perform the conversion.

$$756 \text{ mL} = \underline{\hspace{1cm}} \text{ cm}^3$$

2. Find a convenient unit of measure and add.

$$230 \text{ L} + 2.4 \text{ kL} + 150{,}000 \text{ mL} = \underline{\hspace{1cm}} \text{ kL}$$

3. Perform the conversion.

$$43{,}000 \text{ mg} = \underline{\hspace{1cm}} \text{ kg}$$

4. Change to a convenient unit of measure and add.

$$320 \text{ g} + 8 \text{ kg} + 870 \text{ mg} = \underline{\hspace{1cm}} \text{ g}$$

Concept Check
Explain how you would convert 54 kilograms to milligrams.

Chapter 6 Measurements
6.4 Converting Units

Vocabulary
Fahrenheit system   •   approximations   •   Celsius scale   •   unit fraction

1. To convert between American units and metric units, it is helpful to have equivalent values. Most of these are _____.

2. To convert from one unit to another you multiply by a(n) _____.

3. In the _____ water boils at $212°$ and freezes at $32°$.

4. In the _____ water boils at $100°$ and freezes at $0°$

Example	**Student Practice**
1. Convert 3 feet to meters.	**2.** Convert 85 feet to meters.
The unit fraction is $\dfrac{0.305 \text{ meter}}{1 \text{ foot}}$.	
$3 \text{ feet} \times \dfrac{0.305 \text{ meter}}{1 \text{ foot}} = 0.915 \text{ meter}$	
3. Answer parts **(a)** and **(b)**.	**4.** Answer parts **(a)** and **(b)**.
(a) Convert 26 m to yd.	**(a)** Convert 1.5 gal to L.
$26 \text{ m} \times \dfrac{1.09 \text{ yd}}{1 \text{ m}} = 28.34 \text{ yd}$	
(b) Convert 1.9 km to mi.	**(b)** Convert 0.9 L to qt.
$1.9 \text{ km} \times \dfrac{0.62 \text{ mi}}{1 \text{ km}} = 1.178 \text{ mi}$	

Example	Student Practice
5. Convert 235 cm to ft. Round to the nearest hundredth of a foot.	**6.** Convert 47 cm to ft. Round to the nearest hundredth of a foot.

5. (continued)

Our first fraction converts centimeters to inches. Our second fraction converts inches to feet.

$$235 \text{ cm} \times \frac{0.394 \text{ in.}}{1 \text{ cm}} \times \frac{1 \text{ ft}}{12 \text{ in.}} = \frac{92.59 \text{ ft}}{12}$$
$$\approx 7.72 \text{ ft}$$

---

7. Convert 100 km/hr to mi/hr.

We need to multiply by a unit fraction. The fraction we multiply by must have kilometers in the denominator.

$$\frac{100 \text{ km}}{\text{hr}} \times \frac{0.62 \text{ mi}}{1 \text{ km}} = 62 \text{ mi/hr}$$

Thus 100 km/hr is approximately equal to 62 mi/hr.

8. Convert 120 km/hr to mi/hr.

---

9. A rocket carrying a communication satellite is launched from a rocket launch pad. It is traveling at 700 miles per hour. How many feet per second is the rocket traveling? Round to the nearest whole number.

Because the conversion is from miles per hour to feet per second it requires two unit fractions.

$$\frac{700 \text{ miles}}{\text{hr}} \times \frac{5280 \text{ ft}}{1 \text{ mile}} \times \frac{1 \text{ hr}}{60 \text{ min}} \times \frac{1 \text{ min}}{60 \text{ sec}}$$
$$= \frac{700 \times 5280 \text{ ft}}{60 \times 60 \text{ sec}}$$
$$= \frac{3,696,000 \text{ ft}}{3600 \text{ sec}}$$
$$\approx 1027 \text{ ft/sec}$$

10. A baseball pitcher throws a pitch that is clocked at 95 miles per hour. What is the speed of the baseball in feet per second? Round to the nearest whole number.

Example	Student Practice
11. When the temperature is $35°\,C$, what is the Fahrenheit reading?	**12.** When the temperature is $15°\,C$, what is the Fahrenheit reading?

Use the formula for converting Celsius to Fahrenheit, $F = 1.8 \times C + 32$.

$$F = 1.8 \times C + 32$$
$$= 1.8 \times 35 + 32$$
$$= 63 + 32$$
$$= 95$$

The temperature is $95°\,F$.

---

13. When the temperature is $50°\,F$, what is the Celsius reading?	**14.** When the temperature is $41°\,F$, what is the Celsius reading?

Use the formula for converting Fahrenheit to Celsius, $C = \dfrac{5 \times F - 160}{9}$.

$$C = \frac{5 \times F - 160}{9}$$
$$= \frac{5 \times 50 - 160}{9}$$
$$= \frac{250 - 160}{9}$$
$$= \frac{90}{9}$$
$$= 10$$

The temperature is $10°\,C$.

Extra Practice

1. Perform the conversion. Round to the nearest hundredth if necessary.

   22 m to yd

2. Perform the conversion. Round to the nearest hundredth if necessary.

   8 qt to L

3. Perform the conversion. Round to the nearest hundredth if necessary.

   10 lb to kg

4. Perform the conversion. Round to the nearest hundredth if necessary.

   $125°C$ to Fahrenheit

Concept Check

Explain how you would convert a speed of 65 miles per hour to a speed in kilometers per hour.

Chapter 6 Measurements
6.5 Solving Applied Measurement Problems

Vocabulary

perimeter   •   unit fraction   •   estimated   •   units

1. To convert from one unit to another unit, multiply by a(n) _____.

2. In solving applied problems with fractions, we can check our work by using _____ values.

3. The _____ is the sum of the length of the sides of a geometric figure.

4. In order to make calculations and solve applied measurement problems with fractions it is helpful to get all the measurements in the same _____.

Example	**Student Practice**
1. A triangular support piece holds a solar panel. The sketch shows the dimensions of the triangle. Find the perimeter of this triangle. Express the answer in feet.	2. Sylvia is having a fence built in her yard to keep her goats from straying. The dimensions of the rectangular area she is going to have fenced in are $12\frac{1}{8}$ meters by $6\frac{5}{8}$ meters. What is the length of the fencing that she must buy? Express the answer in yards.

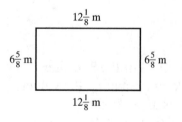

Read the problem carefully and create a Mathematics Blueprint. The perimeter is the sum of the length of the sides.

$$2\frac{1}{4}+1\frac{1}{4}+2\frac{3}{4}=5\frac{5}{4}=6\frac{1}{4} \text{ yards}$$

Convert $6\frac{1}{4}$ or $\frac{25}{4}$ yards to feet.

$$\frac{25}{4}\text{ yards}\times\frac{3\text{ feet}}{1\text{ yards}}=\frac{75}{4}\text{ feet}=18\frac{3}{4}\text{ feet}$$

Example	Student Practice

3. How many 210- liter gasoline barrels can be filled from a tank of 5.04 kiloliters of gasoline?

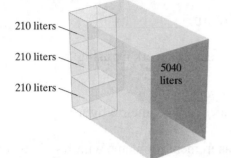

Read the problem carefully and create a Mathematics Blueprint. We are going to divide the tank size of 5.04 kiloliters by the 210-liter barrels to find out how many barrels can be filled.

First we convert 5.04 kiloliters to liters.

5.04 kiloliters = 5040 liters

Now we find out how many 210-liter barrels can be filled by 5040 liters. We need to divide.

$$\frac{5040 \text{ liters}}{210 \text{ liters}} = 24$$

Thus, we can fill 24 of the 210-liter barrels. Check by estimating the values. 5.04 kiloliters is approximately 5 kiloliters or 5000 liters. 210-liter barrels hold approximately 200 liters.

$$\frac{5000}{200} = 25$$

The estimate is close to our calculated value, so our value of 24 is reasonable.

4. Brad has made 36.9 liters of tomato sauce from the tomatoes in his garden. He wants to bottle the sauce in containers with a volume of 450 milliliters each. How many containers will he fill?

164

Extra Practice

1. A rectangular window measures 28 inches × 54 inches. Window insulation is applied along all four sides. The insulation costs $6.00 per meter. What will it cost to insulate the window?

2. A powdered sauce company has 400 kilograms of powdered barbecue sauce mix that is to be packaged into 50-g packets. How many packets can be made from this powdered barbecue sauce?

3. The temperature in London, England was 18°C, while on the same day the temperature in Boston, Massachusetts was 68°F. How much hotter was Boston than London that day?

4. A leaky faucet drips at a rate of 1 cup every 15 hours. How many gallons of water will be lost in 30 days? Round to the nearest hundredth.

Concept Check

Suppose you are reading a map that has a scale showing that 3 inches represents 6.5 miles on the ground. Explain how you would find out how many miles there are between two cities that are 5 inches apart on the map.

MATH COACH

Mastering the skills you need to do well on the test.

Watch the MATH COACH videos in MyMathLab® or on YouTube while you work the problems below. These helpful hints will help you avoid making common errors on test problems.

Convert American Units—Problem 6   3 cups = _____ qt

> **Helpful Hint:** Recall that there are four cups in one quart. Make sure to round your answer to the nearest hundredth as requested.

Did you use the correct unit fraction for this problem, $\frac{1 \text{ quart}}{4 \text{ cups}}$?

Yes _____   No _____

If you answered No, stop and consider why you should use this unit fraction.

Did you multiply $3 \text{ cups} \times \frac{1 \text{ quart}}{4 \text{ cups}}$?

Yes _____   No _____

If you answered No, go back and perform this operation.

Now go back and rework Problem 6 using these suggestions.

Convert Metric Units—Problem 16   983 g = _____ kg

> **Helpful Hint:** Remember that a kilogram is much larger than a gram so there will be fewer kilograms than grams when the conversion is made.

Did you remember that when converting from grams (smaller unit) to kilograms (larger unit), you must move the decimal point to the LEFT?

If you answered No, stop and make this adjustment in your work.

Yes _____   No _____

If you answered No, go back and perform this step.

Did you remember to move the decimal point three places?

Yes _____   No _____

Convert between Metric and American Units—Problem 22   30 km = _____ mi

Helpful Hint: Recall that 1 kilometer is approximately equal to 0.62 mile.

Did you remember to choose a unit fraction with kilometers in the denominator and miles in the numerator?

If you answered No, stop and perform this operation.

Yes ____   No ____

If you answered No, stop and consider which unit fraction is best for this problem.

Did you multiply $30 \text{ km} \times \dfrac{0.62 \text{ mi}}{1 \text{ km}}$?

Yes ____   No ____

Convert Temperatures—Problem 32   The warmest day this year in Acapulco, Mexico was $40°C$. What was the Fahrenheit temperature?

Helpful Hint: When looking for a Fahrenheit temperature, we use the formula $F = 1.8 \times C + 32$, where F is the number of Fahrenheit degrees and C is the number of Celsius degrees.

Did you choose the correct formula, $F = 1.8 \times C + 32$?

If you answered Problem 32 incorrectly, please rework the problem using these suggestions.

Yes ____   No ____

If you answered No, go back and think about why this is the correct formula for this problem.

Did you substitute for C correctly by writing
$F = 1.8 \times 40 + 32$?

Yes ____   No ____

If you answered No, stop and make this correction to your work.

Did you perform the multiplication of 1.8×40 before adding 32?

Yes ____   No ____

If you answered No, go back and follow the correct order of operations.

Chapter 7 Geometry
7.1 Angles

Vocabulary
geometry • line • line segment • ray • angle • sides • vertex • degrees • right angle
perpendicular • straight angle • acute angle • obtuse angle • complementary angles
complement • supplementary angles • supplement • adjacent • vertical angles
parallel lines • transversal • alternate interior angles • corresponding angles

1. A 90° angle is called a(n) _____.

2. _____ never meet.

3. Angles are commonly measured in _____.

4. _____ is a branch of mathematics that deals with the properties of and
   relationships between figures and space.

Example	**Student Practice**
1. In the following sketch, determine which angles are acute, obtuse, right, or straight angles.	**2.** In the following sketch, determine which angles are acute, obtuse, right, or straight angles.

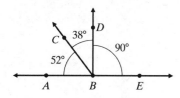

∠ABC and ∠CBD are acute angles.

∠CBE is an obtuse angle.

∠ABD and ∠DBE are right angles.

∠ABE is a straight angle.

Example	Student Practice

3. Angle A measures $39°$.

   (a) Find the complement of angle A.

      Complementary angles have a sum of $90°$. So the complement of angle A measures $90° - 39° = 51°$.

   (b) Find the supplement of angle A.

      Supplementary angles have a sum of $180°$. So the supplement of angle A measures $180° - 39° = 141°$.

4. Angle B measures $42°$.

   (a) Find the complement of angle B.

   (b) Find the supplement of angle B.

---

5. In the following sketch, two lines intersect, forming four angles. The measure of angle a is $55°$. Find the measure of all the other angles.

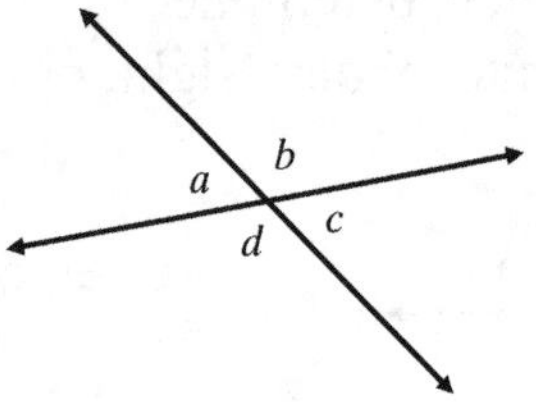

Since $\angle a$ and $\angle c$ are vertical angles, we know that they have the same measure. Thus, we know that $\angle c$ measures $55°$.

Since $\angle a$ and $\angle b$ are adjacent angles of intersecting lines, we know that they are supplementary angles. Thus, we know that $\angle b$ measures $180° - 55° = 125°$.

Finally, $\angle b$ and $\angle d$ are vertical angles, so we know that they have the same measure. Thus, we know that $\angle d$ measures $125°$.

6. In the following sketch, two lines intersect, forming four angles. The measure of angle h is $116°$. Find the measure of all the other angles.

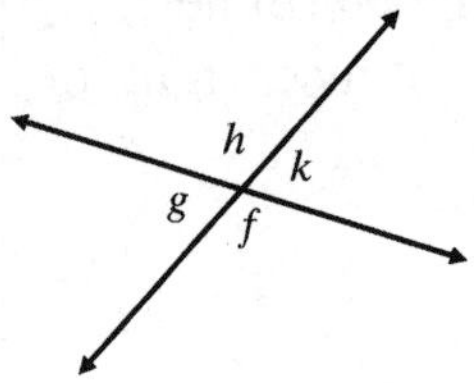

Example	Student Practice
7. In the following figure, $m \parallel n$ and the measure of $\angle a$ is $64°$. Find the measures of $\angle b$, $\angle c$, $\angle d$, and $\angle e$. 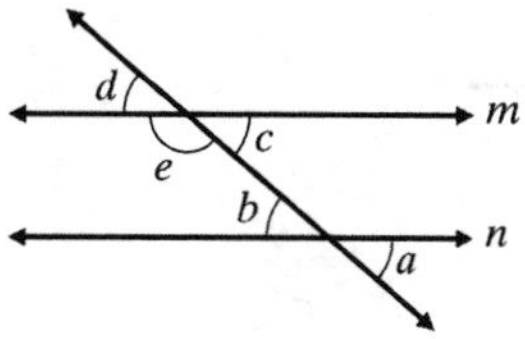	**8.** In the following figure, $p \parallel q$ and the measure of $\angle x$ is $125°$. Find the measures of $\angle w$, $\angle y$, $\angle z$, and $\angle v$. 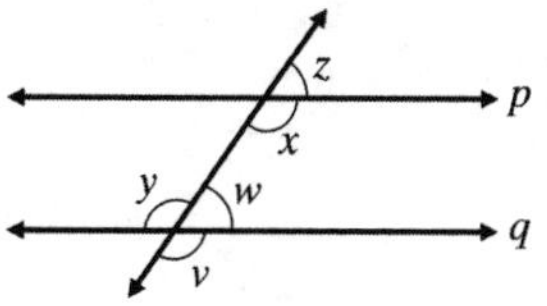

$\angle a$ and $\angle b$ are vertical angles. They have the same measure.

$\angle a = \angle b = 64°$

$\angle b$ and $\angle c$ are alternate interior angles. They have the same measure.

$\angle b = \angle c = 64°$

$\angle b$ and $\angle d$ are corresponding angles. They have the same measure.

$\angle b = \angle d = 64°$

$\angle e$ and $\angle d$ are adjacent angles of intersecting lines. They are supplementary angles.

$\angle e = 180° - 64° = 116°$

1. Given $p \parallel q$ and $\angle a = 142°$, find the measure of $\angle h$.

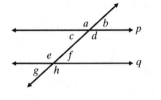

2. Given $p \parallel q$ and $\angle a = 142°$, find the measure of $\angle g$

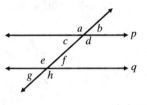

3. Find the measure of $\angle JOM$, as shown in the following sketch below. Assume that angle LOK is a right angle, and angle JOK is a straight angle.

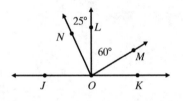

4. Find the complement of an angle that measures $79°$.

Concept Check

In the figure below, lines m and n are parallel. In this figure, explain what the relationship is between $\angle e$ and $\angle a$. If you know the measure of $\angle e$, how can you find the measure of $\angle a$?

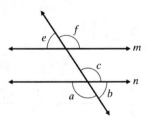

Name: _____ Date: _____

Instructor: _____ Section: _____

Chapter 7 Geometry
7.2 Rectangles and Squares

Vocabulary
rectangle • square • perimeter • area

1. The _____ of a rectangle is the sum of the lengths of all its sides.

2. _____ is the measure of the surface inside a geometric figure.

3. A(n) _____ is a four-sided figure that has four right angles.

4. If all four sides have the same length, then the rectangle is called a(n) _____.

Example	Student Practice
1. A helicopter has a 3-cm by 5.5-cm insulation pad near the control panel that is rectangular. Find the perimeter of the rectangle.	**2.** Find the perimeter of the rectangle.

Example

1. A helicopter has a 3-cm by 5.5-cm insulation pad near the control panel that is rectangular. Find the perimeter of the rectangle.

Length = l = 5.5 cm
Width = w = 3 cm

In the formula for the perimeter of a rectangle, we substitute 5.5 cm for l and 3 cm for w. Remember, $2l$ means 2 times l and $2w$ means 2 times w.

$P = 2l + 2w$

$= (2)(5.5 \text{ cm}) + (2)(3 \text{ cm})$

$= 11 \text{ cm} + 6 \text{ cm} = 17 \text{ cm}$

Thus, the perimeter of the rectangle is 17 cm.

Student Practice

2. Find the perimeter of the rectangle.

Example	Student Practice
3. High Ridge Stables has a new sign at the highway entrance that is in the shape of a square, with each side measuring 8.6 yards. Find the perimeter of the sign.	**4.** Find the perimeter of the square.

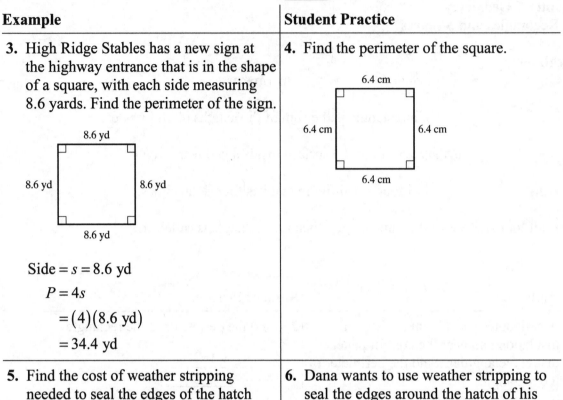

3. (Example continued)

Side $= s = 8.6$ yd

$P = 4s$

$\quad = (4)(8.6 \text{ yd})$

$\quad = 34.4$ yd

4. 6.4 cm (each side, square)

---

5. Find the cost of weather stripping needed to seal the edges of the hatch shown below. Weather stripping costs $0.12 per foot.

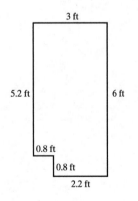

$3.0 + 6.0 + 2.2 + 0.8 + 0.8 + 5.2 = 18.0$ ft

The perimeter is 18 ft. Now calculate the cost.

$18.0 \text{ ft} \times \dfrac{0.12 \text{ dollar}}{\text{ft}} = \$2.16$

Thus the cost for the weather stripping materials is $2.16.

6. Dana wants to use weather stripping to seal the edges around the hatch of his boat. The weather stripping Dana wants to use costs $0.17 per foot. Find the cost of the weather stripping required.

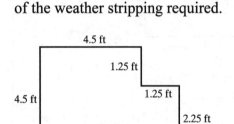

Example	Student Practice
7. Find the area of the rectangle shown below.	**8.** Find the area of the rectangle shown below.

7. Find the area of the rectangle shown below.

19 ft

7 ft

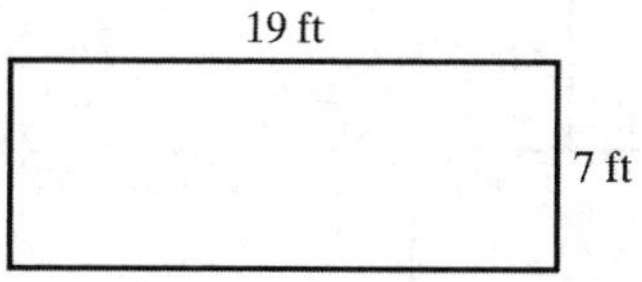

Our answer must be in square feet because the measures of the length and width are in feet.

The length of the rectangle is $l = 19$ ft. The width of the rectangle is $w = 7$ ft.

$A = (l)(w)$

$\quad = (19 \text{ ft})(7 \text{ ft})$

$\quad = 133 \text{ ft}^2$

The area is 133 square feet.

8. Find the area of the rectangle shown below.

27 m

13 m   13 m

27 m

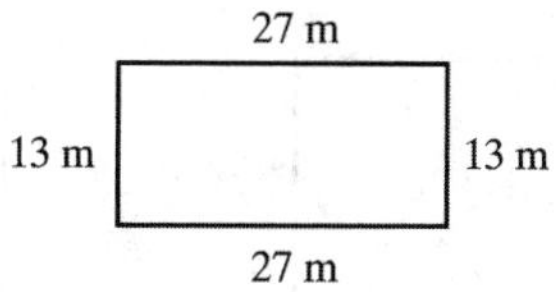

9. A square measures 9.6 in. on each side. Find its area.

We know our answer will be measured in square inches. We will write this as in.^2.

$A = s^2$

$\quad = (9.6 \text{ in.})^2$

$\quad = (9.6 \text{ in.})(9.6 \text{ in.})$

$\quad = 92.16 \text{ in.}^2$

10. A square measures 4.8 m on each side. Find its area.

Example	Student Practice
11. Consider the shape shown below, which is made up of a rectangle and a square. Find the area of the region. 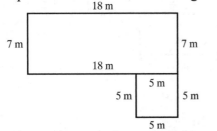	**12.** Find the area of the region shown in the figure below. 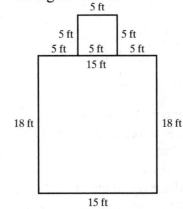

The total area is the sum of the two separate areas. The area of the rectangle is $(7 \text{ m})(18 \text{ m}) = 126 \text{ m}^2$.

The area of the square is $(5 \text{ m})^2 = 25 \text{ m}^2$.

$126 \text{ m}^2 + 25 \text{ m}^2 = 151 \text{ m}^2$

Thus, the area of the region is 151 m^2.

Extra Practice

1. Find the perimeter of the rectangle.
   length $= 12\dfrac{1}{3}$ m, width $= 7\dfrac{1}{3}$ m

2. Find the area of the square.
   length = width = 42.38 cm

3. The rectangular swimming pool at the health club measures 30 feet wide and 60 feet long. A new pool cover is needed and costs $0.75 per square foot. If the manager purchases exactly what is needed, how much will the pool cover cost?

4. Find the perimeter and area of the shape made up of rectangles and squares.
   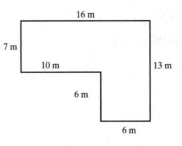

Concept Check

If a rectangle that measures 12 feet by 15 feet is attached to a square that measures 15 feet on a side, explain how you would find the area of the entire region.

Chapter 7 Geometry
7.3 Parallelograms, Trapezoids, and Rhombuses

Vocabulary
quadrilaterals • parallelogram • perimeter • area • base • height • rhombus • trapezoid

1. A(n) _____ is a four-sided figure in which both pairs of opposite sides are parallel.

2. All four-sided figures are _____.

3. A(n) _____ is a four-sided figure with two parallel sides.

4. A(n) _____ is a parallelogram with all four sides equal.

Example	**Student Practice**
1. Find the perimeter.	**2.** Find the perimeter.
2.6 m 1.2 m / / 1.2 m 2.6 m	6.7 ft 5.1 ft / / 5.1 ft 6.7 ft
The perimeter of a parallelogram is the distance around the parallelogram. $P = (2)(1.2 \text{ meters}) + (2)(2.6 \text{ meters})$ $= 2.4 \text{ meters} + 5.2 \text{ meters}$ $= 7.6 \text{ meters}$	
3. Find the area of a parallelogram with base 7.5 m and height 3.2 m.	**4.** Find the area of a parallelogram with base 9.8 mm and height 2.1 mm.
To find the area of a parallelogram, we multiply the base times the height. $A = bh$ $= (7.5 \text{ m})(3.2 \text{ m})$ $= 24 \text{ m}^2$	

Example	Student Practice
5. A truck is manufactured with an iron brace welded to the truck frame. The brace is shaped like a rhombus. The brace has a base of 9 inches and a height of 5 inches. Find the perimeter and the area of this iron brace. Since all four sides are equal, multiply to find the perimeter. $P = 4(9 \text{ in.}) = 36 \text{ in.}$ Since the rhombus is a special type of parallelogram, use the area formula for a parallelogram. $A = bh = (9 \text{ in.})(5 \text{ in.}) = 45 \text{ in.}^2$	**6.** A tile has the shape of a rhombus. The base is 8.25 cm and the height is 6 cm. Find the perimeter and area of the tile.
7. Find the perimeter of the trapezoid. 12 m 5 m 5 m 18 m $P = 18 \text{ m} + 5 \text{ m} + 12 \text{ m} + 5 \text{ m} = 40 \text{ m}$	**8.** Find the perimeter of the trapezoid. 4 ft 4 ft 8 ft 5 ft
9. A roadside sign is in the shape of a trapezoid. It has a height of 30 ft, and the bases are 60 ft and 75 ft. What is the area of the sign? Use the trapezoid formula with $h = 30$, $b = 60$, and $B = 75$. $$A = \frac{h(b+B)}{2}$$ $$= \frac{(30 \text{ ft})(60 \text{ ft} + 75 \text{ ft})}{2}$$ $$= \frac{4050}{2} \text{ ft}^2 = 2025 \text{ ft}^2$$	**10.** A garden is in the shape of a trapezoid. The trapezoid has a height of 15 m. The bases measure 21 m and 27 m.

Example	Student Practice
11. Find the area of the following piece for inlaid woodwork made by a master carpenter. Since this shape is hard to cut, it is made of one trapezoid and one rectangle laid together.	**12.** Find the area of the following piece for inlaid woodwork. The shape is made of one trapezoid and one rectangle.

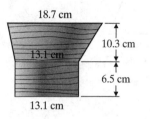

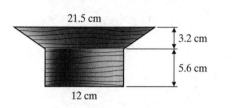

Separate the area into two portions and find the area of each portion.

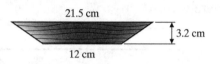

Find the area of the trapezoid.

$$A = \frac{h(b+B)}{2}$$
$$= \frac{(3.2 \text{ cm})(12 \text{ cm} + 21.5 \text{ cm})}{2}$$
$$= \frac{107.2}{2} \text{ cm}^2 = 53.6 \text{ cm}^2$$

Find the area of the rectangle.

$$A = lw$$
$$= (12 \text{ cm})(5.6 \text{ cm}) = 67.2 \text{ cm}^2$$

We now add each area.

$$67.2 \text{ cm}^2 + 53.6 \text{ cm}^2 = 120.8 \text{ cm}^2$$

The total area of the piece for inlaid woodwork is 120.8 cm^2.

Extra Practice

1. Find the perimeter and the area for the figure below.

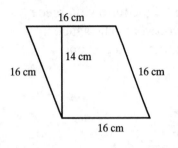

16 cm

14 cm

16 cm   16 cm

16 cm

2. Find the perimeter and the area for the figure below.

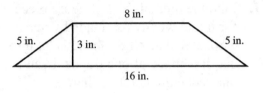

8 in.

5 in.   3 in.   5 in.

16 in.

3. A theater company is building a stage in the shape of a parallelogram. It has a base of 22 yards and a height of 35 yards. Find the area of the stage.

4. Calculate the cost of carpeting the following area, if carpeting costs $18 per square yard.

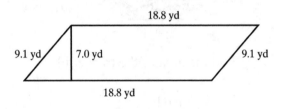

18.8 yd

9.1 yd   7.0 yd   9.1 yd

18.8 yd

Concept Check

The area of a trapezoid with a height of 9 meters and bases of 30 meters and 34 meters is 288 square meters. Explain what would happen to the area of this trapezoid if the height was increased to 16 meters but the length of each base remained the same.

Chapter 7 Geometry
7.4 Triangles

Vocabulary
triangle • isosceles triangle • equilateral triangle • scalene triangle • right triangle
height • base

1. A triangle with one 90° angle is called a(n) _____.

2. A triangle with two equal sides is called a(n) _____.

3. A triangle with three equal sides is called a(n) _____.

4. A(n) _____ has no two sides of equal lengths and no two angles of equal
   measure.

Example	**Student Practice**
1. In the triangle below, angle A measures $35°$ and angle B measures $95°$. Find the measure of angle C.	**2.** In a triangle, angle B measures $100°$ and angle C measures $25°$. What is the measure of angle A?

We will use the fact that the sum of the
measures of a triangle is $180°$.

$$35 + 95 + x = 180$$
$$130 + x = 180$$

What number x when added to 130
equals 180?

Since $130 + 50 = 180$, x must equal 50.

Angle C must measure $50°$.

Example	Student Practice
3. Find the perimeter of a triangular sail whose sides are 12 ft, 14 ft, and 17 ft.	**4.** Find the perimeter of a triangle whose sides are 4.7 cm, 9.3 cm, and 11.6 cm.

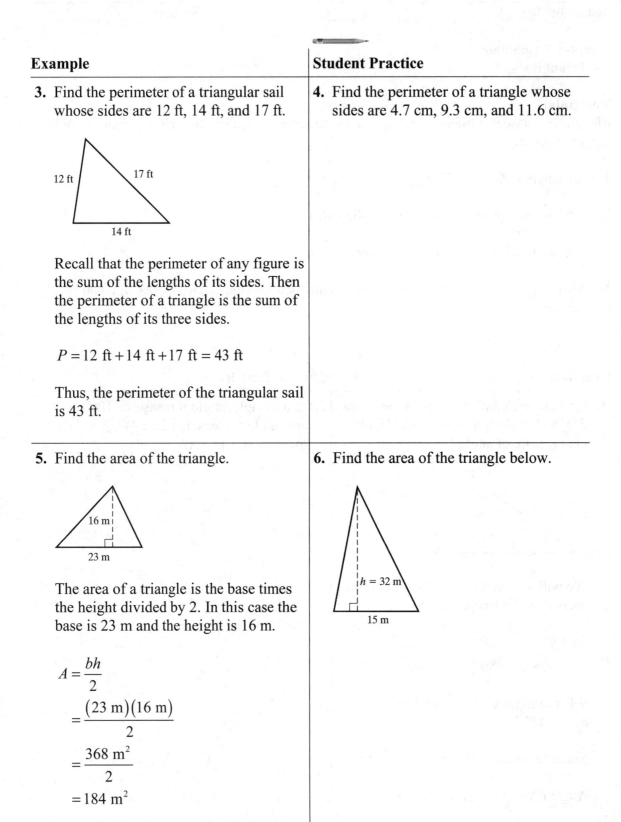

Recall that the perimeter of any figure is the sum of the lengths of its sides. Then the perimeter of a triangle is the sum of the lengths of its three sides.

$P = 12 \text{ ft} + 14 \text{ ft} + 17 \text{ ft} = 43 \text{ ft}$

Thus, the perimeter of the triangular sail is 43 ft.

5. Find the area of the triangle.

6. Find the area of the triangle below.

The area of a triangle is the base times the height divided by 2. In this case the base is 23 m and the height is 16 m.

$$A = \frac{bh}{2}$$
$$= \frac{(23 \text{ m})(16 \text{ m})}{2}$$
$$= \frac{368 \text{ m}^2}{2}$$
$$= 184 \text{ m}^2$$

Thus, the area of the triangle is 184 m^2.

Example	Student Practice

7. Find the area of the side of the house.

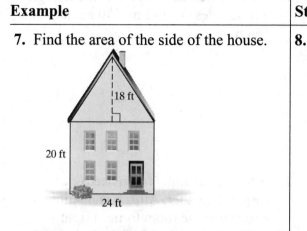

18 ft
20 ft
24 ft

Because the lengths of opposite sides of a rectangle are equal, the triangle has a base of 24 ft. Thus, we can calculate its area.

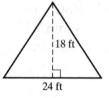

18 ft
24 ft

$$A = \frac{bh}{2}$$

$$= \frac{(24 \text{ ft})(18 \text{ ft})}{2}$$

$$= \frac{432 \text{ ft}^2}{2} = 216 \text{ ft}^2$$

Find the area of the rectangle.

$$A = lw = (24 \text{ ft})(20 \text{ ft}) = 480 \text{ ft}^2$$

Now we find the sum of the two areas.

$$\begin{array}{r} 216 \text{ ft}^2 \\ + 480 \text{ ft}^2 \\ \hline 696 \text{ ft}^2 \end{array}$$

Thus, the area of the side of the house is 696 square feet.

8. Find the area of the figure.

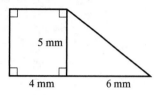

5 mm
4 mm   6 mm

183

Extra Practice

1. A triangle has two angles that measure 101° and 44°. Find the missing angle in the triangle.

2. Find the perimeter of a scalene triangle whose sides are 142 m, 130 m, and 164 m.

3. Find the area of the triangle.

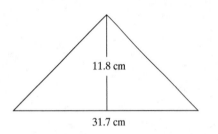

11.8 cm

31.7 cm

4. A triangular study room needs new carpet. Together, Allison and Alicia notice that the room forms a right triangle. The lengths of the walls of the study room measure 27 feet, 27 feet, and 38.2 feet. Allison and Alicia have priced carpet and find that the carpet they want costs $8.99/square *yard*. How much will it cost to carpet the study room? (If necessary, round to the nearest cent.)

Concept Check

A triangle has a base of 20 yards and a height of 20 yards. The triangle is attached to a rectangle that measures 20 yards by 15 yards. Explain how you would find the combined area of the triangle and the rectangle.

Chapter 7 Geometry
7.5 Square Roots

Vocabulary
square root • perfect square • approximation • before • after

1. If square roots are added or subtracted, they must be evaluated _____ being added or subtracted.

2. When we cannot get an exact answer, we use a(n) _____.

3. If a number is a product of two identical factors, than either factor is called a(n) _____.

4. When a whole number is multiplied by itself, the number that is obtained is called a(n) _____.

Example	**Student Practice**
1. Find.	**2.** Find.
(a) $\sqrt{25}$	**(a)** $\sqrt{16}$
Because $(5)(5) = 25$, $\sqrt{25} = 5$.	
(b) $\sqrt{121}$	**(b)** $\sqrt{196}$
Because $(11)(11) = 121$, $\sqrt{121} = 11$.	
3. Find $\sqrt{25} + \sqrt{36}$.	**4.** Find $\sqrt{64} - \sqrt{4}$.
Since we are adding the square roots, they must be evaluated first, then added.	
$\sqrt{25} = 5$ because $(5)(5) = 25$.	
$\sqrt{36} = 6$ because $(6)(6) = 36$.	
Thus, $\sqrt{25} + \sqrt{36} = 5 + 6 = 11$.	

Example	Student Practice
5.	**6.**
(a) Is 81 a perfect square?	**(a)** Is 1 a perfect square?
Yes. 81 is a perfect square because $(9)(9) = 81$.	
(b) If so, find $\sqrt{81}$.	**(b)** If so, find $\sqrt{1}$.
$\sqrt{81} = 9$	
7. Find approximate values using a square root table or a calculator. Round to the nearest thousandth.	**8.** Find approximate values using the square root table or a calculator. Round to the nearest thousandth.
(a) $\sqrt{2}$	**(a)** $\sqrt{5}$
$\sqrt{2} \approx 1.414$	
(b) $\sqrt{12}$	**(b)** $\sqrt{19}$
$\sqrt{12} \approx 3.464$	
(c) $\sqrt{7}$	**(c)** $\sqrt{11}$
$\sqrt{7} \approx 2.646$	
9. Approximate to the nearest thousandth of an inch the length of the side of a square that has an area of 6 in.^2.	**10.** Approximate to the nearest thousandth of a foot the length of the side of a square that has an area of 34 ft^2.
$\sqrt{6 \text{ in.}^2} \approx 2.449 \text{ in.}$	

6 in.²   2.449 in.

2.449 in.

Thus, to the nearest thousandth of an inch, the side measures 2.449 in. | |

Extra Practice

1. Find $\sqrt{100}$. Do not use a calculator. Do not refer to a table of square roots.

2. Evaluate the square roots first, then add or subtract the results. Do not use a calculator or a square root table.

$$\sqrt{144} + \sqrt{0}$$

3. Use a table of square roots or a calculator with a square root key to approximate $\sqrt{83}$ to the nearest thousandth.

4. Use a table of square roots or a calculator with a square root key to approximate $\sqrt{99}$ to the nearest thousandth.

Concept Check

Explain how you would find the length of the side of a tiny square with an area of 0.81 cm^2 without using a calculator.

Name: _____   Date: _____
Instructor: _____   Section: _____

Chapter 7 Geometry
7.6 The Pythagorean Theorem

Vocabulary
Pythagorean Theorem • hypotenuse • leg • 30°-60°-90° triangle • 45°-45°-90° triangle

1. In a right triangle, the side opposite the right angle is called the _____.

2. The _____ states that for any right triangle, the square of the hypotenuse equals the sum of the squares of the two legs of the triangle.

3. In a _____, the length of the hypotenuse is equal to $\sqrt{2} \times$ the length of either leg.

4. In a _____, the length of the leg opposite the 30° angle is $\frac{1}{2}$ the length of the hypotenuse.

Example	**Student Practice**
1. Find the hypotenuse of a right triangle with legs of 5 in. and 12 in.	**2.** Find the hypotenuse of a right triangle with legs of 3 ft and 4 ft.

Square each value first.

$$\text{Hypotenuse} = \sqrt{(\text{leg})^2 + (\text{leg})^2}$$
$$= \sqrt{(5)^2 + (12)^2}$$
$$= \sqrt{25 + 144}$$

Add together the two values. Then take the square root.

$$\sqrt{25 + 144} = \sqrt{169} = 13$$

Thus, the hypotenuse is 13 in.

Example	Student Practice
3. A right triangle has a hypotenuse of 15 cm and a leg of 12 cm. Find the length of the other leg.	**4.** A right triangle has a hypotenuse of 20 ft and a leg of 12 ft. Find the length of the other leg.

3. (continued)

It may help to draw a picture.

Use the property for the length of one leg of a right triangle and simplify.

$$\text{Leg} = \sqrt{(\text{hypotenuse})^2 - (\text{leg})^2}$$
$$= \sqrt{(15)^2 - (12)^2}$$
$$= \sqrt{225 - 144}$$
$$= 9 \text{ cm}$$

---

5. A pilot flies 13 mi east from Pennsville to Salem. She then flies 5 mi south from Salem to Elmer. What is the straight-line distance from Pennsville to Elmer? Round to the nearest tenth of a mile.

6. A camper walks 2.5 km east from his campsite. He then walks 3.3 km north to a clearing. What is the straight-line distance from the campsite to the clearing? Round to the nearest tenth of a kilometer.

It may help to draw a picture.

The distance from Pennsville to Elmer is the hypotenuse of the triangle.

$$\text{Hypotenuse} = \sqrt{(\text{leg})^2 + (\text{leg})^2}$$
$$= \sqrt{(13)^2 + (5)^2}$$
$$= \sqrt{169 + 25}$$
$$\approx 13.928$$

To the nearest tenth, the distance is 13.9 mi.

Example	Student Practice
7. Find the requested sides of each special right triangle. Round to the nearest tenth.	**8.** Find the requested sides of each special right triangle. Round to the nearest tenth.

(a) Find the lengths of sides x and y.

16 yd   60° y 30° x

In a 30°-60°-90° triangle, the side opposite the 30° angle is $\frac{1}{2}$ of the hypotenuse, $\frac{1}{2} \times 16 = 8$. Therefore, $y = 8$ yd.

Find the third side using the Pythagorean Theorem.

$$\text{Leg} = \sqrt{16^2 - 8^2} = \sqrt{256 - 64}$$
$$= \sqrt{192} \approx 13.856$$

Thus, rounded to the nearest tenth, $x = 13.9$ yd.

(a) Find the lengths of sides x and y.

20 ft   60° y 30° x

(b) Find the length of hypotenuse z.

z 45°   6 m   45°   6 m

In a 45°-45°-90° triangle, we have the following.

$$\text{Hypotenuse} = \sqrt{2} \times \text{leg}$$
$$\approx 1.414(6)$$
$$= 8.484$$

To the nearest tenth, $z = 8.5$ m.

(b) Find the length of hypotenuse z.

z 45°   7 m   45°   7 m

Extra Practice

1. Find the unknown side of the right triangle. Use a calculator or square root table when necessary and round to the nearest thousandth.

    leg = 8 m , hypotenuse = 12 m

2. Find the unknown sides of the right triangle to the nearest thousandth using the given information.

    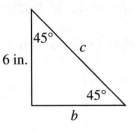

3. Find the unknown side of the right triangle to the nearest thousandth using the given information.

    leg = 15 ft , hypotenuse = 24 ft

4. Find the length of the wheelchair ramp. Round to the nearest tenth.

    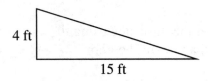

Concept Check

You look up at a plane that is flying at a distance of exactly two miles from your position. The plane is flying at an altitude of exactly 1.5 miles above the land and drops a package of supplies. Explain how you would find the distance from your position to the supplies.

Chapter 7 Geometry
7.7 Circles

Vocabulary
circle • center • radius • diameter • circumference • pi • semicircle

1. A _____ is a line segment across the circle that passes through the center with end-points on the circle.

2. A _____ is a line segment from the center to a point on the circle.

3. A _____ is a two-dimensional flat figure for which all points are at an equal distance from a given point.

4. The distance around the circle is called the _____.

Example	**Student Practice**
1. A bicycle tire has a diameter of 24 in. How many feet does the bicycle travel if the wheel makes one revolution?	2. A bicycle tire has a diameter of 26 in. How many feet does the bicycle travel if the wheel makes four revolutions? Round to the nearest hundredth if needed.

Start   End

1 revolution

Since we are given the diameter, find the circumference using $C = \pi d$ and 3.14 for π.

$$C = \pi d = (3.14)(24 \text{ in.}) = 75.36 \text{ in.}$$

The answer should be in feet. Change 75.36 inches to feet.

$$75.36 \text{ in.} \times \frac{1 \text{ ft}}{12 \text{ in.}} = 6.28 \text{ ft}$$

When the wheel makes 1 revolution, the bicycle travels 6.28 ft.

Example	Student Practice
3. Find the area of a circle whose radius is 6 cm. Use $\pi \approx 3.14$. Round to the nearest tenth.	**4.** Find the area of a circle whose radius is 4 m. Use $\pi \approx 3.14$. Round to the nearest tenth. Estimate to check.

$A = \pi r^2$

$\quad = (3.14)(6 \text{ cm})^2$

$\quad = (3.14)(36 \text{ cm}^2)$

$\quad = 113.04 \text{ cm}^2$

Thus, rounded to the nearest tenth, the area of the circle is 113.0 cm^2.

5. Find the area of the shaded region. Use $\pi \approx 3.14$. Round to the nearest tenth.	**6.** Find the area of the shaded region. Use $\pi \approx 3.14$. Round to the nearest tenth.

Subtract the area of the circle from the area of the square to find the area of the shaded region.

Find the area of the square.
$A = s^2 = (8 \text{ ft})^2 = 64 \text{ ft}^2$

Find the area of the circle.
$A = \pi r^2$

$\quad = (3.14)(3 \text{ ft})^2$

$\quad = (3.14)(9 \text{ ft}^2) = 28.26 \text{ ft}^2$

Now subtract.
$64 \text{ ft}^2 - 28.26 \text{ ft}^2 = 35.74 \text{ ft}^2$

Thus, rounded to the nearest tenth, the area of the shaded region is 35.7 ft^2.

Example	Student Practice
7. Find the area of the shaded region. Use $\pi \approx 3.14$. Round to the nearest tenth.	**8.** Find the area of the region. Use $\pi \approx 3.14$. Round to the nearest tenth.

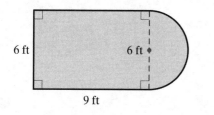

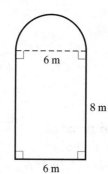

The area of the shaded region is the sum of the area of the rectangle and the area of the semicircle.

To find the area of the semicircle, first find the radius of the semicircle.

$$r = \frac{d}{2} = \frac{6 \text{ ft}}{2} = 3 \text{ ft}$$

The radius is 3 ft. Find the area of a semicircle with radius 3 ft. The area of a semicircle is one-half the area of a circle.

$$A_{\text{semicircle}} = \frac{\pi r^2}{2}$$

$$= \frac{(3.14)(3 \text{ ft})^2}{2} = \frac{(3.14)(9 \text{ ft}^2)}{2}$$

$$= \frac{28.26 \text{ ft}^2}{2} = 14.13 \text{ ft}^2$$

Next find the area of the rectangle.
$$A = lw = (9 \text{ ft})(6 \text{ ft}) = 54 \text{ ft}^2$$

Now add to find the total area.

$$54 \text{ ft}^2 + 14.13 \text{ ft}^2 = 68.13 \text{ ft}^2$$

Thus, rounded to the nearest tenth, the area of the shaded region is 68.1 ft^2.

195

Extra Practice

1. Find the circumference of a circle with radius = 22.5 in. Use $\pi \approx 3.14$. Round to the nearest hundredth.

2. Find the area of a circle with diameter = 24 cm. Use $\pi \approx 3.14$. Round to the nearest hundredth.

3. Raymond bought a circular piece of marble to use as a table in his living room. The marble piece was 4.5 feet in diameter. Find the cost at $85 per square yard of marble. Use $\pi \approx 3.14$. Round to the nearest cent.

4. A circle with a radius of 19.8 m is inscribed in a circle with a radius of 22.4 m. What is the area of the space between the inner and outer circle? Use $\pi \approx 3.14$. Round to the nearest hundredth.

Concept Check

A carpenter makes a semicircle with a radius of 3 feet. Explain how you would find the area of the semicircle.

Chapter 7 Geometry
7.8 Volume

Vocabulary
volume • cube • unit of volume • rectangular solid • cylinder • sphere
cone • pyramid

1. The volume of a _____ is 4 times π times the radius cubed divided by 3.

2. The volume of a _____ is π times the radius of the base squared times the height divided by 3.

3. The volume of a _____ is the length times the width times the height.

4. The volume of a _____ is the area of its circular base, πr^2, times the height.

Example	**Student Practice**
1. Find the volume of a box of width 2 ft, length 3 ft, and height 4 ft.	**2.** Find the volume of a box of width 3 cm, length 7 cm, and height 6 cm.

When we measure volume, we are measuring the space inside an object.

The volume of a rectangular solid, or box, is the length times the width times the height.

$V = lwh$

$\quad = (3\text{ ft})(2\text{ ft})(4\text{ ft})$

$\quad = (6)(4)\text{ ft}^3$

$\quad = 24\text{ ft}^3$

Thus, the volume of the box is 24 in.^3.

Example	Student Practice
3. Find the volume of a cylinder of radius 3 in. and height 7 in. Round to the nearest tenth.	**4.** Find the volume of a cylinder of radius 4 m and height 9 m. Round to the nearest tenth.

7 in.

$r = 3$ in.

Remember, the volume of a cylinder is the area of its circular base, πr^2, times the height h. This gives the formula $V = \pi r^2 h$.

We will continue to use 3.14 as an approximation for π in all volume problems requiring the use of π.

Substitute 3 in. for r and 7 in. for h. Square the value for r, then multiply.

$$V = \pi r^2 h$$
$$= (3.14)(3 \text{ in.})^2 (7 \text{ in.})$$
$$= (3.14)(9 \text{ in.}^2)(7 \text{ in.})$$
$$= (28.26 \text{ in.}^2)(7 \text{ in.})$$
$$= 197.82 \text{ in.}^3$$

Thus, rounded to the nearest tenth, the volume of the cylinder is 197.8 in.^{3}.

Example	Student Practice
5. Find the volume of a sphere with radius 3 m. Round to the nearest tenth.	**6.** Find the volume of a sphere with a radius 9 ft. Round to the nearest tenth.

5. Find the volume of a sphere with radius 3 m. Round to the nearest tenth.

Use $V = \dfrac{4\pi r^3}{3}$ to find the volume of the sphere.

$$V = \frac{4\pi r^3}{3} = \frac{(4)(3.14)(3 \text{ m})^3}{3}$$

$$= \frac{(4)(3.14)(27) \text{ m}^3}{3}$$

$$= (12.56)(9) \text{ m}^3 = 113.04 \text{ m}^3$$

Thus, rounded to the nearest tenth, the volume of the sphere is 113.0 m^3.

6. Find the volume of a sphere with a radius 9 ft. Round to the nearest tenth.

---

7. Find the volume of a cone of radius 7 m and height 9 m. Round to the nearest tenth.

Use $V = \dfrac{\pi r^2 h}{3}$ to find the volume of the cone.

$$V = \frac{\pi r^2 h}{3} = \frac{(3.14)(7 \text{ m})^2 (9 \text{ m})}{3}$$

$$= \frac{(3.14)(49 \text{ m}^2)(9 \text{ m})}{3}$$

$$= (3.14)(49)(3) \text{ m}^3 = 461.58 \text{ m}^3$$

Thus, rounded to the nearest tenth, the volume of the cone is 461.6 m^3.

8. Find the volume of a cone of radius 6 yd and height 14 yd. Round to the nearest tenth.

Example	Student Practice
9. Find the volume of a pyramid with height 6 m, length of base 7 m, and width of base 5 m. 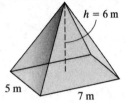 The volume of a pyramid is obtained by multiplying the area B of the base of the pyramid by the height h and dividing by 3. The base of the given pyramid is a rectangle. Find the area of the base. $$\text{Area of base} = (7 \text{ m})(5 \text{ m}) = 35 \text{ m}^2$$ Now find the volume of the pyramid. $$V = \frac{Bh}{3} = \frac{(35 \text{ m}^2)(6 \text{ m})}{3} = 70 \text{ m}^3$$	**10.** Find the volume of a pyramid with height 12 ft, length of base 5 ft, and width of base 5 ft.

Extra Practice

1. Find the volume of a hemisphere with radius 18 in. Use $\pi \approx 3.14$. Round to the nearest tenth.

2. Find the volume of a cone of radius 2.7 ft and height 3 ft. Use $\pi \approx 3.14$. Round to the nearest tenth.

3. Find the volume of a pyramid of height 300 m, length of base 1400 m, and width of base 160 m.

4. A cylindrical pipe has a radius of 3 feet and a length of 25 feet. It is surrounded by lining that is 2 inches thick. Calculate the volume of the lining. Use $\pi \approx 3.14$. Round to the nearest tenth.

Concept Check

A cylinder with a radius of 3 meters and height of 13 meters has a volume of 367.38 cubic meters. Suppose a new cylinder is formed similar to the described cylinder but the new radius is 4 meters while the height is unchanged. Explain how to determine how much larger the volume of the new cylinder is compared to the original cylinder.

Chapter 7 Geometry
7.9 Similar Geometric Figures

Vocabulary
similar • similar triangles • corresponding angles • corresponding sides • ratio
perimeters • areas

1. The _____ of similar figures–whatever the figures–have the same ratio as their corresponding sides.

2. The _____ of similar triangles are equal.

3. The corresponding sides of similar geometric figures have the same _____.

4. _____ are triangles with the same shape but not necessarily the same size.

Example	**Student Practice**
1. These two triangles are similar. Find the length of side n. Round to the nearest tenth.	**2.** The two triangles are similar. Find the length of side n. Round to the nearest tenth.

Example

1. These two triangles are similar. Find the length of side n. Round to the nearest tenth.

The ratio of 12 to 19 is the same as the ratio of 5 to n, $\dfrac{12}{19} = \dfrac{5}{n}$.

Cross-multiply and simplify.
$$12n = (5)(19)$$
$$12n = 95$$

Now solve for n.
$$\frac{12n}{12} = \frac{95}{12}$$
$$n = 7.91\overline{6}$$

Thus, rounded to the nearest tenth, side n has length 7.9.

Student Practice

2. The two triangles are similar. Find the length of side n. Round to the nearest tenth.

Example	Student Practice
3. These two triangles are similar. Name the sides that correspond. 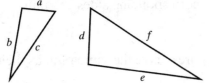 Turn the second triangle such that the shortest side is on the top and the longest side is on the right. 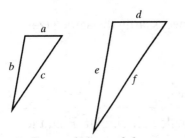 *a* corresponds to *d*. *b* corresponds to *e*. *c* corresponds to *f*.	**4.** The two triangles are similar. Name the sides that correspond. 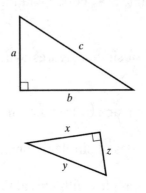
5. A flagpole casts a shadow of 36 ft. At the same time, a nearby tree that is 3 ft tall has a shadow of 5 ft. How tall is the flagpole? 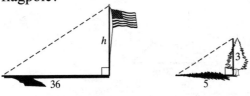 Let h = the height of the flagpole. Thus, h is to 3 as 36 is to 5, $\dfrac{h}{3} = \dfrac{36}{5}$. Cross-multiply and solve for h. $5h = (3)(36)$ $5h = 108$ $h = 21.6$ Thus, the flagpole is 21.6 feet tall.	**6.** A person who is 1.9 m tall casts a shadow of 1.4 m. At the same time, a flagpole that is 6.8 m casts a shadow of length s. What is the length of the flagpole's shadow? Round your answer to the nearest tenth. 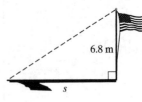

Example	**Student Practice**
7. The two rectangles shown here are similar because the corresponding sides of the two rectangles have the same ratio. Find the width of the larger rectangle.	**8.** The two rectangles are similar. Find the width of the larger rectangle.

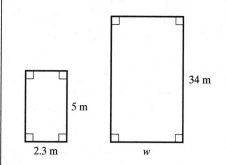

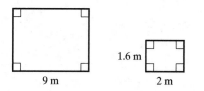

Let $w =$ the width of the larger rectangle. Set up a proportion and cross-multiply.

$$\frac{w}{1.6} = \frac{9}{2}$$

$$2w = (1.6)(9)$$

$$2w = 14.4$$

Solve for w.

$$2w = 14.4$$

$$\frac{2w}{2} = \frac{14.4}{2}$$

$$w = 7.2$$

Thus, the width of the larger rectangle is 7.2 meters.

Extra Practice

1. The two triangles are similar. Find the missing side n. Round to the nearest tenth, if necessary.

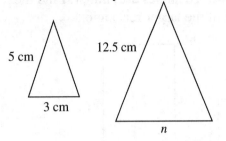

2. The two triangles are similar. Find the missin side n. Round to the nearest tenth, if necessa

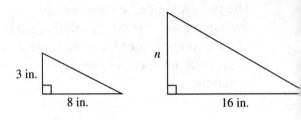

3. The two trapezoids are similar. Find the missing side n. Round to the nearest tenth, if necessary.

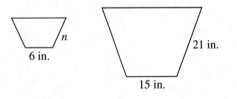

4. An architect draws a scale model of a rectang tower he is building. His drawing is 8.5 in. ta and 3.2 in. wide. The actual tower is 48 ft wid How tall is the tower?

Concept Check

A safari guide conducts tours in a rectangular park in Zambia that measures 5 miles by 9 miles. The drawing in the guide's office is drawn to scale. The smaller sides of the rectangle in the scale drawing are 6 inches. Explain how you would find the larger sides in the scale drawing.

Chapter 7 Geometry
7.10 Solving Applied Problems Involving Geometry

Vocabulary
area   •   volume

1. To find how much a box can hold, find its _____.

2. To find how many square feet of wallpaper is needed for a room, find the total
    _____ of the walls in the room.

Example	Student Practice
1. A professional painter can paint 90 ft^2 of wall space in 20 minutes. How long will it take the painter to paint four walls with the following dimensions: $14 \text{ ft} \times 8 \text{ ft}$, $12 \text{ ft} \times 8 \text{ ft}$, $10 \text{ ft} \times 7 \text{ ft}$, and $8 \text{ ft} \times 7 \text{ ft}$?	**2.** Bob rented an electric floor buffer. It will buff 65 ft^2 of hardwood floor in 10 minutes. He needs to buff the floors in three rooms. The floor dimensions are $25 \text{ ft} \times 12 \text{ ft}$, $11 \text{ ft} \times 8 \text{ ft}$, and $20 \text{ ft} \times 4 \text{ ft}$. How long will it take him to buff the floors in all three rooms?

Read the problem carefully and create a Mathematics Blueprint. Find the total area of the four walls. Each wall is a rectangle. Use $A = lw$ for each wall.

$$A = (14)(8) + (12)(8) + (10)(7) + (8)(7)$$
$$= 112 + 96 + 70 + 56 = 334$$

The total area of the walls is 334 ft^2.
Set up and solve a proportion to find how long it will take to paint the walls.

$$\frac{90 \text{ ft}^2}{20 \text{ minutes}} = \frac{334 \text{ ft}^2}{t \text{ minutes}}$$
$$90(t) = 334(20)$$
$$90t = 6680$$
$$t \approx 74$$

Rounded to the nearest minute, the work will take about 74 minutes.

Example	Student Practice
3. Carlos and Rosetta want to put vinyl siding on the front of their home in West Chicago. The house dimensions are shown in the figure below. The door dimensions are 6 ft×3 ft. The windows measure 2 ft×4 ft. Excluding windows and doors, how many square feet of siding will be needed?	**4.** Suppose Carlos and Rosetta want to put vinyl siding on the front of a similar house but with a length of 30 ft and a width of 23 ft. The dimensions of the door are 6.5 ft×2.5 ft. The windows measure 3 ft×5 ft. Excluding windows and doors, how many square feet of siding will be needed?

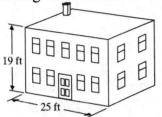

Read the problem carefully and create a Mathematics Blueprint. We will find the area of the front of the house. Then we will subtract the area of the windows and the door to determine the amount of siding needed. Use $A = lw$.

Area of 1 window $= (2 \text{ ft})(4 \text{ ft}) = 8 \text{ ft}^2$

Area of 9 windows $= (9)(8 \text{ ft}^2) = 72 \text{ ft}^2$

Area of 1 door $= (6 \text{ ft})(3 \text{ ft}) = 18 \text{ ft}^2$

The combined area of the door and the windows is $18 \text{ ft}^2 + 72 \text{ ft}^2 = 90 \text{ ft}^2$.

The area of the front of the house is $(19 \text{ ft})(25 \text{ ft}) = 475 \text{ ft}^2$.

Subtract 90 ft^2 from 475 ft^2.

$$\begin{array}{r} 475 \text{ ft}^2 \\ -\ \ 90 \text{ ft}^2 \\ \hline 385 \text{ ft}^2 \end{array}$$

Thus, 385 ft^2 of siding will be needed.

Extra Practice

1. As a fundraiser, the Student Senate on campus sells decorated boxes of candy. Each box measures 6 in. wide by 8 in. long by 4 in. high. The group has projected that they will be able to sell 450 decorated boxes of candy. If one bag of candy costs $1.99 and can fill 240 cubic inches, how much will it cost to purchase candy for the fundraiser?

2. According to a map, Woodbridge is 24 km from Calvington, which is 36 km from Thames, which is 22 km from Ashton, which is 41 km from Colbrook. How long will it take to drive from Woodbridge to Colbrook if the average speed of the car is 45 kilometers per hour? Round to the nearest tenth.

3. The side of a house measures 25 feet wide and 18 feet high. The roof is shaped like a triangle with a height of 6 feet. There are two windows measuring 3 feet by 2 feet on the side of the house. Aluminum siding costs $16 per square yard. What is the cost of putting aluminum siding on this side of the house?

4. A company is giving 80 cylindrical canisters filled with chocolate to their employees. Each canister is 12 in. high with a diameter of 6 in. The canister is filled to 75% of its capacity with chocolate. They purchase boxes of chocolate to fill the canisters. Each box contains 250 in.^{3} of chocolate and costs $5.99. What is the cost of the chocolate required for the canisters? Use $\pi \approx 3.14$ and round to the nearest cent.

Concept Check

Suppose there are two fields at Camp Cherith. The field with the new rope-obstacle course is triangular in shape. The field has a base of 200 meters and a height of 140 meters. It will cost $700 to spray this entire field with a weed killer that costs $0.05 per square meter. The second field has a base of 300 meters and the same height as the other field. The second field needs to be sprayed with weed killer that costs $0.20 per square meter. Explain how you would find out how much more it will cost to spray the second field than the first field.

MATH COACH

Mastering the skills you need to do well on the test.

Watch the **MATH COACH** videos in MyMathLab® or on You▶Tube™ while you work the problems below. These helpful hints will help you avoid making common errors on test problems.

Find the Perimeter of a Parallelogram—Problem 4

Find the perimeter of a parallelogram with sides measuring 6.5 m and 3.5 m.

> **Helpful Hint:** Remember that a parallelogram is a four sided figure with opposite sides that are equal in length. The perimeter is the distance around the figure.

Did you realize that two sides of the parallelogram each measure 6.5 m while the other two each measure 3.5 m? Yes _____ No _____

If you answered No, go back and make this correction in your work.

Did you write the problem as either $6.5 + 6.5 + 3.5 + 3.5$ OR $2(6.5) + 2(3.5)$? Yes _____ No _____

If you answered No, stop and rewrite the problem again. Be careful to align the decimal point correctly when adding or multiplying.

Did you include the correct units, m, with your answer? Yes _____ No _____

If you answered No, then make sure to include these units in your final answer.

If you answered Problem 4 incorrectly, please go back and rework this problem using these suggestions.

Find the Area of a Trapezoid—Problem 10   Find the area of a trapezoid with a height of 9 m and bases of 7 m and 25 m.

> **Helpful Hint:** You can use the formula for the area of a trapezoid, $A = \dfrac{h(b+B)}{2}$, where A equals the area , h is the height, b is the shorter base, and B is the longer base.

When you substituted the height and each base into the formula, did you obtain the equation $A = \dfrac{9(7+25)}{2}$?

Yes _____ No _____

If you answered No, stop and make the necessary corrections to your work.

Did you add the numbers within the parentheses first before multiplying by 9 and dividing by 2? Yes _____ No _____

If you answered No, go back and follow the correct order of operations.

Did you include the correct units, m^2, with your answer? Yes _____ No _____

If you answered No, then make sure to include these units in your final answer.

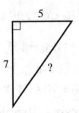

Helpful Hint: When the measure of two legs of a right triangle are known, you can find the hypotenuse by using the formula $\text{Hypotenuse} = \sqrt{(\text{leg})^2 + (\text{leg})^2}$.

Did you substitute correctly to write the expression

$\sqrt{(5)^2 + (7)^2}$ OR $\sqrt{(7)^2 + (5)^2}$?

Yes _____   No _____

If you answered No, stop and look at the problem again to make sure your understand which sides are know and which side is unknown.

Did your problem simplify to $\sqrt{74}$ before approximating the square root?

Yes _____   No _____

If you answered No, go back and perform the necessary calculations again.

Find the Area of a Circle—Problem 24   Find the area of a circle with diameter 12 ft.

Helpful Hint: Try to memorize the formula for the area of a circle, $A = \pi r^2$, where A is the area of a circle, r is the circle's radius, and π is pi. Remember that we typically use the value 3.14 as an approximation for π.

Did you remember to divide the diameter by 2 to find the radius? Yes _____   No _____

If you answered No, go back and complete this step.

Did you remember to square the radius first to obtain the expression 3.14×36 ? Yes _____   No _____

If you answered No, stop and make the necessary corrections to your work. Be sure to round the result to the nearest hundredth and add the correct units to your final answer.

Your final answer should have $A \approx$ written to its left. Note that this is the most accurate way to write your answer because an approximation was used for π.

Now go back and rework the problem using these suggestions.

Chapter 8 Statistics
8.1 Circle Graphs

Vocabulary

statistics   •   graphs   •   circle graphs   •   percent

1. We use _____ to give a visual representation of the data that is easy to read.

2. The entire circle of a circle graph represents 100 _____ .

3. _____ are especially helpful for showing the relationship of parts to a whole.

4. _____ is that branch of mathematics that collects and studies data.

Example	**Student Practice**
1. Use the circle graph below. What is the largest category of students? 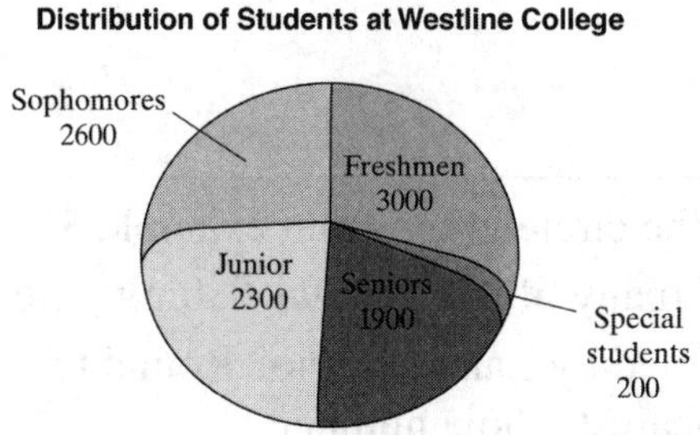 **Distribution of Students at Westline College** The largest pie-shaped section of the circle is labeled "Freshmen." Thus, the largest category is freshman students.	**2.** Use the circle graph from example **1**. What is the second smallest category of students?
3. Use the circle graph from example **1**. What percent of the students are sophomores or juniors? There are 2600 sophomores and 2300 juniors. $2600 + 2300 = 4900$. Thus we see that there are 4900 out of 10,000 students who are sophomores or juniors. $$\frac{4900}{10,000} = 0.49 = 49\%$$	**4.** Use the circle graph from example **1**. What percentage of students are juniors or seniors?

Example	Student Practice

5. Use the circle graph below. What percent of the total area is occupied by Lake Erie or Lake Ontario?

Percentage of Area Occupied by Each of the Great Lakes

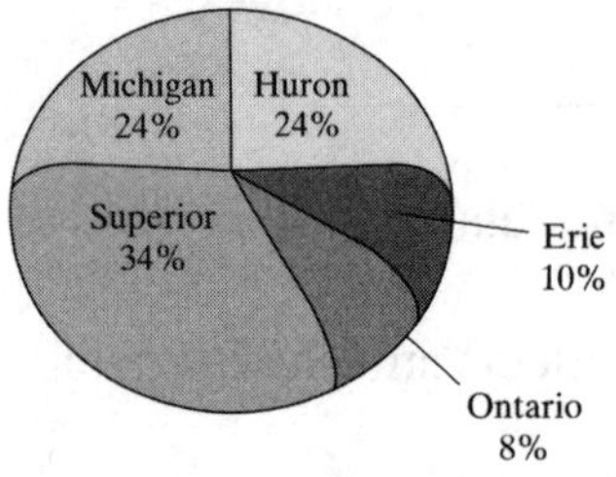

If we add 10% for Lake Erie and 8% for Lake Ontario, we get $10\% + 8\% = 18\%$. Thus, 18% of the area is occupied by Lake Erie or Lake Ontario.

6. Use the circle graph from example **5**. What percent of the total area is occupied by Lake Superior or Lake Ontario?

---

7. Use the circle graph from example **5**. How many of the total $94,680$ mi^2 are occupied by Lake Michigan? Round to the nearest whole number.

Remember that we multiply the percent times the base to obtain the amount. Here, 24% of 94,680 is what is occupied by Lake Michigan.

$$(0.24)(94,680) = n$$
$$22,723.2 = n$$

Rounded to the nearest whole number, $22,723$ mi^2 are occupied by Lake Michigan.

8. Use the circle graph from example **5**. How many of the total $94,680$ mi^2 are occupied by Lake Ontario? Round to the nearest whole number.

Example	Student Practice

Example

9. Use the following circle graph to answer parts **(a)** and **(b)**.

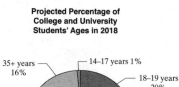

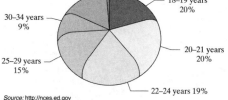

Source: http://nces.ed.gov

(a) What percent of the students represented in this circle graph are projected to be between 20 and 24 years old?

We add 20% to 19% to obtain 39%.

Thus, 39% of the students represented by this graph are projected to be between 20 and 24 years old.

(b) Of the projected 19,882,000 college and university students in 2018, how many will be between 30 and 34 years old?

We take 9% of the 19,882,000 students expected to be enrolled in colleges and universities.

$$(0.09)(19,882,000) = 1,789,380$$

Approximately 1,789,380 college and university students will be between 30 and 34 years old in 2018.

Student Practice

10. Use the circle graph from example **9**. Answer parts **(a)** and **(b)**.

(a) What percent of the students represented in this circle graph are projected to be 25 years or older?

(b) Of the projected 19,882,000 college and university students in 2018, how many will be 25 years or older?

Extra Practice

1. The following circle graph shows the age distribution of students at a college. What percent of students are 18 to 25 years old?

2. Use the circle graph from extra practice 1. What percent of students are 33 and under?

Age Distribution of Students

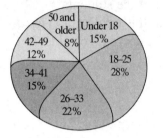

3. Use the circle graph from extra practice 1. If there are 3000 total students, how many students are 50 and older?

4. Use the circle graph from extra practice 1. What is the ratio of the number of students age 25 and younger to the number of students age 26 and older?

Concept Check

Recently a group of car dealers estimated the number of new cars and SUVs sold in New England in 2010 to be 850,000 vehicles. They estimated the distribution by category as show in the circle graph below. Explain how you would find the total number of the vehicles sold that were station wagons, four-door sedans, or minivans.

Sales of Cars and SUVs in New England in 2010

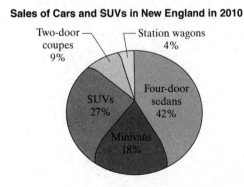

214

Chapter 8 Statistics
8.2 Bar Graphs and Line Graphs

Vocabulary
bar graphs   •   double-bar graphs   •   line graph   •   comparison line graph

1. _____ are useful for making comparisons.

2. A _____ shows two or more line graphs together.

3. _____ are helpful for seeing changes over a period of time.

4. In a _____, only a few points are actually plotted from measured values.

Example	**Student Practice**
1. The graph below shows the approximate population of California from 1950 to 2010. What was the increase in population from 1980 to 1990?	**2.** Use the bar graph from example **1**. What was the increase in population from 1950 to 2000?

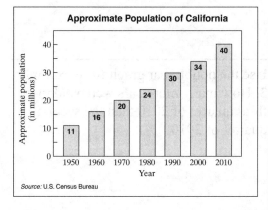

The bar for 1980 rises to 24. Thus the approximate population was 24,000,000. The bar for 1990 rises to 30. Thus the approximate population was 30,000,000.

$$30,000,000 - 24,000,000 = 6,000,000$$

The increase in population was 6 million or 6,000,000.

Example	Student Practice
3. The following double-bar graph illustrates the sales of new cars at a Ford dealership for two different years, 2009 and 2010. The sales are recorded for each quarter of the year. How many cars were sold in the second quarter of 2009?	**4.** Use the double-bar graph from example **3.** How many cars were sold in the third quarter of 2010?

The bar rises to 150 for the second quarter of 2009. Therefore, 150 cars were sold.

Example	Student Practice
5. Use the double-bar graph from example **3.** How many more cars were sold in the third quarter of 2010 than in the third quarter of 2009? From the double-bar graph, we see that 300 cars were sold in the third quarter of 2010 and that 200 cars were sold in the third quarter of 2009.	**6.** Use the double-bar graph from example **3.** How many more cars were sold in the third quarter of 2010 than the second quarter of 2010?

$$\begin{array}{r} 300 \\ -\ 200 \\ \hline 100 \end{array}$$

Thus, 100 more cars were sold.

Example	Student Practice

7. The following line graph shows the number of customers per month coming into a restaurant in a vacation community. Between what two months was the increase in the number of customers the largest?

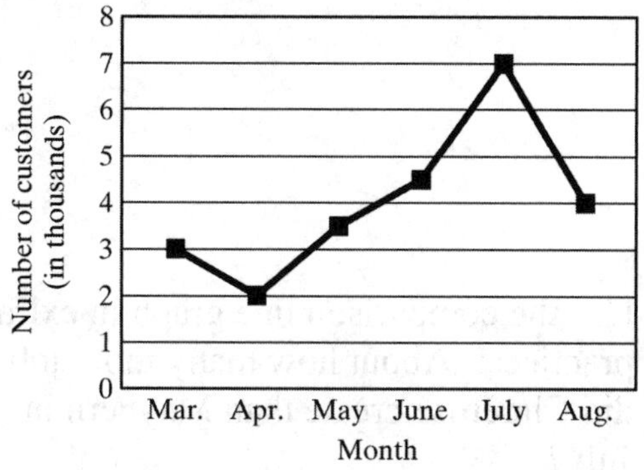

The line from June to July goes upward at the steepest angle. This represents the largest increase. (You can check this by reading the numbers from the left axis.) Thus the greatest increase in customers was between June and July.

8. Use the line graph from example **5**. Was there a greater decrease in customers from March to April or from July to August?

9. Use the comparison line graph below. In what academic year were more degrees awarded in the visual and performing arts than in computer science?

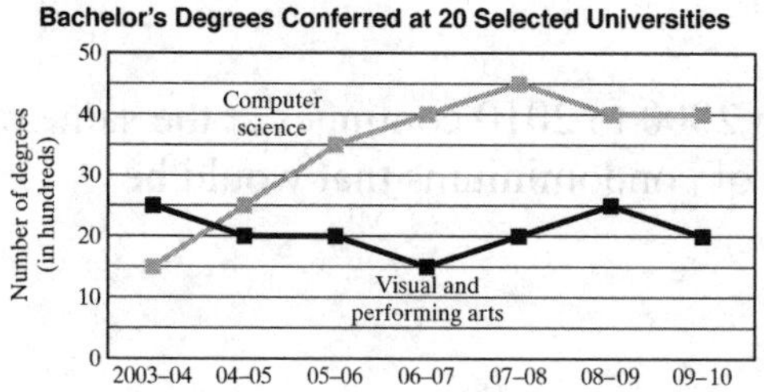

The only year when more bachelor's degrees were awarded in the visual and performing arts was the academic year 2003-2004.

10. Use the comparison line graph from example **9**. In how many academic years were there more computer science degrees awarded than for the visual and performing arts?

Extra Practice

1. The following bar graph shows the number of textbooks sold by Joe and Angela during each quarter of a year. Who sold the most books in any one quarter?

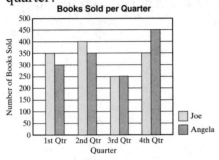

2. Use the double-bar graph in extra practice **1**. How many more books did Angela sell than Joe in the 4th quarter?

3. The comparison line graph below shows the number of new jobs created in the summer in two college towns. How many more jobs were created in Maybern than in Clarerose during the month of June?

4. Use the comparison line graph in extra practice **3**. About how many more jobs did Clarerose create than Maybern in July?

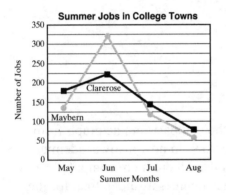

Concept Check

If the increase in the number of condominiums from 2000 to 2010 continues at the same rate until 2020, explain how you would find the number of condominiums that would be constructed in 2020.

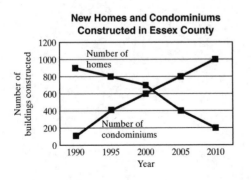

Chapter 8 Statistics
8.3 Histograms

Vocabulary
histogram   •   class interval   •   class frequency   •   raw data

1. Data that has not yet been organized or interpreted is called _____ .

2. A _____ is a special type of bar graph.

3. The width of each bar in a histogram is the same, and this width represents the possible range of scores and is called a _____ .

4. The _____ is the number of times a score occurs in a particular class interval.

Example	Student Practice
1. Use the histogram below. How many students scored a B on the test if the professor considers a test score of 80-89 a B?	**2.** Use the histogram from example **1**. How many students scored an A on the test if the professor considers a test score of 90-99 an A?

Since the 80-89 rises to a height of 8, eight students scored a B on the test

3. Use the histogram from example **1**. How many students scored less than 80 on the test? From the histogram, we see that four tests were scored 50-59, six tests were scored 60-69, and 16 tests were scored 70-79. When we add $4+6+16=26$, we can see that 26 students scored less than 80 on the test.	**4.** Use the histogram from example **1**. How many students scored greater than 59 on the test?

Example	Student Practice
5. Use the histogram below. How many light bulbs lasted between 1400 and 1599 hours?	**6.** Use the histogram from example **5**. How many light bulbs lasted between 400 and 599 hours?

Hours of operation of light bulbs

A total of 10 light bulbs lasted between 1400 and 1599 hours long.

Example	Student Practice
7. Use the histogram from example **5**. How many light bulbs lasted less than 1000 hours? Five bulbs lasted 400-599 hours, 15 bulbs lasted 600-799 hours, and 20 bulbs lasted 800-999 hours. Add $5 + 15 + 20 = 40$. Thus 40 light bulbs lasted less than 1000 hours.	**8.** Use the histogram from example **5**. How many light bulbs lasted more than 1000 hours?

Example	Student Practice
9. Each of the following numbers represents the number of kilowatt-hours of electricity used in 15 homes during a one-month period. Create a set of class intervals for this data and then determine the frequency for each class interval. 770, 520, 850, 900, 1100, 1200, 1150, 730, 680, 900, 1160, 590, 670, 1230, 980	**10.** Each of the following numbers represents the grades from a recent math test for one class. Create a set of class intervals for this data and then determine the frequency for each class interval. 90, 84, 58, 87, 98, 74, 65, 88, 52, 82, 76, 77, 60, 94, 81.

Choose intervals of 200. Make a table.

Kilowatt-Hours Used (Class Interval)	Tally	Frequency
500–699	||||	4
700–899	|||	3
900–1099	|||	3
1100–1299	⦚⦚	5

Example	Student Practice

11. Draw a histogram from the table in example **9**.

Number of kilowatt-hours of electricity used in a home in one month

12. Draw a histogram from the table in Student Practice **10**.

13. Draw a histogram for the following table of recent data showing the number of people in the United States, in each of the five age categories.
Source: U.S. Census Bureau

Age Category	Number of People in the United States
19 or younger	84,151,000
20-34	63,597,000
35-54	85,982,000
55-64	36,275,000
65 or older	40,229,000

Number of People in the U.S. by Age

Age (in years)

14. Draw a histogram for the following table of data showing the number of people in a small town in each of the five categories.

Age Category	Number of People in the town
19 or younger	350
20-34	550
35-54	1500
55-64	700
65 or older	200

Extra Practice

1. Use the histogram below. How many households paid rents of $2000 and above?

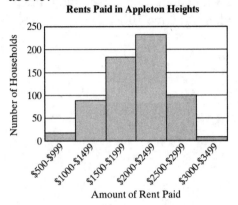

Rents Paid in Appleton Heights

2. Use the histogram in extra practice **1**. How many households paid rents between $1000 and $1999?

3. Use the histogram in extra practice **1**. How many households paid rents less than $3000?

4. A city in the southern United States had the following high temperatures in degrees Fahrenheit for June 1st through June 28th. Construct a histogram.

85	87	82	91	90	76	78
81	85	86	86	88	90	92
93	95	98	98	100	101	97
95	94	96	99	98	94	97

Concept Check

During a promotion month last summer, nonmembers were allowed to visit the gym for free. The number of people who visited the gym 1-4 times a month tripled. The number of people who visited the gym 5-8 times a month doubled. Explain how you would find how many people visited the gym between one and eight times per month.

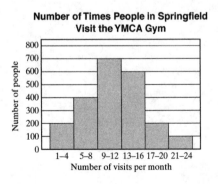

Number of Times People in Springfield Visit the YMCA Gym

Chapter 8 Statistics
8.4 Mean, Median, and Mode

Vocabulary

mean   •   average   •   median   •   mode   •   no mode

1. The _____ of a set of data is the number or numbers that occur most often.

2. If a set of numbers is arranged in order from smallest to largest, the _____ is that value that has the same number of values above it as below it.

3. The mean is often called the _____.

4. The _____ of a set of values is the sum of the values divided by the number of values.

Example	Student Practice
1. Carl recorded the miles per gallon achieved by his car for the last two months. His results were as follows:	**2.** Sally and Megan kept records of their electric bills for the last seven months. Their bills were $50.12, $75.89, $67.42, $84.86, $61.64, $72.46, and $58.34. Find the mean monthly bill, rounded to the nearest cent.

Week	1	2	3	4	5	6	7	8
Miles per Gallon	26	24	28	29	27	25	24	23

What is the mean miles-per-gallon figure for the last eight weeks? Round to the nearest whole number.

The mean is the sum of the values divided by the number of values.

$$\frac{26+24+28+29+27+25+24+23}{8}$$

$$=\frac{206}{8}=25.75\approx 26$$

The mean miles-per-gallon figure to the nearest whole number is 26.

Example	Student Practice
3. Find the median value of the following costs for microwave ovens: $100, $60, $120, $200, $190, $120, $320, $290, $180.	**4.** Find the median value of the following bi-weekly salaries: $800, $1000, $300, $900, $640.

3. (continued)

We must arrange the numbers in order from the smallest to largest (or largest to smallest).

$60, $100, $120, $120, $\underbrace{\quad\quad\quad\quad\quad\quad}_{\text{four numbers}}$ $\underbrace{\$180,}_{\text{middle number}}$

$190, $200, $290, $320 $\underbrace{\quad\quad\quad\quad\quad\quad}_{\text{four numbers}}$

Thus $180 is the median cost.

Example	Student Practice
5. Find the median of the following numbers: 26, 31, 39, 33, 13, 16, 18, 38.	**6.** Find the median value of the following numbers: 125, 163, 182, 146, 184, 113.

5. (continued)

First we place the numbers in order from smallest to largest.

$\underbrace{13,\ 16,\ 18}_{\substack{\text{three}\\\text{numbers}}}$ $\underbrace{26,\ 31}_{\substack{\text{two middle}\\\text{numbers}}}$ $\underbrace{33,\ 38,\ 39}_{\substack{\text{three}\\\text{numbers}}}$

The average (mean) of 26 and 31 is as follows.

$$\frac{26+31}{2}=\frac{57}{2}=28.5$$

Thus the median value is 28.5.

Example	Student Practice
7. The following numbers are the weights of automobiles measured in pounds: 2345, 2567, 2967, 3105, 3105, 3245, 3546. Find the mode of these weights. The value 3105 occurs twice 2345, 2567, 2967, ⬚3105⬚, ⬚3105⬚, 3245, 3546 Each of the other values occurs just once. The mode is 3105 pounds.	**8.** The following numbers are the grades of a recent Mathematics test: 64, 64, 68, 73, 77, 80, 82, 85, 86, 88, 90, 92, 93, 95. Find the mode of these grades.
9. The following numbers are finish times for 12 high school students who ran a distance of one mile. The finish times are measures in seconds. 290, 272, 268, 260, 290, 272, 330, 355, 368, 290, 370, 272 Find the mode of these times. First we need to arrange the numbers in order from smallest to largest and include all repeats. 260, 268, 272, 272, 272, 290, 290, 290, 330, 355, 368, 370 Now we can see that the value 272 occurs three times, as does the value 290. 260, 268, ⬚272⬚, ⬚272⬚, ⬚272⬚, ⬚290⬚, ⬚290⬚, ⬚290⬚, 330, 355, 368, 370 Thus, the modes for these finish times are 272 seconds and 290 seconds.	**10.** The following numbers are distances in miles that 15 faculty travels to teach classes at a community college each day. 3, 6, 2, 7, 23, 6, 4, 7, 4, 23, 4, 15, 3, 16, 20 Find the mode of these distances.

Extra Practice

1. The numbers of miles jogged by a distance runner over seven days were 14, 16, 12, 8, 15, 13, and 13. Find the mean. Round to the nearest tenth if necessary.

2. The cellular phone bills for Alicia over the last six months were as follows: $114, $91, $118, $142, $122, and $107. Find the median value.

3. The numbers of people attending an art show during the last ten days were 245, 325, 250, 245, 190, 275, 250, 185, 245, and 250. Find the mode.

4. Freshmen at a local college take an average of 15 credit hours per semester. Sophomores average 14 credit hours, juniors average 12 credit hours, seniors average 12 credit hours, and graduate students average 7 credit hours. Find the mean of the average number of credit hours taken each semester.

Concept Check

Professor Blair wants to conduct a survey of her students to determine how many times a year they order Chinese food. Would you select the mean, the median, or the mode for this survey? Explain your reasoning.

MATH COACH

Mastering the skills you need to do well on the test.

Watch the **MATH COACH** videos in MyMathLab®or on You Tube™ while you work the problems below. These helpful hints will help you avoid making common errors on test problems.

Interpreting a Double Bar Graph—Problem 11

Please refer to the double bar graph on page 554 in order to answer this question. How much more did a year at private college cost than a year at public college in 2011-2012?

Helpful Hint: If every bar height is not labeled, then you must examine the distance between two labeled values and determine what the bar height represents.

Did you determine that the bar height for public college costs in 2011-2012 was 6? Yes _____ No _____

If you answered No, look carefully at the first bar for 2011-2012. Use a sheet of paper to align horizontally so you can accurately read the scale. Do you see that it aligns to the line that appears halfway between 4 and 8?

Did you determine that the bar height for private college costs in 2011-2012 was 22?
Yes _____ No _____

If you answered No, look carefully at the second bar for 2011-2012. Use a sheet of paper to align horizontally so that you can accurately read the scale. Do you see that it aligns to the line that appears halfway between 20 and 24?

Now go back and rework the problem using these suggestions. Remember to write your answer in thousands of dollars.

Interpreting a Comparison Line Graph—Problem 14

Please refer to the comparison line graph on page 555 in order to answer this question. According to this graph, approximately how much longer is a 25-year-old nonsmoker expected to live than a 25-year-old smoker?

Helpful Hint: If every tic mark on the vertical scale is not labeled, then you must examine the distance between two labeled values closest to the desired value and determine the value of each tic mark between those two labeled values.

Did you determine that the life expectancy for nonsmokers at age 25 is 48? Yes _____ No _____

If you answered No, look carefully at the first point of the top line on the graph. Use a sheet of paper to align horizontally. Notice that this first point appears in between the labeled values 40 and 50. There are four tic marks between 40 and 50. They represent 42, 44, 46, and 48. The first point aligns with the tic mark for 48.

Did you determine that the life expectancy for smokers at age 25 is 36? Yes _____ No _____

If you answered No, look carefully at the first point of the lower line on the graph. Use a sheet of paper to align horizontally. Notice that the first point appears in between the labeled values 30 and 40. There are four tic marks between 30 and 40. They represent 32, 34, 36, and 38. The first point aligns with the tic mark for 36.

Now go back and rework the problem using these suggestions. Remember to include the unit "years" with your answer.

Interpreting a Histogram—Problem 20

Please refer to the histogram on page 555 in order to answer this question. How many color television sets lasted 9-14 years?

> **Helpful Hint:** Be sure to identify which bars are required for the specific question. Often you will have to determine two or three values and then add them together.

Did you realize that the number of color television sets that lasted 9-14 years includes the bar for 9-11 years added to the bar 12-14 years?

Yes _____ No _____

If you answered No, reread the question and examine the graph more closely.

Did you realize that the value for the 9-11 years bar is 45 and the value for the 12-14 years bar is 15?

Yes _____ No _____

If you answered No, examine the graph and interpret the scale more carefully.

If you answered Problem 20 incorrectly, go back and rework the problem using these suggestions.

Finding the Median—Problem 22

A chemistry student had the following scores on ten quizzes in her chemistry class: 10, 16, 15, 12, 18, 17, 14, 10, 20. Find the median quiz score.

> **Helpful Hint:** If there is an even number of values, then the median is the average of the two middle values when they are arranged in order from smallest to largest.

Did you arrange the quiz scores from smallest to largest to obtain 10, 10, 12, 13, 14, 15, 16, 17, 18, 20?
Yes _____ No _____

If you answered No, go back and perform this step.
Remember that any duplicate numbers should be placed next to each other in the list.

Did you identify the two middle values as 14 and 15?
Yes _____ No _____

If you answered No, stop and count to see which values are in fifth and sixth place within the ordered list of ten numbers.

Did you calculate the average of the two middle values?
Yes _____ No _____

If you answered No, go back and perform this step.

Chapter 9 Signed numbers
9.1 Adding Signed numbers

Vocabulary

negative number   •   positive number   •   number line
order   •   signed numbers   •   absolute value

1. A(n) _____ is to the right of zero on the number line.

2. The _____ of a number is the distance between that number and zero on the number line.

3. A(n) _____ is a line on which each number is associated with a point.

4. A(n) _____ is to the left of zero on the number line.

Example	**Student Practice**
1. In each case, replace the ? with $<$ or $>$.	**2.** In each case, replace the ? with $<$ or $>$.
(a) -8 ? -4	**(a)** -3 ? -1
Since -8 lies to the left of -4 on the number line, we know that $-8 < -4$.	
(b) 7 ? 1	**(b)** 4 ? -3
Since 7 lies to the right of 1, we know that $7 > 1$.	
(c) -2 ? 0	**(c)** 0 ? -4
-2 lies to the left of 0, so $-2 < 0$.	
(d) -6 ? 3	**(d)** -4 ? 9
-6 lies to the left of 3, so $-6 < 3$.	
(e) 2 ? -5	**(e)** -2 ? 3
2 lies to the right of -5, so $2 > -5$.	

Example	Student Practice
3. Add.	**4.** Add.
(a) $7 + 5$	**(a)** $11 + 3$
To add two numbers with the same sign, add the absolute values of the numbers, and use the common sign in the answer.	
$$\begin{array}{r} 7 \\ +\ 5 \\ \hline 12 \end{array}$$	
(b) $-3.2 + (-5.6)$	**(b)** $-12.4 + (-1.7)$
Add the absolute values of the two numbers, 3.2 and 5.6. Use a negative sign in the answer.	
$$\begin{array}{r} -3.2 \\ +\ -5.6 \\ \hline -8.8 \end{array}$$	
5. Add. $-\dfrac{1}{7} + \left(-\dfrac{3}{5}\right)$	**6.** Add. $-\dfrac{2}{9} + \left(-\dfrac{3}{7}\right)$
The LCD is 35. Thus, $-\dfrac{1}{7} = -\dfrac{5}{35}$, and $-\dfrac{3}{5} = -\dfrac{21}{35}$. We add two negative numbers, so the answer is negative.	
$$\begin{array}{r} -\dfrac{5}{35} \\ +\ -\dfrac{21}{35} \\ \hline -\dfrac{26}{35} \end{array}$$	

Example	Student Practice
7. Add.	**8.** Add.

7. Add.

(a) $8+(-10)$

The signs are different, so we find the difference, $10-8=2$. The sign of the number with the larger absolute value is negative, so the answer is negative.

$$\begin{array}{r} 8 \\ +\ -10 \\ \hline -2 \end{array}$$

(b) $-16.6+12.3$

$$\begin{array}{r} -16.6 \\ +\ \ 12.3 \\ \hline -4.3 \end{array}$$

8. Add.

(a) $9+(-12)$

(b) $-19.2+11.1$

---

9. Last night the temperature dropped to $-14°F$. From that low, today the temperature rose $34°F$. What was the high temperature today?

We want to add $-14°F$ to $34°F$. Because addition is commutative, it does not matter whether we add $-14+34$ or $34+(-14)$.

$$\begin{array}{r} 34°F \\ +\ -14°F \\ \hline 20°F \end{array}$$

The 34 is larger than 14. The difference between 34 and 14 is 20. The number with the larger absolute value is positive, so the answer is positive.

10. Two nights ago the temperature dropped to $-11°F$. From that low, yesterday the temperature rose $26°F$. What was the high temperature yesterday?

Example	Student Practice
11. Add. $24 + (-16) + (-10)$	**12.** Add. $7 + (-13) + (-5)$

We can go from left to right and start with 24 or we can start with -16.

Step 1   24   Step 1 -16

$$\begin{array}{r} 24 \\ +\ -16 \\ \hline 8 \end{array} \quad \text{or} \quad \begin{array}{r} -16 \\ +\ -10 \\ \hline -26 \end{array}$$

Step 2   8   Step 2 -26

$$\begin{array}{r} 8 \\ +\ -10 \\ \hline -2 \end{array} \quad \text{or} \quad \begin{array}{r} -26 \\ +\ \ 24 \\ \hline -2 \end{array}$$

We obtain the same answer either way.

Extra Practice

1. Add the pair of signed numbers that have the same sign.

$$\left(-1\frac{6}{11}\right) + \left(-2\frac{3}{22}\right)$$

2. Add the pair of signed numbers that have different signs.

$$14.2 + (-36.3)$$

3. Add.

$$-\frac{5}{12} + (-4)$$

4. Add.

$$\left(-\frac{4}{15}\right) + \left(\frac{9}{10}\right) + \left(-\frac{5}{6}\right)$$

Concept Check

In calculating an addition problem such as $4 + (-12) + 23 + (-15)$ some students add from left to right. Other students first add the positive numbers, then add the negative numbers, and then add the two results. Explain which method you prefer and why.

Chapter 9 Signed Numbers
9.2 Subtracting Signed Numbers

Vocabulary

opposite   •   add the opposite   •   zero   •   equal

1. To subtract signed numbers, _____ of the second number to the first number.

2. The _____ of a positive number is a negative number with the same absolute value.

3. If a number is the opposite of another number, these two numbers are at a(n) _____ distance from zero on the number line.

4. The sum of a number and its opposite is _____.

Example	Student Practice
1. Subtract $-8-(-2)$. Write the opposite of -2, which is 2, and change subtraction to addition, $-8+2$. Now we use the rules of addition for two numbers with opposite signs. $-8+2=-6$	**2.** Subtract $-15-(-7)$.
3. Subtract. **(a)** $7-8$ $\quad 7-8=7+(-8)$ $\qquad =-1$ **(b)** $-12-16$ $\quad -12-16=-12+(-16)$ $\qquad =-28$	**4.** Subtract. **(a)** $-11-2$ **(b)** $7-19$

Example	Student Practice
5. Subtract.	**6.** Subtract.
(a) $5.6 - (-8.1)$	**(a)** $7.3 - (-8.9)$

5. (a)

Change the subtraction to adding the opposite. Then add.

$$5.6 - (-8.1) = 5.6 + 8.1$$
$$= 13.7$$

(b) $-\dfrac{6}{11} - \left(-\dfrac{1}{22}\right)$

6. (b) $-\dfrac{5}{9} - \left(-\dfrac{5}{36}\right)$

Change the subtraction to adding the opposite.

$$-\frac{6}{11} - \left(-\frac{1}{22}\right) = -\frac{6}{11} + \frac{1}{22}$$

We see that the $LCD = 22$. So change $\dfrac{6}{11}$ to a fraction with a denominator of 22. Then add.

$$-\frac{6}{11} + \frac{1}{22} = -\frac{6}{11} \times \frac{2}{2} + \frac{1}{22}$$
$$= -\frac{12}{22} + \frac{1}{22}$$
$$= -\frac{11}{22}$$
$$= -\frac{1}{2}$$

Example	Student Practice
7. Subtract.	**8.** Subtract.

(a) $6-(+3)$

$$6-(+3)=6+(-3)$$
$$=3$$

(b) $-\dfrac{1}{2}-\left(-\dfrac{1}{3}\right)$

$$-\dfrac{1}{2}-\left(-\dfrac{1}{3}\right)=-\dfrac{1}{2}+\dfrac{1}{3}$$
$$=-\dfrac{3}{6}+\dfrac{2}{6}$$
$$=-\dfrac{1}{6}$$

(c) $2.7-(-5.2)$

$$2.7-(-5.2)=2.7+5.2$$
$$=7.9$$

(a) $9-(+7)$

(b) $-\dfrac{1}{5}-\left(-\dfrac{1}{7}\right)$

(c) $4.2-(-7.5)$

---

9. Perform the following set of operations, working from left to right.

$$-8-(-3)+(-5)$$

First we change subtracting a -3 to adding 3.

$$-8-(-3)+(-5)=-8+3+(-5)$$
$$=-5+(-5)$$
$$=-10$$

10. Perform the following set of operations, working from left to right.

$$-7-(-12)+(-23)$$

Example	Student Practice
11. Find the difference in temperature between 38°F during the day in Anchorage, Alaska, and −26°F at night.	**12.** Find the difference in temperature between 35°F during the day in Fairbanks, Alaska, and −42°F at night.

Subtract the lower temperature from the higher temperature.

$$38 - (-26) = 38 + 26$$
$$= 64$$

The difference is 64°F.

Extra Practice

1. Subtract the signed numbers by adding the opposite of the second number to the first number.

$$-20.1 - (-3.2)$$

2. Subtract the signed numbers by adding the opposite of the second number to the first number.

$$-3\frac{2}{9} - \left(-1\frac{7}{12}\right)$$

3. Perform each set of operations, working from left to right.

$$0.6 - (-1.9) - 1.0 + (-3.5)$$

4. On a cold morning in northern Minnesota, the temperature was −15°F. That day, the temperature rose 5°F, and that night, the temperature fell a total of 27°F. What was the low temperature that night?

Concept Check

Explain how you would perform the indicated operations to evaluate $8 - (-13) + (-5)$.

Chapter 9 Signed Numbers
9.3 Multiplying and Dividing Signed Numbers

Vocabulary

multiplication   •   commutative   •   negative   •   positive

1. To multiply or divide two numbers with different signs, multiply or divide the absolute values. The result is _____ .

2. To multiply or divide two numbers with the same sign, multiply or divide the absolute values. The result is _____ .

3. Because multiplication is _____ , when multiplying more than two numbers, you can multiply any two numbers first.

4. It is common to use parentheses to mean _____ .

Example	**Student Practice**
1. Evaluate $(7)(8)$. $(7)(8) = 56$	**2.** Evaluate $5(14)$.
3. Multiply. **(a)** $2(-8)$ Note that we are multiplying two signed numbers with different signs. We will always get a negative number for an answer. $2(-8) = -16$ **(b)** $(-3)(25)$ The factors have different signs. The answer will be negative. $(-3)(25) = -75$	**4.** Multiply. **(a)** $(-9)(4)$ **(b)** $(6)(-70)$

Example	Student Practice
5. Divide.	**6.** Divide.
(a) $-20 \div 5$	**(a)** $-72 \div 8$
The dividend and the divisor have different signs. The quotient will be negative.	
	(b) $91 \div (-13)$
$-20 \div 5 = -4$	
(b) $36 \div (-18)$	
$36 \div (-18) = -2$	
7. Multiply.	**8.** Multiply.
(a) $-5(-6)$	**(a)** $-12(-6)$
Note that we are multiplying two numbers with the same sign. We will always obtain a positive number.	
$-5(-6) = 30$	
(b) $-\dfrac{1}{2}\left(-\dfrac{3}{5}\right)$	**(b)** $-\dfrac{1}{5}\left(-\dfrac{4}{7}\right)$
$-\dfrac{1}{2}\left(-\dfrac{3}{5}\right) = \dfrac{3}{10}$	
9. Divide.	**10.** Divide.
$(-50) \div (-2)$	$(-4.8) \div (-4)$
The dividend and the divisor have the same sign. The quotient will be positive.	
$(-50) \div (-2) = 25$	

Example	Student Practice
11. Multiply $5(-2)(-3)$.	**12.** Multiply $7(-4)(-6)$.

11. When multiplying more than two numbers, multiply any two numbers first, then multiply the result by another number. Continue until each factor has been used. First multiply $5(-2) = -10$.

$$5(-2)(-3) = -10(-3)$$

Then multiply $-10(-3) = 30$.

$$5(-2)(-3) = 30$$

---

13. Chemists have determined that a phosphate ion has an electrical charge of -3. If 10 phosphate ions are removed from a substance, what is the change in the charge of the remaining substance?

Removing 10 ions from a substance can be represented by the number -10.

We can use multiplication to determine the result of removing 10 ions each having an electrical charge of -3.

$$(-3)(-10) = 30$$

Thus the change in charge would be $+30$.

14. An oxide ion has an electrical charge of -2. If 13 oxide ions are removed from a substance, what is the change in the charge of the new substance?

Example	Student Practice
15. Travis Tobey went outside his house to measure the temperature at 4:00 P.M. for seven days in October in Copper Center, Alaska. His temperature readings in degrees Fahrenheit were $-11°$, $-8°$, $-15°$, $-3°$, $5°$, $12°$, and $-1°$. Find the average temperature for this seven-day period.	**16.** In March, Mark measured the morning temperature at his farm in North Dakota at 7:00 A.M. each day. For a six-day period the temperatures in degrees Fahrenheit were $18°$, $15°$, $5°$, $-6°$, $-7°$, and $-17°$. What was the average temperature at his farm over this six-day period? Round to the nearest tenth.

To find the average, we take the sum of the seven days of temperature readings and divide by seven.

$$\frac{-11+(-8)+(-15)+(-3)+5+12+(-1)}{7}$$

$$=\frac{-21}{7}=-3$$

The average temperature was $-3°F$.

Extra Practice

1. Multiply $(-120)(13)$.

2. Multiply $\left(-\frac{4}{7}\right)\left(-\frac{21}{32}\right)$.

3. Divide $(100)\div(-25)$.

4. Divide $\dfrac{-\dfrac{20}{27}}{-\dfrac{30}{81}}$.

Concept Check

A student was doing the problem $(-4)+(-8)=-12$ and comparing it to the problem $(-4)(-8)=+32$. The student commented, "In the first problem two negative numbers give you a negative number. In the second problem two negative numbers give you a positive answer. This is confusing!" Explain how the student could keep from being confused with these two types of problems.

Chapter 9 Signed Numbers
9.4 Order of Operations with Signed Numbers

Vocabulary

order of operations   •   exponents   •   multiply or divide   •   parentheses

1. The _____ for whole numbers applies to signed numbers as well.

2. The first step in the order of operations is to perform operations inside _____.

3. Steps three and four of the order of operations are to _____ from left to right and to add and subtract from left to right, respectively.

4. The second step in the order of operations is to simplify any expressions with _____, and find any square roots.

Example	**Student Practice**
1. Perform the indicated operations in the proper order.	**2.** Perform the indicated operations in the proper order.
(a) $7 + 6(-2)$	**(a)** $36 \div (-6) + 81 \div (-9)$
Multiplication and division must be done first. We begin with $6(-2)$.	
$7 + 6(-2) = 7 + (-12)$ $= -5$	
(b) $9 \div 3 - 16 \div (-2)$	**(b)** $37 + 75 \div (-5)$
There is no multiplication, but there is division, and we do that first.	
$9 \div 3 - 16 \div (-2) = 3 - (-8)$	
Transform subtraction to adding the opposite.	
$3 - (-8) = 3 + 8 = 11$	

Example	Student Practice
3. Perform the indicated operations in the proper order. $$\dfrac{7(-2)+4}{8\div(-2)(5)}$$ We perform the multiplication and division first in the numerator and the denominator, respectively. $$\dfrac{7(-2)+4}{8\div(-2)(5)} = \dfrac{-14+4}{(-4)(5)}$$ We perform the addition in the numerator and the multiplication in the denominator. Then simplify the fraction. $$\dfrac{-14+4}{(-4)(5)} = \dfrac{-10}{-20} = \dfrac{1}{2}$$	**4.** Perform the indicated operations in the proper order. $$\dfrac{8(-3)-6}{4(-5)\div(-2)}$$
5. Perform the indicated operations in the proper order. $$4(6-9)+(-2)^3+3(-5)$$ Combine numbers inside parentheses. $$4(-3)+(-2)^3+3(-5)$$ Next, simplify the exponent. $$4(-3)+(-8)+3(-5)$$ Next, perform each multiplication. $$-12+(-8)+(-15)$$ Finally, add three numbers. $$-12+(-8)+(-15)=-35$$	**6.** Perform the indicated operations in the proper order. $$-3(-18+13)+(-3)^4+5(-7)$$

Example	Student Practice

7. Perform the indicated operations in the proper order.

$$\left(\frac{1}{2}\right)^3 + 2\left(\frac{3}{4} - \frac{3}{8}\right) \div \left(-\frac{3}{5}\right)$$

First find the LCD and combine the two fractions inside the parentheses.

$$\left(\frac{1}{2}\right)^3 + 2\left(\frac{3}{4} - \frac{3}{8}\right) \div \left(-\frac{3}{5}\right)$$

$$= \left(\frac{1}{2}\right)^3 + 2\left(\frac{6}{8} - \frac{3}{8}\right) \div \left(-\frac{3}{5}\right)$$

$$= \left(\frac{1}{2}\right)^3 + 2\left(\frac{3}{8}\right) \div \left(-\frac{3}{5}\right)$$

Next simplify $\left(\frac{1}{2}\right)^3 = \left(\frac{1}{2}\right)\left(\frac{1}{2}\right)\left(\frac{1}{2}\right) = \frac{1}{8}$

$$\frac{1}{8} + 2\left(\frac{3}{8}\right) \div \left(-\frac{3}{5}\right)$$

Next multiply and divide from left to right.

$$\frac{1}{8} + 2\left(\frac{3}{8}\right) \div \left(-\frac{3}{5}\right) = \frac{1}{8} + \frac{3}{4} \div \left(-\frac{3}{5}\right)$$

$$= \frac{1}{8} + \frac{3}{4} \times \left(-\frac{5}{3}\right)$$

$$= \frac{1}{8} + \left(-\frac{5}{4}\right)$$

Next, add the two fractions.

$$\frac{1}{8} + \left(-\frac{5}{4}\right) = \frac{1}{8} + \left(-\frac{10}{8}\right) = -\frac{9}{8} \text{ or } -1\frac{1}{8}$$

8. Perform the indicated operations in the proper order.

$$\left(\frac{1}{6}\right)^2 + 8\left(\frac{1}{5} - \frac{7}{10}\right) \div \frac{2}{3}$$

Extra Practice

1. Perform the indicated operations in the proper order. Reduce fractions.

$$(-48) \div 6 + (90)\left(-\frac{3}{10}\right)$$

2. Perform the indicated operations in the proper order. Simplify the numerator and denominator first. Reduce fractions.

$$\frac{60 \div 6 + (8)(-2)}{72 \div 9 + (-2)}$$

3. Perform the indicated operations in the proper order. Reduce fractions.

$$-90 \div 3^2 + (5-3)^3$$

4. Perform the indicated operations in the proper order.

$$1.69 - (-0.7)(3)$$

Concept Check

Explain in what order to perform the operations in the expression $4(3-9)+5^2$.

Chapter 9 Signed Numbers
9.5 Scientific Notation

Vocabulary
scientific notation • power of 10 • positive exponent • negative exponent

1. Numbers in scientific notation may be added or subtracted if they have the same _____.

2. A number that is larger than or equal to 10 and written in scientific notation will always have a _____ as the power of 10.

3. A positive number is in _____ if it is in the form $a \times 10^n$, where a is a number greater than (or equal to) 1 and less than 10, and n is a whole number or the negative of a whole number.

4. A positive number that is smaller than 1 and written in scientific notation will always have a _____ as the power of 10.

Example	Student Practice
1. Write in scientific notation.	**2.** Write in scientific notation.
(a) 9826	**(a)** 5789
$9826. = 9.826 \times 10^{\boxed{\text{What power?}}}$	
The decimal point is moved 3 places to the left. Therefore, use 3 for the power of 10, $9826 = 9.826 \times 10^3$.	
(b) 163,457	**(b)** 423,598
$163,457. = 1.63457 \times 10^{\boxed{\text{What power?}}}$	
The decimal point is moved 5 places to the left.	
$163,457 = 1.63457 \times 10^5$	

Example	Student Practice
3. Write in scientific notation.	**4.** Write in scientific notation.
(a) 0.036	**(a)** 0.092
Change the given number to a number greater than or equal to 1 and less than 10. Thus we change 0.036 to 3.6.	
$0.036 = 3.6 \times 10^{\boxed{\text{What power?}}}$	
The decimal point is moved 2 places to the right. Since the original number is less than 1, use -2 for the power of 10.	
$0.036 = 3.6 \times 10^{-2}$	
	(b) 0.579
(b) 0.72	
$0.72 = 7.2 \times 10^{\boxed{\text{What power?}}}$	
The decimal point is moved 1 place to the right. Because the original number is less than 1, use -1 for the power of 10.	
$0.72 = 7.2 \times 10^{-1}$	
5. Write in standard notation.	**6.** Write in standard notation.
5.8671×10^{4}	3.78915×10^{5}
Move the decimal point four places to the right.	
$5.8671 \times 10^{4} = 58,671$	

Example	Student Practice
7. Write in standard notation.	**8.** Write in standard notation.
(a) 9.8×10^5	**(a)** 9.3×10^5
Move the decimal point five places to the right. Then add four zeros.	
$9.8 \times 10^5 = 980,000$	
(b) 3×10^3	**(b)** 4×10^4
Move the decimal point three places to the right. Then add three zeros.	
$3 \times 10^3 = 3000$	
9. Write in standard notation.	**10.** Write in standard notation.
(a) 2.48×10^{-3}	**(a)** 9.95×10^{-3}
Move the decimal point three places to the left. Add two zeros between the decimal point and 2.	
$2.48 \times 10^{-3} = 0.00248$	
(b) 1.2×10^{-4}	**(b)** 3.5×10^{-5}
Move the decimal point four places to the left. Add three zeros between the decimal point and 1.	
$1.2 \times 10^{-4} = 0.00012$	
11. Add 5.89×10^{-20} meters $+ 3.04 \times 10^{-20}$ meters.	**12.** Add 6.95×10^{12} miles $+ 2.12 \times 10^{12}$ miles.
5.89×10^{-20} meters	
$\underline{+\ 3.04 \times 10^{-20}}$ meters	
8.93×10^{-20} meters	

Example	Student Practice
13. Add $7.2 \times 10^6 + 5.2 \times 10^5$.	**14.** Subtract $9.47 \times 10^5 - 3.5 \times 10^4$.

Note that these two numbers have different powers of ten, so we cannot add them as written. Therefore, change one number from scientific notation to another form.

Now $5.2 \times 10^5 = 520,000$ can be written as 0.52×10^6. Now we can add.

$$\begin{array}{r} 7.2 \times 10^6 \\ + \ 0.52 \times 10^6 \\ \hline 7.72 \times 10^6 \end{array}$$

Extra Practice

1. Write in scientific notation.

$30,000,000$

2. Write in scientific notation.

0.00546

3. Write in standard notation.

3.72×10^{12}

4. Add.

$4 \times 10^8 + 2.65 \times 10^7$

Concept Check

Explain how to place the zeros and commas in the correct locations to write the number 5.398×10^8 in standard notation.

MATH COACH

Mastering the skills you need to do well on the test.

Watch the **MATH COACH** videos in MyMathLab® or on You[Tube]™ while you work the problems below. These helpful hints will help you avoid making common errors on test problems.

Adding Signed Numbers—Problem 6 Add $-\dfrac{1}{4}+\left(-\dfrac{5}{8}\right)$.

> **Helpful Hint:** Be sure to find the least common denominator (LCD) for the two fractions before adding.

Did you identify the LCD as 8?

Yes _____ No _____

If you answered No, go back and make this change.

Did you change $-\dfrac{1}{4}$ to $-\dfrac{2}{8}$ before adding the fractions?

Yes _____ No _____

If you answered No, stop and make this conversion.

Is your final answer negative?

Yes _____ No _____

If you answered No, review the sign rules and perform the addition again.

If you answered Problem 6 incorrectly, please go back and rework the problem using these suggestions.

Subtracting Signed Numbers—Problem 11 Subtract $-2.5-(-6.5)$.

> **Helpful Hint:** Remember to rewrite the problem as adding the first number plus the opposite of the second number.

Did you rewrite the problem correctly as $-2.5+(6.5)$?

Yes _____ No _____

If you answered No, stop and make this correction to your work.

Did you realize that your answer would be a positive number?

Yes _____ No _____

If you answered No, review the sign rules and do the problem again.

Using the Order of Operations with Signed Numbers—Problem 26 $\quad -6(-3)-4(3-7)^2$

Helpful Hint: First, combine numbers inside the parentheses. Next, raise numbers to a power. Then perform multiplication from left to right. Finally, perform subtraction from left to right.

Did you first combine $3-7$ inside the parentheses to obtain -4?

Yes _____ No _____

If you answered No, please review the Helpful Hint above and try that step again.

Did you next calculate $(-4)^2 = 16$?

Yes _____ No _____

If you answered No, please review the Helpful Hint above and try that step again.

Remember to follow the order of operations and use the Helpful Hint for the last two steps of the problem.

Changing Numbers from Scientific Notation to Standard Notation—Problem 33

$$9.36 \times 10^{-5}$$

Helpful Hint: When a number in scientific notation has a negative power of 10, then the number is less than 1. We move the decimal point to the left and insert zeros as needed.

Did you move the decimal point five places to the LEFT?

Yes _____ No _____

If you answered No, please review the Helpful Hint above and try that step again.

Did you remember to insert four zeros to the left of the digit 9?

Yes _____ No _____

If you answered No, remember that for numbers smaller than one, there must be a zero for every place to the right of the point until you have a nonzero digit.

Now go back and rework the problem using these suggestions.

Chapter 10 Introduction to Algebra
10.1 Variables and Like Terms

Vocabulary

variable   •   term   •   like terms   •   numerical coefficients

1. _____ have identical variables and identical exponents.

2. A _____ is a symbol, usually a letter of the alphabet, that stands for a number.

3. To combine like terms, you combine the numbers, called the _____, that are directly in front of the terms by using the rules for adding signed numbers.

4. A _____ is a number, a variable, or a product of a number and one or more variables separated from other terms in an expression by a + sign or a − sign.

Example	**Student Practice**
1. Name the variables in each equation.	**2.** Name the variables in each equation.
(a) $A = lw$	**(a)** $A = 2\pi rh$
The variables are A, l, and w.	
(b) $V = \dfrac{4\pi r^3}{3}$	**(b)** $V = \dfrac{1}{3}\pi r^2 h$
The variables are V and r.	
3. Write the formula without a multiplication sign.	**4.** Write the formula without a multiplication sign.
(a) $V = \dfrac{B \times h}{3}$ $V = \dfrac{Bh}{3}$	**(a)** $V = \pi \times r^2 \times h$
(b) $A = \dfrac{h \times (B+b)}{2}$ $A = \dfrac{h(B+b)}{2}$	**(b)** $A = 2 \times \pi \times r \times h$

Example	Student Practice
5. Combine like terms.	**6.** Combine like terms.

5. Combine like terms.

(a) $3x + 7x - 15x$

$3x + 7x - 15x$

$= \boxed{3x + 7x + (-15x)}$

$= 10x + (-15x)$

$= -5x$

(b) $9x - 12x - 11x$

$9x - 12x - 11x$

$= \boxed{9x + (-12x) + (-11x)}$

$= -3x + (-11x)$

$= -14x$

6. Combine like terms.

(a) $5x - 17x - 25x$

(b) $5x + 7x - 48x$

7. Combine like terms.

(a) $3x - 8x + x$

$3x - 8x + x = 3x - 8x + 1x$

$= -5x + 1x$

$= -4x$

(b) $12x - x - 20.5x$

$12x - x - 20.5x = 12x - 1x - 20.5x$

$= 11.0x - 20.5x$

$= -9.5x$

8. Combine like terms.

(a) $7x - 15x + x$

(b) $7.5x - 9x - x$

Example	Student Practice
9. Combine like terms.	**10.** Combine like terms.

9. Combine like terms.

$$7.8 - 2.3x + 9.6x - 10.8$$

In each case, we combine the numbers separately and the variable terms separately. It may help to use the commutative and associative properties first.

$$7.8 - 2.3x + 9.6x - 10.8$$
$$= 7.8 - 10.8 - 2.3x + 9.6x$$
$$= -3 + 7.3x$$

10. Combine like terms.

$$9.3 - 5.2x + 8.5x - 12.5$$

---

11. Combine like terms.

(a) $5x + 2y + 8 - 6x + 3y - 4$

For convenience, we will rearrange the problem to place like terms next to each other. This is an optional step.

$$5x - 6x + 2y + 3y + 8 - 4 = -1x + 5y + 4$$
$$= -x + 5y + 4$$

(b) $\dfrac{3}{4}x - 12 + \dfrac{1}{6}x + \dfrac{2}{3}$

$$\dfrac{3}{4}x + \dfrac{1}{6}x - 12 + \dfrac{2}{3}$$
$$= \dfrac{9}{12}x + \dfrac{2}{12}x - \dfrac{36}{3} + \dfrac{2}{3}$$
$$= \dfrac{11}{12}x - \dfrac{34}{3}$$

The order of the terms in an answer is not important in this type of problem.

12. Combine like terms.

(a) $5m + 8n - 15 - 9m - n - 24$

(b) $\dfrac{5}{7}a + 9 - \dfrac{3}{11}a - \dfrac{1}{4}$

Extra Practice

1. Name the variables in the given equation.

$$P = 2 \times l + 2 \times w$$

2. Write the given equation without multiplication signs.

$$p = 4 \times a \times b \times c$$

3. Combine like terms.

$$2.4x - 3.6y + 2 - 1.2x - 1.7y$$

4. Combine like terms.

$$\frac{1}{3}x - \frac{1}{4}y - \frac{2}{3}x + \frac{3}{4}y$$

Concept Check

Explain how you would combine like terms in the following expression without making any sign errors.

$$-8.2x - 3.4y + 6.7z - 3.1x + 5.6y - 9.8z$$

Chapter 10 Introduction to Algebra
10.2 The Distributive Property

Vocabulary
distributive property of multiplication over addition   •   property
simplify   •   distributive property of multiplication over subtraction

1. The direction _____ means remove parentheses, combine like terms, and leave the answer in the most basic form.

2. The _____ states that if a, b, and c are signed numbers, then $a(b-c)=ab-ac$ and $(b-c)a=ba-ca$.

3. A _____ is an essential characteristic.

4. The _____ states that if a, b, and c are signed numbers, then $a(b+c)=ab+ac$ and $(b+c)a=ba+ca$.

Example	**Student Practice**
1. Simplify.	**2.** Simplify.
(a) $8(x+5)$	**(a)** $4(a+9)$
$8(x+5)=8x+8(5)$ $\qquad =8x+40$	
(b) $-3(x+3y)$	**(b)** $-6(x+4y)$
$-3(x+3y)=-3x+(-3)(3y)$ $\qquad =-3x+(-9y)$ $\qquad =-3x-9y$	
(c) $6(3a-7b)$	**(c)** $7(5m-8n)$
$6(3a-7b)=6(3a)-6(7b)$ $\qquad =18a-42b$	

Example	Student Practice
3. Simplify $(5x+y)(2)$.	**4.** Simplify $(6x+9y)(3)$.

3. (continued)

$$(5x+y)(2)=(5x)(2)+(y)(2)$$
$$=10x+2y$$

Notice that we write our final answer with the numerical coefficient to the left of the variable. We would not leave $(y)(2)$ as an answer but would write $2y$.

5. Simplify.

(a) $-5(x+2y-8)$

$$-5(x+2y-8)$$
$$=-5\left[x+2y+(-8)\right]$$
$$=-5(x)+(-5)(2y)+(-5)(-8)$$
$$=-5x+(-10y)+40$$
$$=-5x-10y+40$$

(b) $(1.5x+3y+7)(2)$

$$(1.5x+3y+7)(2)$$
$$=(1.5x)(2)+(3y)(2)+(7)(2)$$
$$=3x+6y+14$$

6. Simplify.

(a) $-7(a+8b+6)$

(b) $(3.4x+7.5y+8)(4)$

7. Simplify $\dfrac{2}{3}\left(x+\dfrac{1}{2}y-\dfrac{1}{4}z+\dfrac{1}{5}\right)$.

$$\dfrac{2}{3}\left(x+\dfrac{1}{2}y-\dfrac{1}{4}z+\dfrac{1}{5}\right)$$
$$=\dfrac{2}{3}(x)+\dfrac{2}{3}\left(\dfrac{1}{2}y\right)+\dfrac{2}{3}\left(-\dfrac{1}{4}z\right)+\dfrac{2}{3}\left(\dfrac{1}{5}\right)$$
$$=\dfrac{2}{3}x+\dfrac{1}{3}y-\dfrac{1}{6}z+\dfrac{2}{15}$$

8. Simplify $\dfrac{4}{5}\left(\dfrac{1}{5}x-\dfrac{1}{4}y+6z-\dfrac{1}{8}\right)$.

Example	Student Practice
9. Simplify. $2(x+3y)+3(4x+2y)$ Use the distributive property. $2(x+3y)+3(4x+2y)$ $=2x+6y+12x+6y$ Combine like terms. $2x+6y+12x+6y=14x+12y$	**10.** Simplify. $5(4x+6y)+8(7x+y)$
11. Simplify. $2(x-3y)-5(2x+6)$ Use the distributive property. $2(x-3y)-5(2x+6)$ $=2x-6y-10x-30$ Combine like terms. $2x-6y-10x-30=-8x-6y-30$ Notice that in the final step, only the x terms could be combined. There are no other like terms.	**12.** Simplify. $-6(x-7)+3(-5y+4x)$

Extra Practice

1. Simplify. Be sure to combine like terms.

$$(-10)(2x - 3y - 4z)$$

2. Simplify. Be sure to combine like terms.

$$\frac{1}{5}\left(3x - 2y - \frac{5}{7}z + \frac{1}{3}\right)$$

3. Simplify. Be sure to combine like terms.

$$0.2(-x - 2.1y + 5)$$

4. Simplify. Be sure to combine like terms.

$$2(3y - 4) + 8(y - 7)$$

Concept Check

Explain the steps that are needed to simplify $-3(2x + 5y) + 4(5x - 1)$.

Chapter 10 Introduction to Algebra
10.3 Solving Equations Using the Addition Property

Vocabulary
solving an equation • equation • solution • addition property of equations

1. A(n) _____ is a mathematical statement that says that two expressions are equal.

2. The _____ of an equation is that number which makes the equation true.

3. The _____ states that you may add the same number to each side of an equation to obtain an equivalent equation.

4. One of the most important skills in algebra is that of _____.

Example	Student Practice
1. Solve.	**2.** Solve.

1. Solve.

$x - 9 = 3$

We want to isolate the variable x.

Think: "Add the opposite of −9 to both sides of the equation."

The opposite of −9 is 9.

$x - 9 = 3$
$x - 9 + 9 = 3 + 9$

Now, simplify on each side.

$x + 0 = 12$
$x = 12$

Thus, the solution is $x = 12$.

2. Solve.

$x + 5 = -9$

Example	Student Practice
3. Solve. Check your solution to **(a)**.	**4.** Solve. Check your solution to **(a)**.

(a) $x + 1.5 = 4$

Use the addition property to solve this equation since it involves only addition. Isolate the variable.
Add the opposite of 1.5 to both sides of the equation.

$$x + 1.5 = 4$$
$$x + 1.5 + (-1.5) = 4 + (-1.5)$$
$$x = 2.5$$

Check by substituting 2.5 for x in the original equation.

$$x + 1.5 = 4$$
$$2.5 + 1.5 \overset{?}{=} 4$$
$$4 = 4$$

(b) $\dfrac{3}{8} = x - \dfrac{3}{4}$

Note that here the variable is on the right-hand side.

$$\frac{3}{8} + \frac{3}{4} = x - \frac{3}{4} + \frac{3}{4}$$

Now change $\dfrac{3}{4}$ to $\dfrac{6}{8}$ and simplify.

$$\frac{3}{8} + \frac{6}{8} = x + 0$$
$$\frac{9}{8} \text{ or } 1\frac{1}{8} = x$$

(a) $y - 5.3 = 8$

(b) $\dfrac{5}{6} = x + \dfrac{1}{3}$

Example	Student Practice
5. Solve $2x + 7 = x + 9$.	**6.** Solve $5x - 7 = 3x + 11$.

We want to remove the $+7$ on the left side of the equation. First, add -7 to both sides of the equation.

$$2x + 7 = x + 9$$
$$2x + 7 + (-7) = x + 9 + (-7)$$

Now we need to remove the x on the right-hand side of the equation.

$$2x + 7 + (-7) = x + 9 + (-7)$$
$$2x = x + 2$$

Finally, add $-x$ to both sides of the equation.

$$2x = x + 2$$
$$2x + (-x) = x + (-x) + 2$$
$$x = 2$$

Extra Practice

1. Solve $-12 + a = -4$.

2. Solve $-2.5 + x = -9$.

3. Solve $y - \dfrac{4}{7} = \dfrac{6}{7}$.

4. Solve $16x + 52 = 15x - 13$.

Concept Check
Explain the steps that are needed to solve the following equation: $-7x + 5 = -8x - 13$.

Chapter 10 Introduction to Algebra
10.4 Solving Equations Using the Division or Multiplication Property

Vocabulary
mixed numbers   •   equation   •   division property of equations
multiplication property of equations

1. Whatever you do to one side of a(n) _____, you must do to the other side.

2. The _____ states that you may multiply each side of an equation by the same nonzero number to obtain an equivalent equation.

3. The _____ states that you may divide each side of an equation by the same nonzero number to obtain an equivalent equation.

4. Remember to write all _____ as improper fractions or decimals before solving linear equations.

Example	**Student Practice**
1. Solve for n.	**2.** Solve for n.
$6n = 72$	$5n = 85$
Note that the variable n is multiplied by 6.	
$6n = 72$	
Divide each side by 6.	
$\dfrac{6n}{6} = \dfrac{72}{6}$	
Since $72 \div 6 = 12$, we have the following.	
$1 \cdot n = 12$	
$n = 12$	

Example	Student Practice
3. Solve for the variable. $-3n = 20$	**4.** Solve for the variable. $-9n = 55$

3. Solve for the variable. $-3n = 20$

Note that the coefficient of n is -3.

Now, divide each side of the equation by -3.

$$\frac{-3n}{-3} = \frac{20}{-3}$$

Next, simplify. Watch your signs.

$$n = -\frac{20}{3}$$

---

5. Solve for the variable. $2.5y = 20$

Note that the coefficient of y is 2.5.
Now, divide each side of the equation by 2.5.

$$\frac{2.5y}{2.5} = \frac{20}{2.5}$$

Perform the division from the right side of the equation.

$$2.5_\wedge \overline{)20.0_\wedge} \quad \overset{8}{}$$

Simplify both sides.

$$\frac{2.5y}{2.5} = \frac{20}{2.5}$$
$$y = 8$$

Check by substituting 8 for y in the original equation. It is always best to check the solution to equations involving decimals or fractions.

6. Solve for the variable. $5.2a = 26$

Example	Student Practice
7. Solve for the variable and check your solution.	**8.** Solve for the variable and check your solution.

7.

$$\frac{5}{8}y = 1\frac{1}{4}$$

First, change the mixed number to a fraction. It will be easier to work with.

$$\frac{5}{8}y = \frac{5}{4}$$

Now, multiply both side of the equation by $\frac{8}{5}$ because $\frac{8}{5} \cdot \frac{5}{8} = 1$.

$$\frac{8}{5} \cdot \frac{5}{8} y = \frac{5}{4} \cdot \frac{8}{5}$$
$$1 \cdot y = 2$$
$$y = 2$$

Check by substituting 2 for y in the original equation.

$$\frac{5}{8}y = 1\frac{1}{4}$$
$$\frac{5}{8}(2) \overset{?}{=} 1\frac{1}{4}$$
$$\frac{10}{8} \overset{?}{=} 1\frac{1}{4}$$
$$1\frac{2}{8} \overset{?}{=} 1\frac{1}{4}$$
$$1\frac{1}{4} = 1\frac{1}{4}$$

8.

$$2\frac{3}{5}z = 39$$

Extra Practice

1. Solve for the variable.

$$-8y = -48$$

2. Solve for the variable.

$$3.3z = 9.9$$

3. Solve for the variable.

$$15 = -\frac{3}{4}y$$

4. Solve for the variable.

$$-2\frac{1}{3}x = 21$$

Concept Check

Explain the steps you would take to solve the equation $12 - 5 = -14x$.

Chapter 10 Introduction to Algebra
10.5 Solving Equations Using Two Properties

Vocabulary
equivalent equation   •   distributive property   •   like terms   •   signs

1. The first step in the procedure of solving equations is removing any parentheses by using the _____.

2. To solve an equation, we take logical steps to change the equation to a simpler _____.

3. The chance of making a simple error with _____ is quite high. Checking gives you a chance to detect this type of error.

4. If there are _____ on either side of the equation, these should be combined first.

Example	Student Practice
1. Solve $3x + 18 = 27$. Check your solution.	**2.** Solve $-70 = 8x - 6$. Check your solution.

We want only x terms on the left and only numbers on the right. Begin by removing 18 from the left side of the equation. Add the opposite of 18 to both sides of the equation.

$$3x + 18 + (-18) = 27 + (-18)$$
$$3x = 9$$

Divide both sides of the equation by 3 so that x stands alone.

$$\frac{3x}{3} = \frac{9}{3}$$
$$x = 3$$

The solution to the equation is $x = 3$.

Check by substituting 3 for x in the original equation.

Example	Student Practice
3. Solve $8x = 5x - 21$.	**4.** Solve $10x = 3x - 28$.

Move $5x$ to the left side so that all the variables are on one side of the equation and all of the numbers are on the other side of the equation.

$$8x + (-5x) = 5x + (-5x) - 21$$
$$3x = -21$$

Divide both sides of the equation by 3.

$$3x = -21$$
$$x = -7$$

Check by substituting -7 for x in the original equation.

---

5. Solve for the variable.	**6.** Solve for the variable.
$-5 + 2y + 8 = 7y + 23$	$-5 + 4y + 9 = 8y + 20$

Begin by combining the like terms on the left side of the equation. Then, move the variables to the left side and the numbers to the right side. Finally, divide both sides of the equation by -5.

$$-5 + 2y + 8 = 7y + 23$$
$$2y + 3 = 7y + 23$$
$$2y + (-7y) + 3 = 7y + (-7y) + 23$$
$$-5y + 3 = 23$$
$$-5y + 3 + (-3) = 23 + (-3)$$
$$-5y = 20$$
$$y = -4$$

Check by substituting -4 for y in the original equation.

Example	Student Practice

7. Isolate the variable on the right-hand side. Then solve for x.

$$7x - 3(x - 4) = 9(x + 2)$$

Remove the parentheses by using the distributive property.

$$7x - 3x + 12 = 9x + 18$$

Add like terms.

$$4x + 12 = 9x + 18$$

Now, add -18 to each side.

$$4x + 12 + (-18) = 9x + 18 + (-18)$$

Simplify.

$$4x + (-6) = 9x$$

Add $-4x$ to each side. This isolates the variable on the right-hand side.

$$4x + (-4x) - 6 = 9x + (-4x)$$

Simplify.

$$-6 = 5x$$

Divide each side of the equation by 5.

$$\frac{-6}{5} = \frac{5x}{5}$$

$$-\frac{6}{5} = x$$

Check by substituting $-\dfrac{6}{5}$ for x in the original equation.

8. Isolate the variable on the right-hand side. Then solve for x.

$$9(x - 4) + 3 = 17(x - 6)$$

Extra Practice

1. Solve.

$$13x - 12 = 27$$

2. Solve.

$$-17 = 5x + 8$$

3. Solve.

$$-25 - 2y + 15 = 3y - 20$$

4. Solve.

$$14x - 3(x + 6) = 3(x + 2)$$

Concept Check

Explain the steps you would take to solve the following equation:

$$7x - 3(x - 6) = 2(x - 3) + 8.$$

Chapter 10 Introduction to Algebra
10.6 Translating English to Algebra

Vocabulary
addition • subtraction • multiplication • division • algebraic expressions

1. To translate English to algebra means to translate English expressions into

   _____.

2. The English phrases "greater than," "sum of," and "added to" represent _____.

3. The English phrases "divided by," "ratio of," and "quotient of" represent _____.

4. The English phrases "times," "of," and "the product of" represent _____.

Example	Student Practice
1. Translate the English sentence into an equation using variables. Use r to represent Roberto's weight and j to represent Juan's weight. Roberto's weight is 42 pounds more than Juan's weight. r is 42 pounds more than j ↓ ↓ ↓ ↓ ↓ r = 42 + j	**2.** Translate the English sentence into an equation using variables. Use a to represent Abby's height and p to represent Peter's height. Abby's height is 9 inches more than Peter's height.
3. Translate the English sentence into an equation using variables. Use c to represent the cost of the chair in dollars and s to represent the cost of the sofa in dollars. The chair costs $200 less than the sofa. When translating "less than," be sure that the number before the phrase is the value that is subtracted. $c = s - 200$	**4.** Translate the English sentence into an equation using variables. Use n to represent the number of students in the noon class and m to represent the number of students in the morning class. The noon class has 15 fewer students than the morning class.

Example	Student Practice
5. Translate the English sentence into an equation using variables. Use f for the daily cost of a 14-foot truck and e for the daily cost of an 11-foor truck. *The daily cost of a 14-foot truck is 20 dollars more than the daily cost of an 11-foot truck.* The 14-foot truck daily cost is 20 more than the 11-foot truck daily cost. $f = 20 + e.$	**6.** Translate the English sentence into an equation using variables. Use m to represent the number of boxes carried on Monday and w to represent the number of boxes carried on Wednesday. *On Monday John carried seven more boxes into the apartment than he did on Wednesday.*
7. Translate the following English sentence into an equation using the variables indicated. Use l for the length and w for the width of the rectangle. *The length of a rectangle is 3 feet shorter than double the width.* The length of the rectangle is compared to the width. Therefore, we begin with the width. We have $w =$ the width of the rectangle. Now the length of the rectangle is 3 feet shorter than double the width. Double the width is $2w$. If it is 3 feet shorter than double the width, we will have to take away 3 from the $2w$. Therefore, $2w - 3 =$ the length of the rectangle. So we have $l = 2w - 3$.	**8.** Translate the following English sentence into an equation using the variables indicated. Use l for the length and w for the width of the rectangle. *The length of a rectangle is 9 feet longer than double the width.*

Example	Student Practice
9. Write algebraic expressions for Bob's salary and Fred's salary. Use the letter b. *Fred's salary is $150 more than Bob's salary.* Let $b =$ Bob's salary. Now, Fred's salary is $150 more than Bob's salary. Therefore, $b + 150 =$ Fred's salary.	**10.** Write algebraic expressions for Jim's trip and Robert's trip. Use the letter r. *Roberts's trip is 450 miles longer than Jim's trip.*
11. Write algebraic expressions for the size of each of two angles of a triangle. Use the letter A. *Angle B of the triangle is 34° less than angle A.* Let $A =$ the number of degrees in angle A. Now, angle B of the triangle is 34° less than angle A. So therefore $A - 34 =$ the number of degrees in angle B.	**12.** Write algebraic expressions for the size of each of two angles of a triangle. Use the letter C. *Angle D of the triangle is 55° less than angle C.*
13. Write the algebraic expression for the length of each of three sides of a triangle. Use the letter x. *The second side is 4 inches longer than the first. The third side is 7 inches shorter than triple the length of the first side.* Since the other two sides are described in terms of the first side, start by writing an expression for the first side. Let $x =$ the length of the first side. Therefore, $x + 4 =$ the length of the second side. Therefore, $3x - 7 =$ the length of the third side.	**14.** Write the algebraic expression for the length of each of three sides of a triangle. Use the letter y. *The second side is triple the length of the first side. The third side is 8 inches longer than the first side.*

Extra Practice

1. Translate the English sentence into an equation using the variable indicated.

   The combined number of hours Amy and Bill study per school day is 9 hours. Use a for the number of hours Amy studies and b for the number of hours Bill studies.

2. Translate the English sentence into an equation using the variable indicated.

   The product of your hourly wage and the number of hours worked is $300. Let $h =$ the hourly wage and $t =$ number of hours worked.

3. Write algebraic expressions for each quantity using the given variable.

   Anastasia, Nicholas, and Elias were raising money for the Diabetes Foundation. Anastasia collected $540 more in donations than Elias. Nicholas raised $60 less than Elias. Use the letter e.

4. Write algebraic expressions for each quantity using the given variable.

   In the Jones family there are three children. The oldest is 6 years older than the middle child. The youngest child's age is 11 less than twice the middle child's age. Use the letter m.

Concept Check

In Dr. Tobey's Basic Mathematics class, 12 more students have part-time jobs than full-time jobs. The students wanted to describe this relationship with algebraic expressions. One student said let $p =$ the number of students with part-time jobs and let $p - 12 =$ the number of students with full-time jobs. Another student said let $f =$ the number of students with full-time jobs and let $f + 12 =$ the number of students with part-time jobs. What student is right? Are both right? Explain you answer.

Chapter 10 Introduction to Algebra
10.7 Solving Applied Problems

Vocabulary
variable • equations • perimeter

1. You can use _____ to solve problems that involve rates and percents.

2. Sometimes three items are compared. Let a(n) _____ represent the quantity to which the other two quantities are compared.

3. The _____ of a rectangle is given by $P = 2w + 2l$.

Example	Student Practice
1. A 12-foot piece of wood is cut into two pieces. The longer piece is 3.5 feet longer than the shorter piece. What is the length of each piece?	**2.** A 19-foot board is cut into two pieces. The longer piece is 2.7 feet longer than the shorter piece. What is the length of each piece?

Read the problem carefully and create a Mathematics Blueprint.

Let $x =$ the length of the shorter piece. The longer piece is 3.5 feet longer than the shorter piece. So, $x + 3.5 =$ the length of the longer piece. The sum of the lengths of the two pieces is 12 feet. Write an equation, $x + (x + 3.5) = 12$.

Solve the equation for x.

$2x + 3.5 = 12$

$\qquad 2x = 8.5$

$\qquad\quad x = 4.25$

The shorter piece is 4.25 feet long. The longer piece is $4.25 + 3.5 = 7.75$ feet long. Verify the solutions by making sure they meet the original conditions.

Example	Student Practice
3. A farmer wishes to fence in a rectangular field with 804 feet of fence. The length is to be 3 feet longer than double the width. How long and how wide is the field?	**4.** What are the length and the width of a rectangular field that has a perimeter of 992 feet and a length that is 10 feet longer than double the width?

3. A farmer wishes to fence in a rectangular field with 804 feet of fence. The length is to be 3 feet longer than double the width. How long and how wide is the field?

Read the problem carefully and create a Mathematics Blueprint.

The perimeter of a rectangle is given by $P = 2w + 2l$.

Let w = the width. The length is 3 feet longer than double the width. Therefore, Length $= 3 + 2w$. You may wish to draw a diagram and label it with the given facts.

Substitute these facts into the perimeter formula, $2w + 2(2w + 3) = 804$.

Now use the distributive property and solve the equation for w.

$$2w + 2(2w + 3) = 804$$
$$2w + 4w + 6 = 804$$
$$6w + 6 = 804$$
$$6w + 6 + (-6) = 804 + (-6)$$
$$6w = 798$$
$$\frac{6w}{6} = \frac{798}{6}$$
$$w = 133$$

The width is 133 feet. The length $= 2w + 3$. When $w = 133$, we have $2(133) + 3 = 266 + 3 = 269$. Thus, the length is 269 feet. Verify the solutions by making sure they meet the original conditions.

4. What are the length and the width of a rectangular field that has a perimeter of 992 feet and a length that is 10 feet longer than double the width?

Example	Student Practice

5. A triangle has three angles, A, B, and C. The measure of angle C is triple the measure of angle B. The measure of angle A is $105°$ larger than the measure of angle B. Find the measure of each angle. Check your answer.

Read the problem carefully and create a Mathematics Blueprint.

Let $x =$ the number of degrees in angle B. Therefore, $3x =$ the number of degrees in angle C and $x + 105 =$ the number of degrees in angle A.

The sum of the interior angles of a triangle is $180°$. Thus, we can write the equation $x + 3x + (x + 105) = 180$.

Solve the equation for x.

$$x + 3x + x + 105 = 180$$
$$5x + 105 = 180$$
$$5x + 105 + (-105) = 180 + (-105)$$
$$5x = 75$$
$$\frac{5x}{5} = \frac{75}{5}$$
$$x = 15$$

Thus, angle B measures $15°$. Angle C measures $3x = (3)(15) = 45°$. Angle A measures $x + 105 = 15 + 105 = 120°$.

Verify the solutions by making sure they meet the original conditions.

6. The measure of angle C of a triangle is triple the measure of angle A. The measure of angle B is $20°$ less than the measure of angle A. Find the measure of each angle.

Example	Student Practice
7. This month's salary for an appliance saleswoman was $3000. This includes her base monthly salary of $1800 plus a 5% commission on total sales. Find her total sales for the month.	**8.** A salesperson at a boat dealership earns $4000 a month plus a 4% commission on the total sales of the boats he sells. Last month he earned $4250. What was the total sales of the boats he sold?

Read the problem carefully and create a Mathematics Blueprint. Let $s =$ the amount of total sales. Then $0.05s =$ the amount of commission earned from the sales. This month's salary is represented by the equation $3000 = 1800 + 0.05s$. Solve the equation for s.

$$3000 = 1800 + 0.05s$$
$$1200 = 0.05s$$
$$24,000 = s$$

She sold $24,000 worth of appliances.

Extra Practice

1. A 45-foot rope is cut into two pieces. The longer piece is 15 feet longer than the shorter piece. What is the length of each piece?

2. The perimeter of a rectangle is 450 inches. The length is 25 inches more than 3 times the width. What are the dimensions of the rectangle?

3. The perimeter of a square is 64 feet. If each side is tripled, what will be the perimeter of the larger square?

4. During a baseball game, the Astros scored 4 fewer runs than the Giants. A total of 20 runs were scored. How many runs did each team score?

Concept Check

The first angle of a triangle is twice as large as the second angle. The third angle is 10 degrees less than the second angle. Explain how you would write an expression for each of the three angles. How would you set up an equation to find the measures of the angles? Explain how you would solve the equation.

MATH COACH

Mastering the skills you need to do well on the test.

Watch the **MATH COACH** videos in MyMathLab®or on You[Tube]™ while you work the problems below. These helpful hints will help you avoid making common errors on test problems.

Combining Like Terms with Variables and Fractional Coefficients—Problem 2 $\quad \frac{1}{3}x + \frac{5}{8}y - \frac{1}{5}x + \frac{1}{2}y$

> **Helpful Hint:** First identify the like terms. Then, find the LCD for the coefficients of the x terms and find the LCD for the coefficients of the y terms.

Did you correctly identify $\frac{1}{3}x$ and $-\frac{1}{5}x$ as like terms and $\frac{5}{8}y$ and $\frac{1}{2}y$ as like terms? Yes ____ No ____

If you answered No, stop and review the definition of like terms.

Was 15 your LCD for the coefficients of the x terms? Yes ____ No ____

If you answered No, go back and review how to find the LCD of two fractions.

Did you transform $\frac{1}{3}$ to $\frac{5}{15}$ as the coefficient of the first term? Did you transform $-\frac{1}{5}$ to $-\frac{3}{15}$ as the coefficient of the third term? Yes ____ No ____

If you answered No, stop and carefully review how to write equivalent fractions using the LCD. Now follow the same procedure to find the LCD for the y terms and rewrite those fractions before combining life terms.

If you answered Problem 2 incorrectly, go back and rework the problem using these suggestions.

Simplifying Expressions Containing Parentheses—Problem 10
Simplify $2(-3a+2b)-5(a-2b)$.

> **Helpful Hint:** First, use the distributive property to remove parentheses before doing any other operations. Double check your work and be careful to avoid sign errors.

After removing the parentheses, did you obtain $-6a+4b-5a+10b$? Yes ____ No ____

If you answered No, stop and redo the distributive property. Remember to use extra care when multiplying by -5 as you remove the second set of parentheses.
Was $-11a$ your answer when adding $-6a$ and $-5a$? Yes ____ No ____

If you answered No, go back and perform the addition again. Be careful of your signs. Recall that adding two negative numbers always results in a negative number. Now go back and rework the problem using these suggestions.

Solving Equations with Variables on Both Sides—Problem 14   Solve for the variable.

$8x - 2 - x = 3x - 9 - 10x$

Helpful Hint: Be sure to collect and combine like terms on each side of the equation before performing any other steps.

Did you combine like terms on each side to obtain
$7x - 2 = -7x - 9$?

Yes _____   No _____

If you answered No, go back and identify like terms. Then combine the x terms on the left side of the equation and combine the x terms on the right side of the equation.

Did you add $7x$ to each side of the equation and then add 2 to each side of the equation, resulting in $14x = -7$?

Yes _____   No _____

If you answered No, stop and perform these steps.

Consider what the last step would be to solve the equation for x. Now go back and complete this problem again.

Writing an Algebraic Expression for Each Quantity—Problem 20   The length of a rectangle is 5 inches shorter than double the width. Use the letter w.

Helpful Hint: Since length is compared to width, let $w =$ the width.

Did you notice that "double the width" could be written as $2w$?

Yes _____   No _____

If you answered No, reread the problem and write this expression down first.

Did you realize that "5 inches shorter than double the width" translates to $2w - 5$?

Yes _____   No _____

If you answered No, go back and perform this step in the translation.

Remember to write an expression for length and an expression for width in your final answer.

If you answered Problem 20 incorrectly, then go back and rework the problem using these suggestions.

Worksheet Answers Chapter 1

Section 1.1

Vocabulary

1. digits
2. period
3. place-value system
4. decimal system

Student Practice

2. $2,000,000 + 30,000 + 2,000 + 400 + 10$
4. $104,047$
6. (a) 1
   (b) The hundreds place and the ones place
8. thirty million, five hundred ninety-seven thousand, eight hundred twelve
10. ten million, eight hundred ninety-seven thousand, nine hundred eight
12. $8,000,967,000,241$
14. $10,000$

Extra Practice

1. $4000 + 900 + 60 + 5$
2. 837
3. four thousand, six hundred twenty-seven
4. 5000

Concept Check

Answers may vary. Possible solution: The zeros act as placeholders so the place value of each digit is in the right place: three hundred sixty-eight million, five hundred twenty-two $= 368,000,522$.

Section 1.2

Vocabulary

1. identity property of zero
2. commutative property of addition
3. addends
4. associative property of addition

Student Practice

2. 16

4. 25
6. 22
8. 8449
10. $10,791$
12. $77,681$

Extra Practice

1. 183
2. $113,811$
3. 1251
4. $47,862$ square miles

Concept Check

Answers may vary. Possible solution:

$$
\begin{array}{ccccc}
 & \overset{1}{4} & \overset{2}{5} & \overset{2}{6} & 7 \\
 & & 3 & 1 & 8 & 9 \\
+ & & & 8 & 9 & 5 \\
\hline
 & & 8 & 6 & 5 & 1 \\
\end{array}
$$

$7 + 9 + 5 = 21$ (carry the 2)

$2 + 6 + 8 + 9 = 25$ (carry the 2)

$2 + 5 + 1 + 8 = 16$ (carry the 1)

$1 + 4 + 3 = 8$

Section 1.3

Vocabulary

1. subtrahend
2. minuend
3. equation
4. difference

Student Practice

2. (a) 2
   (b) 5
4. 3123
6. 3869
8. 6
10. (a) 40 homes
    (b) 2009 and 2010

Extra Practice

1. 1045
2. $77,632$
3. $x = 35$
4. 391 votes

Concept Check

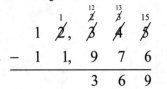

$$\begin{array}{r} 1\ \overset{1}{\cancel{2}},\ \overset{\overset{12}{\cancel{3}}}{\cancel{3}}\ \overset{\overset{13}{\cancel{3}}}{\cancel{4}}\ \overset{15}{\cancel{5}} \\ -\ 1\ 1,\ 9\ 7\ 6 \\ \hline 3\ 6\ 9 \end{array}$$

Borrow 10 ones from the 4. Change 5 ones to 15 ones. Borrow 10 tens from the 3. Change 3 tens to 13 tens. Borrow 10 hundreds from the 2. Change 2 hundreds to 12 hundreds. Then subtract.

Section 1.4

Vocabulary
1. associative property of multiplication
2. factors
3. commutative property of multiplication
4. multiplication property of zero

Student Practice
2. 32
4. 9639
6. 2056
8. (a) 55,000
   (b) 13,500
   (c) 182,000
10. 366,548
12. 2700
14. $1,357,800

Extra Practice
1. 20,609,000
2. 119,322
3. 2665
4. $8890

Concept Check
Answers may vary. Possible solution:

$$\begin{array}{r} 3457 \\ \times\ \ \ 2008 \\ \hline 27\ 656 \\ 6\ 914\ 000 \\ \hline 6,941,656 \end{array}$$

The zeros act as placeholders. Think of 3457 multiplied by 2000.

Section 1.5

Vocabulary
1. undefined
2. dividend
3. remainder
4. quotient

Student Practice
2. 12
4. (a) 0
   (b) undefined
6. 536 R 4
8. 47 R 124
10. 365 miles per hour

Extra Practice
1. 0
2. 137 R 2
3. 98 R 5
4. 211 R 20

Concept Check
Answers may vary. Possible solution: Consider how many times the first digit of the divisor, 4, can be divided into the first two digits of the dividend, 29. Try 7 first. You will find this to be too large, so try 6. This works.

Section 1.6

Vocabulary
1. evaluate
2. order of operations
3. exponent
4. base

Student Practice
2. 19^5
4. (a) 625
   (b) 1
6. (a) 661
   (b) 32,769
   (c) 1028
8. 13
10. 1272
12. 11
14. 106

Extra Practice

1. 10^4
2. 136
3. 29
4. 69

Concept Check

Answers may vary. Possible solution:

$7 \times 6 \div 3 \times 4^2 - 2$

$= 7 \times 6 \div 3 \times 16 - 2$ Exponents

$= 42 \div 3 \times 16 - 2$ Multiply.

$= 14 \times 16 - 2$ Divide.

$= 224 - 2$ Multiply.

$= 222$ Subtract.

Section 1.7

Vocabulary

1. number line
2. estimate
3. approximately equal to
4. round

Student Practice

2. 5,400,000
4. (a) 452,320
   (b) 65,000,000
6. 150,000,000 kilometers
8. $3400
10. 25,000,000
12. $17 \text{ R } 20 \approx 18$ miles per gallon

Extra Practice

1. 209,000
2. 10,000
3. 60,000,000
4. $1000

Concept Check

Answers may vary. Possible solution:
Since the digit to the right of the
millions place is less than 5 (it is 4),
round down to 682,000,000.

Section 1.8

Vocabulary

1. solve and state the answer
2. mathematics blueprint
3. understand the problem

4. check

Student Practice

2. $909
4. 4730
6. 21 miles per gallon

Extra Practice

1. 464 miles
2. $33
3. $77
4. 5 hours or 300 minutes

Concept Check

Answers may vary. Possible solution:
Round each number so there is one
nonzero digit. Then divide the total cost
by the number of cars.

$$40 \overline{)800,000} = 20,000$$

The estimate is $20,000 each.

Section 2.1

Vocabulary

1. denominator
2. fractions
3. undefined
4. numerator

Student Practice

2. $\dfrac{3}{8}$

4. (a)

(b)

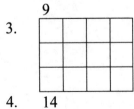

6. (a) $\dfrac{27}{88}$

(b) $\dfrac{2}{7}$

(c) $\dfrac{11}{18}$

8. $\dfrac{17}{33}$

Extra Practice

1. $\dfrac{3}{10}$

2. $\dfrac{8}{9}$

3.

4. $\dfrac{14}{62}$

Concept Check

Answers may vary. Possible solution: There were 120 new businesses. If 30 of the new restaurants went out of business and 25 of the new businesses that were not restaurants went out of business, then

$120 - (30 + 25) = 120 - 55 = 65$ new

businesses did not go out of business:

$$\dfrac{\text{not go out of business}}{\text{new business}} = \dfrac{65}{120}$$

Section 2.2

Vocabulary

1. composite number
2. prime factorization
3. equivalent fractions
4. prime number

Student Practice

2. $2^2 \times 3 \times 7$

4. $\dfrac{2}{3}$

6. $\dfrac{5}{8}$

8. (a) Yes
   (b) No

Extra Practice

1. $2^2 \times 7$

2. $\dfrac{3}{5}$

3. $\dfrac{2}{3}$

4. No

Concept Check

Answers may vary. Possible solution: Consider the factors of 195.

$195 = 3 \times 5 \times 13$

Now find whether 231 has a factor of 3 or 5 or 13. It has a factor of 3.

$$\dfrac{195}{231} = \dfrac{3 \times 5 \times 13}{3 \times 7 \times 11} = \dfrac{65}{77}$$

Vocabulary
1. improper fraction
2. mixed number
3. proper fraction
4. prime

Student Practice
2.
   (a) $\dfrac{65}{9}$

   (b) $\dfrac{46}{11}$

4. $3\dfrac{4}{9}$

6. $\dfrac{14}{5}$

8. $7\dfrac{2}{3}$

10. $1\dfrac{2}{5}$

Extra Practice
1. $\dfrac{53}{14}$

2. $5\dfrac{1}{3}$

3. 19

4. $2\dfrac{5}{8}$

Concept Check
Answers may vary. Possible solution:
Multiply the whole number, 5, by the
denominator of the fraction, 13. Add
the result to the numerator of the
fraction. Write this sum over the
denominator of the fraction.
$$5\dfrac{6}{13}=\dfrac{5\times13+6}{13}=\dfrac{65+6}{13}=\dfrac{71}{13}$$

Section 2.4

Vocabulary
1. numerators
2. proper fraction
3. improper fraction
4. denominator

Student Practice
2.
   (a) $\dfrac{14}{33}$

   (b) $\dfrac{65}{138}$

4. $\dfrac{2}{9}$

6.
   (a) $\dfrac{48}{5}$ or $9\dfrac{3}{5}$

   (b) $\dfrac{7}{5}$ or $1\dfrac{2}{5}$

8. 480

10.
   (a) $\dfrac{154}{5}$ or $30\dfrac{4}{5}$

   (b) $\dfrac{1025}{27}$ or $37\dfrac{26}{27}$

12. $x=\dfrac{4}{7}$

Extra Practice
1. $\dfrac{8}{3}$ or $2\dfrac{2}{3}$

2. 4

3. $\dfrac{2}{3}$

4. $\dfrac{21}{2}$ or $10\dfrac{1}{2}$

Concept Check
Answers may vary. Possible solution:
Write 6 as $\dfrac{6}{1}$. Change the mixed number
to an improper fraction, $4\dfrac{3}{5}=\dfrac{23}{5}$. Then
multiply and simplify.
$$\dfrac{6}{1}\cdot\dfrac{23}{5}=\dfrac{138}{5}\text{ or }27\dfrac{3}{5}$$

Section 2.5

Vocabulary
1. reciprocal
2. invert
3. fractions
4. mixed numbers

Student Practice

2. $\dfrac{14}{5}$ or $2\dfrac{4}{5}$

4. $\dfrac{5}{27}$

6. (a) $\dfrac{7}{12}$

    (b) 0

8. $\dfrac{40}{21}$ or $1\dfrac{19}{21}$

10. $\dfrac{17}{44}$

12. $x = \dfrac{5}{9}$

14. $\dfrac{428}{11}$ or $38\dfrac{10}{11}$ miles per hour

Extra Practice

1. $\dfrac{8}{5}$ or $1\dfrac{3}{5}$

2. $\dfrac{3}{2}$ or $1\dfrac{1}{2}$

3. $\dfrac{3}{2}$ or $1\dfrac{1}{2}$

4. $\dfrac{1}{225}$

Concept Check

Answers may vary. Possible solution:

Write 7 as $\dfrac{7}{1}$. Change the mixed number

to an improper fraction, $3\dfrac{3}{5} = \dfrac{18}{5}$. Then

write the division problem as a
multiplication problem and simplify.

$7 \div 3\dfrac{3}{5} = \dfrac{7}{1} \div \dfrac{18}{5} = \dfrac{7}{1} \times \dfrac{5}{18} = \dfrac{35}{18}$ or $1\dfrac{17}{18}$

Section 2.6
Vocabulary

1. least common multiple
2. least common denominator
3. multiples
4. building fraction property

Student Practice

2. 36
4. 44
6. 21
8. 21
10. 70
12. 126
14. (a) $\dfrac{25}{35}$

    (b) $\dfrac{11}{22}$ and $\dfrac{16}{22}$

16. (a) 70

    (b) $\dfrac{25}{70}$ and $\dfrac{4}{70}$

Extra Practice

1. 90
2. 42
3. $\dfrac{64}{72}$

4. $\dfrac{40}{75}$ and $\dfrac{18}{75}$

Concept Check

Answers may vary. Possible solution:
Write each denominator as a product of
prime factors. Form a product of all
those prime factors, using each factor
the greatest number of times it appears
in any one denominator.

$6 = 2 \times 3$
$14 = 2 \times 7$
$15 = 3 \times 5$
$LCD = 2 \times 3 \times 5 \times 7 = 210$

Section 2.7
Vocabulary

1. common denominators
2. denominator
3. least common denominator (LCD)
4. reduce

Student Practice

2. $\dfrac{8}{15}$

4. $\dfrac{2}{3}$

6. $\dfrac{2}{13}$

8. $\dfrac{41}{72}$

10. $\dfrac{29}{42}$

12. $\dfrac{27}{40}$ mile

14. $x = \dfrac{5}{24}$

16. $\dfrac{184}{225}$

Extra Practice

1. $\dfrac{41}{75}$

2. $\dfrac{3}{4}$

3. $\dfrac{5}{84}$

4. $\dfrac{6}{7}$

Concept Check

Answers may vary. Possible solution:
Find the LCD.
$9 = 3 \times 3$
$7 = 7$
$LCD = 3 \times 3 \times 7 = 63$
Rewrite the fractions with the same denominator and then subtract.
$\dfrac{8}{9} - \dfrac{3}{7} = \dfrac{56}{63} - \dfrac{27}{63} = \dfrac{29}{63}$

Section 2.8

Vocabulary

1. common denominator
2. order of operations
3. mixed numbers
4. borrow

Student Practice

2. $11\dfrac{2}{3}$

4. $7\dfrac{17}{18}$

6. $16\dfrac{9}{20}$

8. $9\dfrac{5}{9}$

10. $2\dfrac{19}{30}$

12. $\dfrac{11}{3}$ or $3\dfrac{2}{3}$

Extra Practice

1. $1\dfrac{9}{10}$

2. $12\dfrac{7}{24}$

3. 3

4. $\dfrac{3}{40}$

Concept Check

Answers may vary. Possible solution:
First, multiply. Then, rewrite fractions with common denominators. Finally, subtract.
$\dfrac{4}{5} - \dfrac{1}{4} \times \dfrac{2}{3} = \dfrac{4}{5} - \dfrac{1}{6} = \dfrac{24}{30} - \dfrac{5}{30} = \dfrac{19}{30}$

Section 2.9

Vocabulary

1. estimation
2. diameter

Student Practice

2. $20\dfrac{1}{5}$ inches

4. $158\dfrac{1}{8}$ cups

6. $50\dfrac{1}{2}$ knots

Extra Practice

1. $76\dfrac{5}{12}$ pounds

2. $1\dfrac{7}{8}$ pounds

3. $4\dfrac{7}{12}$ feet

4. $16\dfrac{2}{3}$ feet

Concept Check

Answers may vary. Possible solution:
Change each mixed number to an
improper fraction, rewrite with a
common denominator, subtract, and
then simplify.

$$3\dfrac{3}{5}-1\dfrac{7}{8}=\dfrac{18}{5}-\dfrac{15}{8}=\dfrac{144}{40}-\dfrac{75}{40}$$

$$=\dfrac{69}{40}=1\dfrac{29}{40}$$

He still has to go $1\dfrac{29}{40}$ miles.

Section 3.1

Vocabulary
1. tropical year
2. decimal fractions
3. decimal point
4. decimal part

Student Practice
2. four hundred thirty-five and six hundred seven thousandths
4. one thousand, six hundred forty-three and $\dfrac{99}{100}$ dollars
6. 12.061
8. $165\dfrac{33}{1000}$
10. $45\dfrac{12}{125}$
12. $\dfrac{1}{625{,}000}$

Extra Practice
1. twenty-four and forty-nine ten-thousandths
2. one thousand, two hundred thirty-eight and $\dfrac{9}{100}$ dollars
3. 18.045
4. $\dfrac{32}{125}$

Concept Check
Answers may vary. Possible solution:
Since the denominator has 5 zeros, the decimal will have 5 decimal places. The numerator only has 3 digits, so 2 zeros will need to be added.
$\dfrac{953}{100{,}000} = 0.00953$

Section 3.2

Vocabulary
1. inequality symbols
2. order
3. round

4. number line

Student Practice
2. $8.217 < 8.219$
4. $2.0031 > 2.003$
6. 3.2, 3.21, 3.213, 3.31, 3.312
8. 6.8
10. 45.237
12. 760.00 kilowatt-hours
14. $2954, $431, $153

Extra Practice
1. $23.0097 > 23.009$
2. 13.574
3. 12.887, 12.88889, 12.909, 12.987, 12.99
4. $2485, $390, $105

Concept Check
Answers may vary. Possible solution:
Find the ten-thousandths place:
34.958$\underline{3}$65
Look at the digit to the right of the ten-thousandths place. If it is less than 5, round down. If it is 5 or greater, round up. In either case, drop the digits to the right of the ten-thousandths place.
34.958365 rounds to 34.9584.

Section 3.3

Vocabulary
1. decimals
2. place value
3. decimal point
4. zeros

Student Practice
2. 403.438
4. 763,410.1 miles
6. $922.71
8. 6278.646
10. 0.605
12. 1366.9 miles
14. $x = 13.1$

Extra Practice
1. 19.357
2. 248.55

3. $5.17

4. 59,421.6 miles

Concept Check

Answers may vary. Possible solution:
First, add two zeros to the right of the
first number, so when they are aligned
vertically, they have the same number
of decimal places. Then subtract all
digits with the same place value,
starting with the right column and
moving to the left. Borrow when
necessary. The math is shown below.

$$
\begin{array}{r}
5\ 6\ \overset{6}{\cancel{7}}.\ \overset{\overset{13}{\cancel{3}}}{\cancel{4}}\ \overset{\overset{14}{\cancel{4}}}{\cancel{3}}\ \overset{\overset{9}{\cancel{10}}}{\cancel{0}}\ \overset{10}{\cancel{0}} \\
-\ 3\ 4\ 5.\ 9\ 8\ 7\ 2 \\
\hline
2\ 2\ 1.\ 4\ 6\ 2\ 8
\end{array}
$$

Section 3.4

Vocabulary

1. product
2. decimal places
3. power of 10
4. factor

Student Practice

2. 0.032
4. 28.67872
6. 15.3811
8. 167.44 square yards
10. 30.42
12. 70,846,000
14. 59,810
16. 92,040 meters

Extra Practice

1. 40.3711
2. 78,904,560
3. 723,407
4. 1968.5 inches

Concept Check

Answers may vary. Possible solution:
Count the number of decimal places and
then add, $2+3=5$. Multiply and then
move the decimal point 5 places to the
left in the product.
$3.45 \times 9.236 = 31.86420$ or 31.8642

Vocabulary

1. divisor
2. dividend
3. round
4. quotient

Student Practice

2. 0.847
4. 0.314
6. 87
8. 0.025
10. 0.149
12. $n = 2.7$

Extra Practice

1. 11.7
2. 44.5
3. 177.65
4. $n = 72$

Concept Check

Answers may vary. Possible solution:
Move the decimal point three places to
the right in the divisor and dividend.
Then divide as whole numbers, adding
zeros to the right of the dividend as
needed.

$$
\begin{array}{r}
.2993 \\
0.578_\wedge \overline{)0.173_\wedge 0000} \\
\underline{115\ 6} \\
57\ 40 \\
\underline{52\ 02} \\
5\ 380 \\
\underline{5\ 202} \\
1780 \\
\underline{1734} \\
46
\end{array}
$$

So, $0.173 \div 0.578 \approx 0.299$.

Section 3.6

Vocabulary

1. equivalent
2. repeating decimals
3. decimal form
4. terminating decimal

290

Student Practice

2. 0.425
4. $1.0\overline{3}$
6. $5.\overline{63}$
8. 1.036
10. $\dfrac{7}{8} < 0.876$
12. 6.25

Extra Practice

1. 4.975
2. 1.571
3. $0.\overline{36}$
4. 44.067

Concept Check

Answers may vary. Possible solution:
Parentheses, exponent, multiplication, subtraction.

$$45.78 - (3.42 - 2.09)^2 \times 0.4$$

$$= 45.78 - 1.33^2 \times 0.4$$

$$= 45.78 - 1.7689 \times 0.4$$

$$= 45.78 - 0.70756$$

$$= 45.07244$$

Section 3.7

Vocabulary

1. check
2. estimate
3. reasonable
4. solve and state the answer

Student Practice

2. (a) 0.03
    (b) 180
4. $630.23
6. (a) 13
    (b) $14.04

Extra Practice

1. 7.5
2. No
3. 3408.08 square feet
4. About 206.8 minutes

Concept Check

Answers may vary. Possible solution:
Divide 24.7 by 1.3.
$24.7 \div 1.3 = 19$
They will need 19 boxes.

Section 4.1

Vocabulary
1. simplest form
2. rate
3. unit rate
4. ratio

Student Practice

2.
   (a) $\dfrac{1}{2}$

   (b) $\dfrac{3}{4}$

   (c) $\dfrac{7}{9}$

4.
   (a) $\dfrac{11}{80}$

   (b) $\dfrac{37}{55}$

6. $\dfrac{59,500 \text{ dollars}}{3 \text{ seconds}}$

8. 41 miles/hour
10. $0.05 per pound
12. $3.29/pound for large package
    $2.51/pound for extra large package

Extra Practice

1. $\dfrac{19 \text{ yards}}{1 \text{ yard}}$

2. $\dfrac{1 \text{ pizza}}{8 \text{ people}}$

3. $145 per washing machine

4. $\dfrac{7}{25}$

Concept Check
Answers may vary. Possible solution:
Divide number of cans by number of people. Then simplify.

$$\frac{663 \text{ cans}}{231 \text{ people}} = \frac{221 \text{ cans}}{77 \text{ people}}$$

Section 4.2

Vocabulary
1. equality test for fractions
2. proportion
3. cross products
4. rates

Student Practice

2. $\dfrac{3 \text{ gallons}}{200 \text{ square feet}} = \dfrac{6 \text{ gallons}}{400 \text{ square feet}}$

4. Not a proportion
6. (a) Proportion
   (b) Not a proportion

8. (a) No
   (b) Yes

Extra Practice

1. $\dfrac{60}{12} = \dfrac{95}{19}$

2. Proportion
3. Not a proportion
4. Yes.

Concept Check
Answers may vary. Possible solution:
Cross multiply. If the products are equal, it is a proportion.

$$33 \times 225 \overset{?}{=} 45 \times 165$$

$$7425 = 7425$$

Yes, it is a proportion.

Section 4.3

Vocabulary
1. equation
2. variable
3. proportion
4. units of measure

Student Practice

2. $n = 30$
4. $53 = n$
6. $9 = n$
8. $n = 7.5$
10. $n = 5$
12. 10.5 grams

14. $n \approx \$2.4$

Extra Practice

1. $n = 2.3$
2. $3 = n$
3. $0.6 \approx n$
4. 27 inches

Concept Check

Answers may vary. Possible solution:
Cross multiply. Divide both sides by
$2\frac{1}{2}$. Then simplify.

The math is shown below.

$$\frac{2\frac{1}{2}}{3\frac{3}{4}} = \frac{16\frac{1}{2}}{n}$$

$$2\frac{1}{2} \times n = 3\frac{3}{4} \times 16\frac{1}{2}$$

$$2\frac{1}{2} \times n = \frac{15}{4} \times \frac{33}{2}$$

$$\frac{2\frac{1}{2} \times n}{2\frac{1}{2}} = \frac{\frac{495}{8}}{2\frac{1}{2}}$$

$$n = \frac{495}{8} \div \frac{5}{2}$$

$$n = \frac{495}{8} \times \frac{2}{5}$$

$$n = \frac{99}{4} \text{ or } 24\frac{3}{4}$$

4. $534.04

Concept Check

Answers may vary. Possible solution:
Write a proportion and solve for n.

$$\frac{70 \text{ euros}}{104 \text{ dollars}} = \frac{n \text{ euros}}{400 \text{ dollars}}$$

$$70 \times 400 = 104 \times n$$

$$28,000 = 104 \times n$$

$$\frac{28,000}{104} = \frac{104 \times n}{104}$$

$$269.23 \approx n$$

It is worth about 269 euros.

Section 4.4

Vocabulary

1. units
2. estimate

Student Practice

2. 15 gallons
4. 135.8 times
6. 160 fish

Extra Practice

1. 495 miles
2. 195 pizzas
3. 375 flowers

Section 5.1

Vocabulary
1. multiplying by 100
2. parts per 100
3. dividing by 100
4. percent

Student Practice
2. (a) 42%
   (b) 26%
4. (a) 465%
   (b) The present cost of two truck tires is 355% of the cost 15 years ago.
6. (a) 0.12%
   (b) 0.00005%
8. (a) 0.52
   (b) 0.07
10. (a) 0.925
    (b) 0.085
    (c) 0.0047
    (d) 5.51
12. (a) 0.5%
    (b) 582%
    (c) 0.72%
    (d) 0.08%

Extra Practice
1. 41%
2. 0.013%
3. 0.0083
4. 0.076%

Concept Check
Answers may vary. Possible solution:
Move the decimal place two places to the left and drop the % symbol.
$0.00072\% = 0.0000072$

Section 5.2

Vocabulary
1. decimal
2. mixed number
3. percent
4. fraction

Student Practice
2. $\dfrac{23}{25}$
4. $\dfrac{1057}{2000}$
6. $4\dfrac{12}{25}$
8. $\dfrac{21}{130}$
10. 18.75%
12. 33.33%
14. $35\dfrac{5}{7}\%$
16.

Fraction	Decimal	Percent
$\dfrac{91}{800}$	0.11375	$11\dfrac{3}{8}\%$

Extra Practice
1. $\dfrac{391}{500}$
2. $\dfrac{9}{200}$
3. 160%
4. 383.33%

Concept Check
Answers may vary. Possible solution:
First change $\dfrac{3}{8}$ to a decimal by dividing 3 by 8. Then add 8. Finally, move the decimal point two places to the left and drop the % symbol.

$$\frac{3}{8} = 0.375$$

$$8\frac{3}{8}\% = 8.375\% = 0.08375$$

Section 5.3A

Vocabulary
1. of
2. equation

3.   is

4.   amount = percent × base

Student Practice

2. $n = 0.7\% \times 800$

4.   (a) $52\% \times n = 650$

     (b) $5.4 = 145\% \times n$

6. $1400

8.   20 people

10. $n = 0.0525\%$

12. 72.2%

Extra Practice

1. $0.30 \times n = 27$

2.   7000

3.   40

4.   76.5%

Concept Check

Answers may vary. Possible solution:
Find "What is 85% of 120?"

$n = 85\% \times 120$

$n = 0.85 \times 120$

$n = 102$

102 people were previous mustang owners.

Section 5.3B

Vocabulary

1.   percent number

2.   base

3.   amount

4.   percent proportion

Student Practice

2. p is the unknown percent number.

4.   (a) $b = 480$; $a = 192$

     (b) base $= b$; $a = 18$

6. $p = 75$; $b = 550$; $a =$ the amount

8.   216

10. $20,000

12. 9%

Extra Practice

1.   68.75

2.   75

3.   5%

4.   9.25%

Concept Check

Answers may vary. Possible solution:
Find "0.7% of what number is $140?"
Let $a = 140$, $b =$ unknown, and $p = 0.7$.
Substitute in formula and solve for b.

$$\frac{a}{b} = \frac{p}{100}$$

$$\frac{140}{b} = \frac{0.7}{100}$$

$$140 \times 100 = 0.7b$$

$$\frac{14,000}{0.7} = \frac{0.7b}{0.7}$$

$$20,000 = b$$

The stock's value was $20,000.

Section 5.4

Vocabulary

1.   discount

2.   markup problems

3.   base

Student Practice

2.   800 monitors

4. $52

6. $65

8. $1168

Extra Practice

1. $90

2.   15.09%

3. $11.25; $63.75

4. $666.25; $4458.75

Concept Check

Answers may vary. Possible solution:
Add the three percents
$(23\% + 14\% + 17\% = 54\%)$.

Then find 54% of $80,000
$(0.54 \times \$80,000 = \$43,200)$.

Since $43,200 is less than $48,000, he did not stay within his budget.

Section 5.5

Vocabulary

1.   interest

2.   commission

3. principal
4. percent of decrease

Student Practice
2. $8750
4. 18%
6. $1380
8. (a) $2080
   (b) $208

Extra Practice
1. $544
2. 32%
3. 2%
4. $100.80; $2900.80

Concept Check
Answers may vary. Possible solution:
Use the formula $I = P \times R \times T$. Let
$P = 5800$, $R = 16\%$ or 0.16, and

$T = \dfrac{3}{12} = \dfrac{1}{4}$ (since T must be in years).

$I = 5800 \times 0.16 \times \dfrac{1}{4} = 232$

The simple interest is $232.

Worksheet Answers Chapter 6

Section 6.1

Vocabulary

1. proportion
2. American units
3. metric system
4. unit fraction

Student Practice

2. (a) 7
   (b) 16
   (c) 3
   (d) 60
   (e) 4
   (f) 5280
4. 14 hours
6. (a) 625.32 inches
   (b) $2\frac{2}{5}$ gallons
8. 8780 pounds
10. 13.75 feet
12. 1223 quarts
14. $61.80

Extra Practice

1. 9
2. 14,500
3. 510
4. 12.5

Concept Check

Answers may vary. Possible solution: There are 2 pints in 1 quart. Use the unit fraction $\dfrac{1 \text{ quart}}{2 \text{ pints}}$ to convert.

$250 \text{ pints} \times \dfrac{1 \text{ quart}}{2 \text{ pints}} = 125 \text{ quarts}$

Section 6.2

Vocabulary

1. metric system
2. kilometer
3. meter
4. millimeter

Student Practice

2. (a) centi-
   (b) deka-

4. 300,000 centimeters
6. 0.247 kilometers
8. (b) 21 meters
10. (a) 0.038 dm
    (b) 526,000 cm
12. (a) 9,650,000 mm
    (b) 4.67 m
14. 1237 cm

Extra Practice

1. 11,500

2. 1.37
3. 71,000
4. 4818.32

Concept Check

Answers may vary. Possible solution: We are converting from a smaller unit, cm, to a larger unit, km. Therefore, there will be fewer kilometers than centimeters. Move the decimal place 5 units to the left.

5643 cm = 0.05643 km

Section 6.3

Vocabulary

1. mass
2. megagram
3. gram
4. liter

Student Practice

2. (a) 590
   (b) 5300
   (c) 27,000
4. (a) 0.0679
   (b) 821.4
6. (a) 8.14
   (b) 1900
8. (a) 1730
   (b) 390
10. (a) 0.0814
    (b) 0.0617
12. $4.50
14. (b) 8.2 L

Extra Practice

1. 756
2. 2.78
3. 0.043
4. 8320.87

Concept Check

Answers may vary. Possible solution:
We are converting from a larger unit,
kg, to a smaller unit, mg. Therefore,
there will be more mg than km. Move
the decimal place 6 units to the right.
$54 \text{ kg} = 54,000,000 \text{ mg}$

Section 6.4

Vocabulary

1. approximations
2. unit fractions
3. Fahrenheit system
4. Celsius scale

Student Practice

2. 25.925 meters
4. (a) 5.685 L
   (b) 0.954 qt
6. 1.54 ft
8. 74.4 mi/hr
10. 139 ft/sec
12. $59° \text{F}$
14. $5°\text{C}$

Extra Practice

1. 23.98 yd
2. 7.57 L
3. 4.54 kg
4. $257°\text{F}$

Concept Check

Answers may vary. Possible solution:
Multiply by the unit fraction $\dfrac{1 \text{ km}}{0.62 \text{ mi}}$.

$\dfrac{65 \text{ mi}}{1 \text{ hr}} \cdot \dfrac{1 \text{ km}}{0.62 \text{ mi}} \approx 104.8 \text{ km/hr}$

Section 6.5

Vocabulary

1. unit fraction
2. estimated
3. perimeter

4. units

Student Practice

2. $40\dfrac{7}{8}$ yards
4. 82 containers

Extra Practice

1. $24.99
2. 8000 packets
3. $3.6° \text{F}$ or $2°\text{C}$
4. 3 gallons

Concept Check

Answers may vary. Possible solution:
Set up a ratio. Cross-multiply and solve
for the unknown.

$\dfrac{3 \text{ in.}}{6.5 \text{ mi}} = \dfrac{5 \text{ in.}}{x \text{ mi}}$

$3x = 6.5(5)$

$\dfrac{3x}{3} = \dfrac{32.5}{3}$

$x \approx 10.8$

The two cities are 10.8 miles apart.

Section 7.1

Vocabulary
1. right angle
2. parallel lines
3. degrees
4. geometry

Student Practice
2. $\angle MNP$ and $\angle ONQ$ are acute angles. $\angle PNO$ and $\angle MNQ$ are obtuse angles. $\angle PNQ$ is a right angle. $\angle MNO$ is a straight angle.
4. (a) $48°$
   (b) $138°$
6. $\angle f = 116°$, $\angle g = 64°$, $\angle k = 64°$
8. $\angle v = 125°$, $\angle w = 55°$, $\angle y = 55°$, $\angle z = 55°$

Extra Practice
1. $\angle h = 142°$
2. $\angle g = 38°$
3. $\angle JOM = 150°$
4. $11°$

Concept Check
Answers may vary. Possible solution:
$\angle e$ and $\angle a$ are supplementary angles.
measure of $\angle a = 180 -$ measure of $\angle e$

Section 7.2

Vocabulary
1. perimeter
2. area
3. rectangle
4. square

Student Practice
2. 12.5 m
4. 25.6 cm
6. About $3.32
8. 351 m^2
10. 23.04 m^2
12. 295 ft^2

Extra Practice
1. $39\dfrac{1}{3}$ m
2. 1796.0644 cm^2
3. $1350
4. Perimeter $= 58$ m ; Area $= 148 \text{ m}^2$

Concept Check
Answers may vary. Possible solution: Find the area of the rectangle. Find the area of the square. Then add the results.

$$\text{Area} = 12 \text{ ft} \times 15 \text{ ft} + (15 \text{ ft})^2$$
$$= 180 \text{ ft}^2 + 225 \text{ ft}^2$$
$$= 405 \text{ ft}^2$$

Section 7.3

Vocabulary
1. parallelogram
2. quadrilaterals
3. trapezoid
4. rhombus

Student Practice
2. 23.6 ft
4. 20.58 mm^2
6. Perimeter $= 33$ cm.; Area $= 49.5 \text{ cm}^2$
8. 21 ft
10. 360 m^2
12. 248.92 cm^2

Extra Practice
1. Perimeter $= 64$ cm; Area $= 224 \text{ cm}^2$
2. Perimeter $= 34$ in.; Area $= 36 \text{ in.}^2$
3. 770 square yards
4. $2368.80

Concept Check
Answers may vary. Possible solution:
The area would increase by $\dfrac{16}{9}$.

$$A = \frac{16(30+34)}{2} = 512 \text{ m}^2 ;$$

$$288 \times \frac{16}{9} = 512 \text{ m}^2$$

Section 7.4

Vocabulary
1. right triangle
2. isosceles triangle
3. equilateral triangle
4. scalene triangle

Student Practice
2. Angle A measures $55°$.
4. 25.6 cm
6. 240 m^2
8. 35 mm^2

Extra Practice
1. $35°$
2. 436 m
3. 187.03 cm^2
4. About $364.10

Concept Check
Answers may vary. Possible solution:
Find the area of each separately. Then add the results.
Triangle area:
$$A = \frac{1}{2}(20 \text{ yd})(20 \text{ yd}) = 200 \text{ yd}^2$$
Rectangle area:
$$A = 20 \text{ yd} \times 15 \text{ yd} = 300 \text{ yd}^2$$
Combined area:
$$A = 200 \text{ yd}^2 + 300 \text{ yd}^2 = 500 \text{ yd}^2$$

Section 7.5

Vocabulary
1. before
2. approximation
3. square root
4. perfect square

Student Practice
2. (a) 4
   (b) 14
4. 6
6. (a) Yes
   (b) 1
8. (a) $\sqrt{5} \approx 2.236$

(b) $\sqrt{19} \approx 4.359$
(c) $\sqrt{11} \approx 3.317$
10. 5.831 ft

Extra Practice
1. 10
2. 12
3. 9.110
4. 9.950

Concept Check
Answers may vary. Possible solution:
Since the square root of 81 is 9
$\left(\sqrt{81} = 9 \right)$, then the square root of 0.81
is 0.9 $\left(\sqrt{0.81} = 0.9 \right)$.

Section 7.6

Vocabulary
1. hypotenuse
2. Pythagorean Theorem
3. $45°$-$45°$-$90°$ triangle
4. $30°$-$60°$-$90°$ triangle

Student Practice
2. 5 ft
4. 16 ft
6. 4.1 km
8. (a) $x = 17.3$ ft; $y = 10$ ft
   (b) 9.9 m

Extra Practice
1. 8.944 m
2. $b = 6$ in. $c = 8.485$ in.
3. 18.735 ft
4. 15.5 ft

Concept Check
Answers may vary. Possible solution:
The hypotenuse is 2 mi and one leg is 1.5 mi. Find the distance to the supplies by using the Pythagorean Theorem.
$$\text{leg} = \sqrt{2^2 - 1.5^2}$$
$$= \sqrt{4 - 2.25}$$
$$= \sqrt{1.75}$$
$$\approx 1.32 \text{ miles}$$

Section 7.7

Vocabulary
1. diameter
2. radius
3. circle
4. circumference

Student Practice
2. 27.21 ft
4. 50.2 m^2
6. 42.5 cm^2
8. 62.1 m^2

Extra Practice
1. 141.3 in.
2. 452.16 cm^2
3. $150.13
4. 344.52 m^2

Concept Check
Answers may vary. Possible solution:
Find the area of a circle with a radius of 3 feet. Then divide the result by 2.

$$\frac{\pi r^2}{2} = \frac{3.14(3 \text{ ft})^2}{2} = 14.13 \text{ ft}^2$$

Section 7.8

Vocabulary
1. sphere
2. cone
3. rectangular solid
4. cylinder

Student Practice
2. 126 cm^3
4. 452.2 m^3
6. 3052.1 ft^3
8. 527.5 yd^3
10. 100 ft^3

Extra Practice
1. 12,208.3 in.^{3}
2. 22.9 ft^3
3. 22,400,000 m^3
4. 80.7 ft^3

Concept Check
Answers may vary. Possible solution:

The volume is larger by a factor of

$$\frac{4^2}{3^2} = \frac{16}{9}.$$

Section 7.9

Vocabulary
1. perimeters
2. corresponding angles
3. ratio
4. similar triangles

Student Practice
2. $n = 46.2$ m
4. a corresponds to z. b corresponds to x. c corresponds to y.
6. 5.0 m
8. 15.64 m

Extra Practice
1. 7.5 cm
2. 6 in.
3. 8.4 in.
4. 127.5 ft

Concept Check
Answers may vary. Possible solution:
Draw a picture. Set up a proportion and solve for the unknown.

$$\frac{5}{6} = \frac{9}{n}$$

$$5n = 6(9)$$

$$5n = 54$$

$$\frac{5n}{5} = \frac{54}{5}$$

$$n = 10.8$$

The larger sides in the drawing are 10.8 inches.

Section 7.10

Vocabulary
1. volume
2. area

Student Practice
2. 72 minutes
4. 538.75 ft^2

Extra Practice

1. $716.40
2. 2.7 hours
3. $912
4. $487.52

Concept Check

Answers may vary. Possible solution:
Find the area of the second field.
Multiply the area by $0.20 to find the
cost. Then subtract the cost of the first
field from this result.

$$\frac{1}{2}(300)(140) = 21,000 \text{ m}^2$$

$$21,000 \times 0.20 = \$4200$$

The difference in cost is
$4200 - \$700 = \$3500$.

Section 8.1

Vocabulary
1. graphs
2. percent
3. circle graphs
4. statistics

Student Practice
2. senior students
4. 42%
6. 42%
8. 7574 mi^2
10. (a) 40%
    (b) 7,952,800 students

Extra Practice
1. 28%
2. 65%
3. 240 students
4. $\dfrac{43}{57}$

Concept Check
Answers may vary. Possible solution:
Add the percents for the three
categories $(4\% + 42\% + 18\% = 64\%)$.
Then find 64% of 850,000.
$0.64 \times 850,000 = 544,000$ of the
vehicles sold were station wagons,
four-door sedans, or minivans.

Section 8.2

Vocabulary
1. double-bar graphs
2. comparison line graph
3. bar graphs
4. line graph

Student Practice
2. 23 million or 23,000,000
4. 300 cars
6. 200 more cars
8. July to August
10. 6 years

Extra Practice
1. Angela

2. 100 more books
3. 100 more jobs
4. 25 more jobs

Concept Check
Answers may vary. Possible solution:
Find the increase from 2000 to 2010
$(1000 - 600 = 400)$. Since this is a 10-
year period, the same increase would be
expected for the next 10-year period.
$1000 + 400 = 1400$ expected
condominiums in 2020.

Section 8.3

Vocabulary
1. raw data
2. histogram
3. class interval
4. class frequency

Student Practice
2. 6 students
4. 36 students
6. 5 light bulbs
8. 70 light bulbs
10.

Math Test Grades (Class Interval)	Tally	Frequency
50–59	||	2
60–69	||	2
70–79	|||	3
80–89	|||||	5
90–99	|||	3

12.

14.

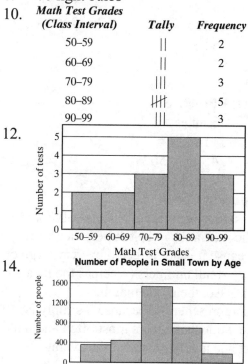

Extra Practice

1. Approximately 340 households
2. Approximately 270 households
3. Approximately 620 households
4.

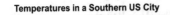
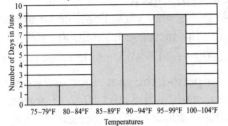

Concept Check

Answers may vary. Possible solution:
Multiply the height of the bar for 1-4
by 3 and the height of the bar for 5-8 by
2 and add the results.

Section 8.4

Vocabulary

1. mode
2. median
3. average
4. mean

Student Practice

2. $67.25
4. $800
6. 154.5
8. 64
10. 4

Extra Practice

1. 13 miles
2. $116
3. 245 and 250
4. 12 credit hours

Concept Check

Answers may vary. Possible solution:
The median may be the best measure
because it is not affected by very large or
very small numbers (whereas the mean
is). Also, the mode may be
misrepresentative because it simply is
the value that occurs most often and may
not accurately represent the average.

Worksheet Answers Chapter 9

Section 9.1

Vocabulary
1. positive number
2. absolute value
3. number line
4. negative number

Student Practice
2. (a) <
   (b) >
   (c) >
   (d) <
   (e) <
4. (a) 14
   (b) −14.1
6. $-\dfrac{41}{63}$
8. (a) −3
   (b) −8.1
10. 15°F
12. −11

Extra Practice
1. $-3\dfrac{15}{22}$
2. −22.1
3. $-4\dfrac{5}{12}$
4. $-\dfrac{1}{5}$

Concept Check
Answers may vary. Possible solution: This solution may have fewer steps if the second method is used.

Section 9.2

Vocabulary
1. add the opposite
2. opposite
3. equal
4. zero

Student Practice
2. −8
4. (a) −13
   (b) −12

6. (a) −16.2
   (b) $-\dfrac{5}{12}$
8. (a) 2
   (b) $-\dfrac{2}{35}$
   (c) 11.7
10. −18
12. 77°F

Extra Practice
1. −16.9
2. $-1\dfrac{23}{36}$
3. −2
4. −37°F

Concept Check
Answers may vary. Possible solution: Change $8-(-13)$ to $8+13$ by adding the opposite. Then evaluate from left to right.

$$8-(-13)+(-5) = 8+13+(-5)$$
$$= 21+(-5)$$
$$= 16$$

Section 9.3

Vocabulary
1. negative
2. positive
3. commutative
4. multiplication

Student Practice
2. 70
4. (a) −36
   (b) −420
6. (a) −9
   (b) −7
8. (a) 72
   (b) $\dfrac{4}{35}$
10. 1.2
12. 168

305

14. +26

16. $1.3°\text{F}$

Extra Practice

1. -1560

2. $\dfrac{3}{8}$

3. -4

4. 2

Concept Check

Answers may vary. Possible solution: Addition of two negative numbers is always negative. Multiplication of two negative numbers is always positive.

Section 9.4

Vocabulary

1. order of operations
2. parentheses
3. multiply or divide
4. exponents

Student Practice

2. (a) -15

    (b) 22

4. -3

6. 61

8. $-\dfrac{215}{36}$

Extra Practice

1. -35

2. -1

3. -2

4. 3.79

Concept Check

Answers may vary. Possible solution: Perform operations inside parentheses. Then simplify expression with exponent. Then multiply and finally add.

$$4(3-9)+5^2 = 4(-6)+5^2$$
$$= 4(-6)+25$$
$$= -24+25$$
$$= 1$$

Section 9.5

Vocabulary

1. power of 10
2. positive exponent
3. scientific notation
4. negative exponent

Student Practice

2. (a) 5.789×10^3

    (b) 4.23598×10^5

4. (a) 9.2×10^{-2}

    (b) 5.79×10^{-1}

6. 378,915

8. (a) 930,000

    (b) 40,000

10. (a) 0.00995

    (b) 0.000035

12. 9.07×10^{12}

14. 9.12×10^5

Extra Practice

1. 3×10^7

2. 5.46×10^{-3}

3. 3,720,000,000,000

4. 4.265×10^8

Concept Check

Answers may vary. Possible solution: Move the decimal point 8 places to the right. You will need to add 5 zeros.

$5.398\times10^8 = 539,800,000$

Section 10.1

Vocabulary
1. like terms
2. variable
3. numerical coefficients
4. term

Student Practice
2. (a) A, r, and h
   (b) V, r, and h
4. (a) $V = \pi r^2 h$
   (b) $A = 2\pi rh$
6. (a) $-37x$
   (b) $-36x$
8. (a) $-7x$
   (b) $-2.5x$
10. $3.3x - 3.2$
12. (a) $-4m + 7n - 39$
    (b) $\dfrac{34}{77}a + \dfrac{35}{4}$

Extra Practice
1. P, l, and w
2. $p = 4abc$
3. $1.2x - 5.3y + 2$
4. $-\dfrac{1}{3}x + \dfrac{1}{2}y$

Concept Check
Answers may vary. Possible solution:
Change all subtraction to addition. Use the commutative property to rearrange terms. Then simplify.
$-8.2x - 3.4y + 6.7z - 3.1x + 5.6y - 9.8z$
$= -8.2x + (-3.4y) + 6.7z + (-3.1x) + 5.6y$
$\quad + (-9.8z)$
$= -8.2x + (-3.1x) + (-3.4y) + 5.6y + 6.7z$
$\quad + (-9.8z)$
$= -11.3x + 2.2y - 3.1z$

Section 10.2

Vocabulary
1. simplify

2. distributive property of multiplication over subtraction
3. property
4. distributive property of multiplication over addition

Student Practice
2. (a) $4a + 36$
   (b) $-6x - 24y$
   (c) $35m - 56n$
4. $18x + 27y$
6. (a) $-7a - 56b - 42$
   (b) $13.6x + 30y + 32$
8. $\dfrac{4}{25}x - \dfrac{1}{5}y + \dfrac{24}{5}z - \dfrac{1}{10}$
10. $76x + 38y$
12. $6x - 15y + 42$

Extra Practice
1. $-20x + 30y + 40z$
2. $\dfrac{3}{5}x - \dfrac{2}{5}y - \dfrac{1}{7}z + \dfrac{1}{15}$
3. $-0.2x - 0.42y + 1$
4. $14y - 64$

Concept Check
Answers may vary. Possible solution:
Use the distributive property to remove parentheses. Then combine like terms to simplify.
$-3(2x + 5y) + 4(5x - 1)$
$= -3(2x) - 3(5y) + 4(5x) - 4(1)$
$= -6x - 15y + 20x - 4$
$= 14x - 15y - 4$

Section 10.3

Vocabulary
1. equation
2. solution
3. addition property of equations
4. solving an equation

Student Practice
2. $x = -14$

4. (a) $y = 13.3$

(b) $x = \dfrac{1}{2}$

6. $x = 9$

Extra Practice

1. $a = 8$
2. $x = -6.5$
3. $y = \dfrac{10}{7}$ or $1\dfrac{3}{7}$
4. $x = -65$

Concept Check

Answers may vary. Possible solution:
Add $8x$ to both sides. Then add -5 to both sides. The math is shown below.

$$-7x + 5 = -8x - 13$$

$$8x - 7x + 5 = 8x - 8x - 13$$

$$x + 5 = -13$$

$$x + 5 + (-5) = -13 + (-5)$$

$$x = -18$$

Section 10.4

Vocabulary

1. equation
2. multiplication property of equations
3. division property of equations
4. mixed numbers

Student Practice

2. $n = 17$
4. $n = -\dfrac{55}{9}$
6. $a = 5$
8. $z = 15$

Extra Practice

1. $y = 6$
2. $z = 3$
3. $y = -20$
4. $x = -9$

Concept Check

Answers may vary. Possible solution:
Simplify the left side. Then divide both sides by -14.

$$12 - 5 = -14x$$

$$7 = -14x$$

$$\dfrac{7}{-14} = \dfrac{-14x}{-14}$$

$$-\dfrac{1}{2} = x$$

Section 10.5

Vocabulary

1. distributive property
2. equivalent equation
3. signs
4. like terms

Student Practice

2. $x = -8$
4. $x = -4$
6. $y = -4$
8. $x = \dfrac{69}{8}$ or $8\dfrac{5}{8}$

Extra Practice

1. $x = 3$
2. $x = -5$
3. $y = 2$
4. $x = 3$

Concept Check

Answers may vary. Possible solution:
Remove parentheses using the distributive property. Combine like terms on each side. Gather the variable terms on one side and the numbers on the other side. Divide both sides of the equation by the numerical coefficient of the variable term.

$$7x - 3(x - 6) = 2(x - 3) + 8$$

$$7x - 3x + 18 = 2x - 6 + 8$$

$$4x + 18 = 2x + 2$$

$$2x = -16$$

$$x = -8$$

Section 10.6

Vocabulary

1. algebraic expressions
2. addition

308

3. division
4. multiplication

Student Practice

2. $a = 9 + p$

4. $n = m - 15$

6. $m = w + 7$

8. $l = 2w + 9$

10. Let $r =$ the length of Robert's trip. So therefore $r - 450 =$ the length of Jim's trip.

12. Let $C =$ the number of degrees in angle C. So therefore $C - 55 =$ the number of degrees in angle D.

14. Let $y =$ the length of the first side. So therefore $3y =$ the length of the second side. So therefore $y + 8 =$ the length of the third side.

Extra Practice

1. $a + b = 9$

2. $ht = 300$

3. Elias: e
   Anastasia: $e + 540$
   Nicholas: $e - 60$

4. Youngest child: $2m - 11$
   Middle child: m
   Oldest child: $m + 6$

Concept Check

Answers may vary. Possible solution: Both are right. In either case, it shows that 12 more students have part-time jobs than full-time jobs.

Section 10.7

Vocabulary

1. equations
2. variable
3. perimeter

Student Practice

2. shorter piece is 8.15 feet; longer piece is 10.85 feet

4. width is 162 feet; length is 334 feet

6. angle A measures $40°$; angle B measures $20°$; angle C measures $120°$

8. $\$6250$

Extra Practice

1. The pieces are 15 feet long and 30 feet long.

2. The rectangle is 175 inches long by 50 inches wide.

3. The perimeter is 192 feet.

4. The Giants scored 12 runs and the Astros scored 8 runs.

Concept Check

Answers may vary. Possible solution: Let $x =$ degrees of 2nd angle. State the degrees of the first and third angles in terms of x (degrees of 2nd angle). Since the sum of the interior angles of a triangle is $180°$, write an equation. Solve the equation by getting all terms with x on one side of the equation and all numbers on the other. Then divide by the coefficient of the x-term.

GUIDED LEARNING
VIDEO WORKSHEETS

Pearson

Name: _____   Date: _____

Instructor: _____   Section: _____

Guided Learning Video Worksheet: Addition with Carrying

Text: *Basic College Mathematics*

Student Learning Objective 1.2.4: Add several-digit numbers when carrying is needed.

Follow along with *Guided Learning Video 1.2.4, Addition with Carrying.*

Understanding the Big Picture: As you listen to the first part of the video, when the pencil icon appears, pause and fill in the blanks, choosing from the words listed.

carrying • vertically • total

1. When you want to find the _____, you add.

2. In the video, the process of adding numbers is discussed when _____ is needed.

3. In the beverage problem, the three numbers are arranged _____ according to place value.

Follow along with the two Guided Learning Video examples, and fill in the blanks on the left as you learn with the instructor. When the pencil icon appears, pause and try the Student Practice on your own.

Guided Learning Video	Pause: Student Practice
1. Add $52 + 19$. Arrange the numbers vertically and line them up according to place value. Add. $\square$ 52 $+19$ $\square\square$	**2.** Add $47 + 26$.

3. Add $698 + 539$.

Arrange the numbers vertically according to place value. Add.

$$
\begin{array}{r}
\boxed{} \\
698 \\
+539 \\
\hline
\boxed{}
\end{array}
$$

4. Add $738 + 576$.

Helpful Hint: Pause the video and write the helpful hint in your own words.

For each Active Video Lesson, when the pencil icon appears, pause and work the problem. Then press play to check your work.

Active Video Lesson 1 | **Active Video Lesson 2**

5. Add $763 + 958$.

6. Add $68 + 74$.

Name: _____   Date: _____

Instructor: _____   Section: _____

Guided Learning Video Worksheet: Subtracting Whole Numbers with Borrowing

Text: *Basic College Mathematics*
Student Learning Objective 1.3.3: Subtract whole numbers when borrowing is necessary.

Follow along with *Guided Learning Video 1.3.3, Subtracting Whole Numbers with Borrowing.*

Understanding the Big Picture: As you listen to the first part of the video, when the pencil icon appears, pause and fill in the blanks, choosing from the words listed.

borrowing • subtrahend • larger • minuend

1. When numbers being subtracted are arranged vertically, the _____ is the upper number.

2. The number being subtracted is called the _____.

3. When subtracting digits of equal place value, _____ is used when the lower digit is _____ than the upper digit.

Follow along with the two Guided Learning Video examples, and fill in the blanks on the left as you learn with the instructor. When the pencil icon appears, pause and try the Student Practice on your own.

Guided Learning Video (▶) GUIDED LEARNING VIDEO	**Pause: Student Practice** ✏️
1. Subtract $724 - 489$. Arrange the numbers vertically and line them up according to place value. Subtract. □ 724 -489 □	**2.** Subtract $836 - 579$.

Guided Learning Video ▶ GUIDED LEARNING VIDEO	**Pause: Student Practice** ✏
3. Subtract $304 - 146$.	**4.** Subtract $503 - 274$.

Arrange the numbers vertically according to place value. Subtract.

$$
\begin{array}{r}
\boxed{} \\
304 \\
-146 \\
\hline
\boxed{}
\end{array}
$$

Helpful Hint: Pause the video and write the helpful hint in your own words.

For each Active Video Lesson, when the pencil icon appears, pause and work the problem. Then press play to check your work.

Active Video Lesson 1 ✏	**Active Video Lesson 2** ✏
5. Subtract $7090 - 4638$.	**6.** Subtract $5621 - 629$.

Guided Learning Video Worksheet: Applying Subtraction to Real-Life Situations

Text: *Basic College Mathematics*
Student Learning Objective 1.3.5: Apply subtraction to real-life situations.

Follow along with *Guided Learning Video 1.3.5, Applying Subtraction to Real-Life Situations.*

Understanding the Big Picture: As you listen to the first part of the video, when the pencil icon appears, pause and fill in the blanks, choosing from the words listed.

total • comparison • difference • subtraction • minus

1. When we want to know *how much more* one amount is than another, _____ is used.

2. When key words like _____ or _____ are used, subtraction is necessary.

3. Subtraction can be required when a problem involves a _____.

Follow along with the two Guided Learning Video examples, and fill in the blanks on the left as you learn with the instructor. When the pencil icon appears, pause and try the Student Practice on your own.

Guided Learning Video ▶ GUIDED LEARNING VIDEO	**Pause: Student Practice**
1. What was the increase in homes sold in Riverside from 2013 to 2014? From the graph, identify the number of homes sold in 2013 and 2014, then subtract. ☐ – ☐ ☐	2. What was the increase in homes sold in Riverside from 2014 to 2015?

3. In 2010, the estimated population of Texas was how much greater than the estimated population of Arizona and New Mexico combined?

    Add the estimated populations of Arizona and New Mexico, Then, subtract the total from the estimated population of Texas.

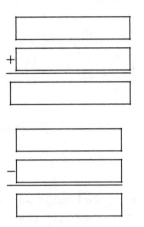

4. In 2010, the estimated population of California was how much greater than the estimated population of Texas and Arizona combined?

Helpful Hint: Pause the video and write the helpful hint in your own words.

For each Active Video Lesson, when the pencil icon appears, pause and work the problem. Then press play to check your work.

Active Video Lesson 1 | **Active Video Lesson 2**

5. In 2010, California's population is how much greater than Texas and New Mexico's population combined?

6. How many more homes were sold in Riverside in 2015 than in 2013?

Name: _____ Date: _____

Instructor: _____ Section: _____

Guided Learning Video Worksheet: Multiplying by a Power of Ten

Text: *Basic College Mathematics*
Student Learning Objective 1.4.3: Multiply a whole number by a power of 10.

Follow along with *Guided Learning Video 1.4.3, Multiplying by a Power of Ten.*

Understanding the Big Picture: As you listen to the first part of the video, when the pencil icon appears, pause and fill in the blanks, choosing from the words listed.

left • end • power • right • zeros

1. A(n) _____ of ten is a whole number beginning with one and ending in one or more zeros.

2. When multiplying by a power of ten, count the number of _____ in the power of ten and attach them to the _____ side of the other whole number.

3. Only zeros on the _____ of a whole number mean it's a multiple of ten.

Follow along with the two Guided Learning Video examples, and fill in the blanks on the left as you learn with the instructor. When the pencil icon appears, pause and try the Student Practice on your own.

Guided Learning Video ▶ GUIDED LEARNING VIDEO	Pause: Student Practice ✎
1. Multiply 2026×100. Count the number of zeros in 100, then multiply. $\boxed{} \times \boxed{} = \boxed{}$	**2.** Multiply 4035×1000.

Guided Learning Video ▶ GUIDED LEARNING VIDEO	Pause: Student Practice ✎
3. Multiply 2300×50. Count the number of zeros at the end of each number, then multiply. $\boxed{} \times \boxed{} = \boxed{}$	**4.** Multiply 6400×30.

Helpful Hint: Pause the video and write the helpful hint in your own words.

7

For each Active Video Lesson, when the pencil icon appears, pause and work the problem. Then press play to check your work.

Active Video Lesson 1	Active Video Lesson 2
5. Multiply 6700×90.	**6.** Multiply 8079×1000.

Name: _____   Date: _____
Instructor: _____   Section: _____

Guided Learning Video Worksheet: Multiplying a Several-Digit Number by a Several-Digit Number

Text: *Basic College Mathematics*
Student Learning Objective 1.4.4: Multiply a several-digit number by a several-digit Number.

Follow along with *Guided Learning Video 1.4.4, Multiplying a Several-Digit Number by a Several-Digit Number.*

Understanding the Big Picture: As you listen to the first part of the video, when the pencil icon appears, pause and fill in the blanks, choosing from the words listed.

add • second • partial products

1. To multiply several-digit numbers, break up the _____ number into smaller numbers.

2. When multiplying the first number times the smaller numbers, _____ are obtained.

3. The last step when multiplying several-digit numbers is to _____ the _____.

Follow along with the two Guided Learning Video examples, and fill in the blanks on the left as you learn with the instructor. When the pencil icon appears, pause and try the Student Practice on your own.

Guided Learning Video ▶ GUIDED LEARNING VIDEO	**Pause: Student Practice**
1. Find the area of the county by multiplying 56 miles times 39 miles. Multiply the first number 56 by each of the smaller numbers that 39 is composed of. Then, add the partial products to obtain the answer. $\begin{array}{r} 56 \\ \times 39 \\ \hline \boxed{} \\ \boxed{} \\ \boxed{} \end{array}$	2. Find the area of a county by multiplying 72 miles times 24 miles.

3. Multiply 897×64.

Multiply the first number 897 by each of the smaller numbers that 64 is composed of. Then, add the partial products to obtain the answer.

$$
\begin{array}{r}
897 \\
\times 64 \\
\hline
\end{array}
$$

☐

☐

☐

4. Multiply 593×72.

Helpful Hint: Pause the video and write the helpful hint in your own words.

For each Active Video Lesson, when the pencil icon appears, pause and work the problem. Then press play to check your work.

Active Video Lesson 1 ✏ | **Active Video Lesson 2** ✏

5. Multiply 968×37.

6. Multiply 74×89.

Name: _____   Date: _____

Instructor: _____   Section: _____

Guided Learning Video Worksheet: Applying Multiplication to Real-Life Situations

Text: *Basic College Mathematics*
Student Learning Objective 1.4.6: Apply multiplication to real-life situations.

Follow along with *Guided Learning Video 1.4.6, Applying Multiplication to Real-Life Situations.*

Understanding the Big Picture: As you listen to the first part of the video, when the pencil icon appears, pause and fill in the blanks, choosing from the words listed.

addition • average • multiplication • operation

1. The operation of _____ can be used when _____ of the same addend occurs.

2. An important step in problem solving is to determine the _____ to perform.

3. The key word _____, indicates the operation of multiplication.

Follow along with the two Guided Learning Video examples, and fill in the blanks on the left as you learn with the instructor. When the pencil icon appears, pause and try the Student Practice on your own.

Guided Learning Video ▶ GUIDED LEARNING VIDEO	**Pause: Student Practice**
1. Steven drove an average speed of 65 miles per hour for 4 hours. How far did he drive? Identify the operation needed to solve the problem. Then, calculate the answer. ☐ ×☐ ☐	2. Rick averaged a speed of 68 miles per hour for 7 hours. How far did he drive?

3. An average of 3152 cars cross over the West River Bridge each weekday. There are 260 weekdays each year. How many cars crossed the bridge last year?

Use the key word in the problem to Identify the necessary operation, then calculate the answer.

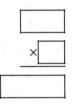

4. Each weekday, an average of 1074 cars pass through each toll booth at a turnpike entrance. There are 8 toll booths at the turnpike entrance. How many cars enter the turnpike each weekday?

Helpful Hint: Pause the video and write the helpful hint in your own words.

For each Active Video Lesson, when the pencil icon appears, pause and work the problem. Then press play to check your work.

Active Video Lesson 1 | **Active Video Lesson 2**

5. Each time Jennifer receives a paycheck, $175 is put into her IRA retirement savings account. If she gets paid twice a month, how much does Jennifer contribute to her IRA in one year?

6. A football player averages 267 yards per game passing. At this average, how many passing yards will be gained in a 9-game season?

Name: _____   Date: _____
Instructor: _____   Section: _____

Guided Learning Video Worksheet: Long Division

Text: *Basic College Mathematics*
Student Learning Objective 1.5.3: Perform division by a two- or three-digit number.

Follow along with *Guided Learning Video 1.5.3, Long Division.*

Understanding the Big Picture: As you listen to the first part of the video, when the pencil icon appears, pause and fill in the blanks, choosing from the words listed.

dividend • long division • quotient • division • equal • divisor

1. Splitting an amount into a(n) _____ number of parts is called _____.

2. The words _____, _____, and _____ are commonly used when referring to division.

3. The process of _____ is used when quantities that aren't basic math facts are being divided.

Follow along with the two Guided Learning Video examples, and fill in the blanks on the left as you learn with the instructor. When the pencil icon appears, pause and try the Student Practice on your own.

Guided Learning Video ▶ GUIDED LEARNING VIDEO	**Pause: Student Practice**
1. Divide: $943 \div 23$	**2.** Divide: $966 \div 23$
Identify the divisor and the dividend, and write the problem in long division form. Then perform long division to calculate the quotient.	

Guided Learning Video GUIDED LEARNING VIDEO	Pause: Student Practice

3. Divide: $6715 \div 197$

Identify the divisor and the dividend, and write the problem in long division form. Then perform long division to calculate the quotient.

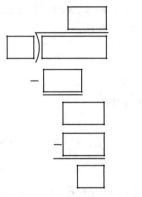

4. Divide: $5701 \div 146$

Helpful Hint: Pause the video and write the helpful hint in your own words.

For each Active Video Lesson, when the pencil icon appears, pause and work the problem. Then press play to check your work.

Active Video Lesson 1	Active Video Lesson 2
5. Divide 5523 by 46.	**6.** Divide 7490 by 214.

14
Copyright © 2017 Pearson Education, Inc.

Guided Learning Video Worksheet: Exponents

Text: *Basic College Mathematics*
Student Learning Objective 1.6.1: Evaluate expressions with whole-number exponents.

Follow along with *Guided Learning Video 1.6.1, Exponents.*

Understanding the Big Picture: As you listen to the first part of the video, when the pencil icon appears, pause and fill in the blanks, choosing from the words listed.

superscript • exponent • shorthand • base

1. An exponent is a(n) _____ number for expressing multiplication of the same number.

2. The _____ is the number being multiplied.

3. The _____ names the number of times the base is multiplied.

4. An exponent is sometimes called a(n) _____.

Follow along with the two Guided Learning Video examples, and fill in the blanks on the left as you learn with the instructor. When the pencil icon appears, pause and try the Student Practice on your own.

Guided Learning Video	**Pause: Student Practice**
1. Find the value of the expression: "Four cubed" Identify the base and the exponent. Then, rewrite the expression and find its value. The exponent is ☐ . The base is ☐ .	2. Find the value of the expression: "Eight cubed"

Guided Learning Video	Pause: Student Practice
3. Write the product in exponent form: "five times five times five times five" Identify the base. To determine the exponent, identify the number of times the base is repeated. Then, write the exponent form. The base is ☐. The exponent is ☐ The exponent form is $\square^{\square}$	**4.** Write the product in exponent form: "six times six times six times six"

Helpful Hint: Pause the video and write the helpful hint in your own words.

For each Active Video Lesson, when the pencil icon appears, pause and work the problem. Then press play to check your work.

Active Video Lesson 1	Active Video Lesson 2
5. Divide 5523 by 46.	**6.** Divide 7490 by 214.

Name: _____   Date: _____

Instructor: _____   Section: _____

Guided Learning Video Worksheet: Order of Operations

Text: *Basic College Mathematics*

Student Learning Objective 1.6.2: Perform several arithmetic operations in the proper order.

Follow along with *Guided Learning Video 1.6.2, Order of Operations*.

Understanding the Big Picture: As you listen to the first part of the video, when the pencil icon appears, pause and fill in the blanks, choosing from the words listed.

dividing • subtracting • priorities • adding • exponents

1. The order of operations is a list of _____ for computing with numbers.

2. The second priority in the order of operations is _____.

3. Multiplying must be completed before _____ and _____.

Follow along with the two Guided Learning Video examples, and fill in the blanks on the left as you learn with the instructor. When the pencil icon appears, pause and try the Student Practice on your own.

Guided Learning Video	Pause: Student Practice
1. Evaluate. $8^2 + 5 \times (7-4)$	**2.** Evaluate. $2^3 + 15 \div (11-6)$

Use the order of operations to evaluate the expression.

$8^2 + 5 \times (7-4)$

= [   ]

= [   ]

= [   ]

= [   ]

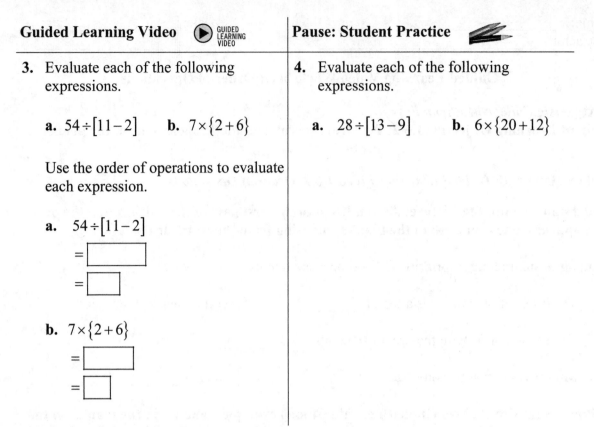
3. Evaluate each of the following expressions.

    a. $54 \div [11-2]$ **b.** $7 \times \{2+6\}$

Use the order of operations to evaluate each expression.

    a. $54 \div [11-2]$

         $= \boxed{}$

         $= \boxed{}$

    b. $7 \times \{2+6\}$

         $= \boxed{}$

         $= \boxed{}$

4. Evaluate each of the following expressions.

    a. $28 \div [13-9]$ **b.** $6 \times \{20+12\}$

Helpful Hint: Pause the video and write the helpful hint in your own words.

For each Active Video Lesson, when the pencil icon appears, pause and work the problem. Then press play to check your work.

Active Video Lesson 1	**Active Video Lesson 2**
5. Evaluate. $(17+7) \div 6 \times 2 + 7 \times 3 - 4$	6. Evaluate. $5 + 6^2 - 2 \times (9-6)^2$

Guided Learning Video Worksheet: Rounding

Text: *Basic College Mathematics*
Student Learning Objective 1.7.1: Round whole numbers.

Follow along with *Guided Learning Video 1.7.1, Rounding*.

Understanding the Big Picture: As you listen to the first part of the video, when the pencil icon appears, pause and fill in the blanks, choosing from the words listed.

place • number line • approximate

1. Large numbers are often rounded because _____ numbers are often "good enough" for certain uses.

2. To round a number, first determine the _____ to round to.

3. A(n) _____ can be used to determine which value a number is closest to.

Follow along with the two Guided Learning Video examples, and fill in the blanks on the left as you learn with the instructor. When the pencil icon appears, pause and try the Student Practice on your own.

Guided Learning Video 🅿 GUIDED LEARNING VIDEO	**Pause: Student Practice** 🖎
1. Round each of the following numbers.	**2.** Round each of the following numbers.
a. Round 368 to the nearest hundred.	**a.** Round 582 to the nearest hundred.
Use a number line to round the number to the nearest hundred.	Use a number line to round the number to the nearest hundred.
300 350 400	
b. Round 129 to the nearest hundred.	**b.** Round 447 to the nearest hundred.
Use a number line to round the number to the nearest hundred.	Use a number line to round the number to the nearest hundred.
100 150 200	

3. Round 25,763 to the nearest hundred.	**4.** Round 67,851 to the nearest hundred.
Use rounding rules to round the number to the nearest hundred.	Use rounding rules to round the number to the nearest hundred.
25,763	67,851
[ ]	[ ]

Helpful Hint: Pause the video and write the helpful hint in your own words.

For each Active Video Lesson, when the pencil icon appears, pause and work the problem. Then press play to check your work.

Active Video Lesson 1	**Active Video Lesson 2**
5. Round 63,286 to the nearest ten.	**6.** Round 627,381 to the nearest thousand.

Guided Learning Video Worksheet: Estimating

Text: *Basic College Mathematics*
Student Learning Objective 1.7.2: Estimate the answer to a problem involving whole numbers.

Follow along with *Guided Learning Video 1.7.2, Estimating.*

Understanding the Big Picture: As you listen to the first part of the video, when the pencil icon appears, pause and fill in the blanks, choosing from the words listed.

rounded • estimate • approximation

1. In mathematics, a(n) _____ is used to determine the approximate value of a calculation.

2. The symbol ≈ is a(n) _____ symbol.

3. Estimating involves calculations with _____ numbers.

Follow along with the two Guided Learning Video examples, and fill in the blanks on the left as you learn with the instructor. When the pencil icon appears, pause and try the Student Practice on your own.

Guided Learning Video (▶) GUIDED LEARNING VIDEO	**Pause: Student Practice** ✏️
1. Estimate the total attendance.	**2.** Estimate the total attendance.
765 men, 837 women, 1153 children	493 men, 524 women, 1923 children
Round each number so it contains one non-zero digit, and add to estimate the total attendance. Compare the estimate with the actual total.	

$765 \approx$ ☐ men
$837 \approx$ ☐ women
$1153 \approx$ ☐ children
Estimated Total:
☐ + ☐ + ☐ ≈ ☐
Actual Total:
☐ + ☐ + ☐ = ☐

☐ ≈ ☐

3. Estimate to see if the sum is correct.

$$35{,}286$$
$$41{,}620$$
$$23{,}589$$
$$+10{,}256$$
$$\overline{71{,}751}$$

Round each number to one non-zero digit, then add to estimate the total. Compare the estimate with the actual total.

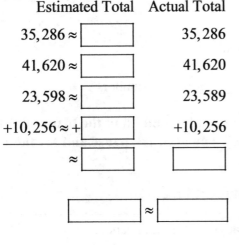

	Estimated Total	Actual Total
35,286 ≈	☐	35,286
41,620 ≈	☐	41,620
23,598 ≈	☐	23,589
+10,256 ≈ +	☐	+10,256
≈	☐	☐
☐	≈	☐

4. Estimate to see if the sum is correct.

$$67{,}392$$
$$23{,}405$$
$$46{,}841$$
$$+12{,}250$$
$$\overline{103{,}888}$$

Helpful Hint: Pause the video and write the helpful hint in your own words.

For each Active Video Lesson, when the pencil icon appears, pause and work the problem. Then press play to check your work.

Active Video Lesson 1 | **Active Video Lesson 2**

5. An amusement park is going to repave their parking lot. The area of the parking lot is 985 feet long by 312 feet wide. Estimate the amount of asphalt in square feet that is needed to do the job.

6. Painting the walls and ceilings of 29 work areas in a large office building required 493 gallons of paint. Estimate the number of gallons required for each of the work areas.

Name: _____   Date: _____
Instructor: _____   Section: _____

Guided Learning Video Worksheet: Solving Problems Using a Mathematics Blueprint

Text: *Basic College Mathematics*
Student Learning Objective 1.8.2: Use the Mathematics Blueprint to solve problems involving more than one operation.

Follow along with *Guided Learning Video 1.8.2, Solving Problems Using a Mathematics Blueprint.*

Understanding the Big Picture: As you listen to the first part of the video, when the pencil icon appears, pause and fill in the blanks, choosing from the words listed.

calculate • plan • understand • check • blueprint • estimate • organize

1. To solve word problems with many facts and steps, it's helpful to _____ the information and _____ the process in a _____.

2. The first two steps of the problem-solving process are to _____ the problem and _____ the answer.

3. After solving a problem, it's important to _____ the answer.

Follow along with the two Guided Learning Video examples, and fill in the blanks on the left as you learn with the instructor. When the pencil icon appears, pause and try the Student Practice on your own.

Guided Learning Video ▶ GUIDED LEARNING VIDEO	**Pause: Student Practice**
1. Trang works 40 hours a week and earns $15 an hour as a payroll clerk. She is considering accepting a job offer to work as an assistant office manager earning a salary of $2400 per month. Which job pays more per year?	2. Bob is evaluating two job offers. The first requires a 40 hour work week and pays $22 per hour. The second offer pays $3900 per month. Which job pays more per year?

1. Organize the information and plan the problem solving process in a mathematics blueprint. Then, perform the calculations to solve the problem.

Current weekly pay: ☐ × ☐ = ☐

Current yearly pay: ☐ × ☐ = ☐

Offered yearly pay: ☐ × ☐ = ☐

☐ is greater than ☐

3. The Hernandez Insurance Agency restocked its office supplies recently. The agency bought 15 reams of paper at $3 each, 2 ink cartridges at $32 each, and 4 boxes of folders at $7 each. How much did the agency pay for the office supplies?

Organize the information and plan the problem solving process in a mathematics blueprint. Then, perform the calculations to solve the problem.

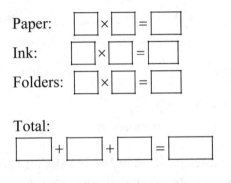

Paper: ☐ × ☐ = ☐

Ink: ☐ × ☐ = ☐

Folders: ☐ × ☐ = ☐

Total:
☐ + ☐ + ☐ = ☐

4. Clarissa's checking account has an initial balance of $1850. This month, she made 3 deposits of $140. She also wrote 2 checks for $74 and 3 checks for $45. What is her final balance?

Helpful Hint: Pause the video and write the helpful hint in your own words.

For each Active Video Lesson, when the pencil icon appears, pause and work the problem. Then press play to check your work.

Active Video Lesson 1 | **Active Video Lesson 2**

5. Juan has investments in the stock market. Last month, his stocks were worth a total of $2347. When he checked his investments this month, 2 stocks had increased in value by $146 and $135. Three stocks had decreased in value by $48, $86, and $93. What is the total value of his stocks this month?

6. Alec drives a taxi. He began his day with a full tank of gas and his odometer read 103,276. At the end of the day, the odometer read 103,591. Alec filled his tank with 12 gallons of gas at noon and filled it again at the end of the day with 9 gallons. How many miles per gallon did the taxi get that day?

Guided Learning Video Worksheet: Using Fractions to Represent Real-Life Situations

Text: *Basic College Mathematics*
Student Learning Objective 2.1.3: Use fractions to represent real-life situations.

Follow along with *Guided Learning Video 2.1.3, Using Fractions to Represent Real-Life Situations.*

Understanding the Big Picture: As you listen to the first part of the video, when the pencil icon appears, pause and fill in the blanks, choosing from the words listed.

numerator • denominator • fractions

1. Oftentimes, _____ are used to describe real-life situations.

2. The _____ specifies the number of equal parts that make up the whole.

3. The _____ specifies the number of parts being talked about.

Follow along with the two Guided Learning Video examples, and fill in the blanks on the left as you learn with the instructor. When the pencil icon appears, pause and try the Student Practice on your own.

Guided Learning Video ⏵ GUIDED LEARNING VIDEO	**Pause: Student Practice**
1. Suppose that there are 11 men and 9 women in your math class. Write a fraction that describes what part of the class consists of women Identify the total number of women and the total number students in the class to describe what portion of the class is women. Number of women: ☐ Total number of students: ☐ + ☐ = ☐ Fraction of the class that is women: ☐/☐	2. A dresser drawer contains 6 pairs of brown socks, 7 pairs of black socks, and 4 pairs of blue socks. What fractional part of the pairs of socks are black pairs?

3. Martha drives 328 miles out of which 217 are highway miles. Tim drives 519 miles out of which 471 are highway miles. Write a fraction that describes for both people together the number of highway miles driven compared with the total number of miles driven.

   Identify the total number of highway miles driven by Martha and Tim. Compare this sum to the total number of miles driven.

   Total highway miles:

   ☐ + ☐ = ☐

   Total miles:

   ☐ + ☐ = ☐

   Fraction that is highway miles:

   ☐ / ☐

4. Jean bakes a batch of 60 cookies of which 24 are chocolate chip. Jocelyn bakes 75 cookies of which 17 are chocolate chip. Write a fraction describing the part of the total number of cookies baked that are chocolate chip.

Helpful Hint: Pause the video and write the helpful hint in your own words.

For each Active Video Lesson, when the pencil icon appears, pause and work the problem. Then press play to check your work.

Active Video Lesson 1 | **Active Video Lesson 2**

5. An inspector found that 1 out of 7 belts were defective. She also found out that 2 out of 9 shirts were defective. Write a fraction that describes what part of all the objects examined was defective.

6. At a dealership, there are 12 blue cars, 17 white cars, 5 black cars and 9 red cars for sale. What fractional part of the cars for sale are red?

Name: _____   Date: _____
Instructor: _____   Section: _____

Guided Learning Video Worksheet: Prime Factoring

Text: *Basic College Mathematics*
Student Learning Objective 2.2.1: Write a number as a product of prime factors.

Follow along with *Guided Learning Video 2.2.1, Prime Factoring.*

Understanding the Big Picture: As you listen to the first part of the video, when the pencil icon appears, pause and fill in the blanks, choosing from the words listed.

terms • factors • factoring • prime

1. Breaking down an amount into a product of smaller numbers is called _____.

2. A factor is _____ if it is divisible only by one and itself.

3. Numbers that are _____ are multiplied together.

Follow along with the two Guided Learning Video examples, and fill in the blanks on the left as you learn with the instructor. When the pencil icon appears, pause and try the Student Practice on your own.

Guided Learning Video	Pause: Student Practice
1. Express the number 42 as the product of prime factors. Use a division ladder to factor the number 42. $\square$ $\square\overline{)\square}$ $\square\overline{)\square}$ $\square = \square \times \square \times \square$	2. Using a division ladder, express the number 90 as the product of prime factors.

3. Factor the number 60 into the product of prime factors.

   Use a factor tree to factor the number.

   60

   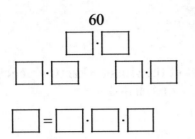

4. Using a factor tree, factor the number 84 into the product of prime factors.

Helpful Hint: Pause the video and write the helpful hint in your own words.

For each Active Video Lesson, when the pencil icon appears, pause and work the problem. Then press play to check your work.

Active Video Lesson 1   **Active Video Lesson 2**

5. Express 56 as the product of prime factors.

6. Express the number 210 as the product of prime factors.

Guided Learning Video Worksheet: Reducing Fractions to Lowest Terms

Text: *Basic College Mathematics*
Student Learning Objective 2.2.2: Reduce a fraction to lowest terms.

Follow along with *Guided Learning Video 2.2.2, Reducing Fractions to Lowest Terms.*

Understanding the Big Picture: As you listen to the first part of the video, when the pencil icon appears, pause and fill in the blanks, choosing from the words listed.

factor • divide • equivalent • reduced

1. Fractions are said to be _____ when the values of the fractions are the same.

2. A fraction is _____ to lowest terms when its numerator and denominator have no common _____ other than one.

3. To reduce a fraction, _____ out any common factor in the numerator and denominator.

Follow along with the two Guided Learning Video examples, and fill in the blanks on the left as you learn with the instructor. When the pencil icon appears, pause and try the Student Practice on your own.

Guided Learning Video	**Pause: Student Practice**
1. a. Reduce the fraction $\frac{6}{14}$. **b.** Reduce the fraction $\frac{32}{40}$.	**2. a.** Reduce the fraction $\frac{12}{20}$. **b.** Reduce the fraction $\frac{30}{54}$.

Find the largest number that is a common factor of both the numerator and denominator. Then, divide the numerator and denominator by that number.

a. $\dfrac{6 \div \square}{14 \div \square} = \dfrac{\square}{\square}$

b. $\dfrac{32 \div \square}{40 \div \square} = \dfrac{\square}{\square}$

3. Simplify $\dfrac{30}{75}$ using the method of prime factors.

    Rewrite the numerator and denominator as the product of prime factors. Then, cancel out common prime factors.

    $$\frac{30}{75} = \frac{\boxed{} \times \boxed{} \times \boxed{}}{\boxed{} \times \boxed{} \times \boxed{}} = \frac{\boxed{}}{\boxed{}}$$

4. Simplify $\dfrac{42}{60}$ using the method of prime factors.

Helpful Hint: Pause the video and write the helpful hint in your own words.

For each Active Video Lesson, when the pencil icon appears, pause and work the problem. Then press play to check your work.

Active Video Lesson 1 ✏ | **Active Video Lesson 2** ✏

5. Reduce $\dfrac{28}{70}$ to lowest terms.

6. Reduce $\dfrac{24}{39}$ to lowest terms.

Name: _____ Date: _____

Instructor: _____ Section: _____

Guided Learning Video Worksheet: Equivalent Fractions

Text: *Basic College Mathematics*

Student Learning Objective 2.2.3: Determine whether two fractions are equal.

Follow along with *Guided Learning Video 2.2.3, Equivalent Fractions*.

Understanding the Big Picture: As you listen to the first part of the video, when the pencil icon appears, pause and fill in the blanks, choosing from the words listed.

identity • equivalent • one

1. Fractions that name the same quantity are called _____ fractions.

2. Multiplying the numerator and denominator of a fraction by the same number is the same as multiplying by _____.

3. The _____ property of one states that if a number is multiplied by one, the value of the number doesn't change.

Follow along with the two Guided Learning Video examples, and fill in the blanks on the left as you learn with the instructor. When the pencil icon appears, pause and try the Student Practice on your own.

Guided Learning Video	Pause: Student Practice
1. Write $\frac{5}{7}$ as an equivalent fraction with a denominator of 21. Determine what number is multiplied times 7 to equal 21. Then multiply both the numerator and the denominator by the same number to calculate the equivalent fraction. $\frac{5}{7} \times \dfrac{\square}{\square} = \dfrac{\square}{21}$	2. Write $\frac{3}{8}$ as an equivalent fraction with a denominator of 32.

Guided Learning Video 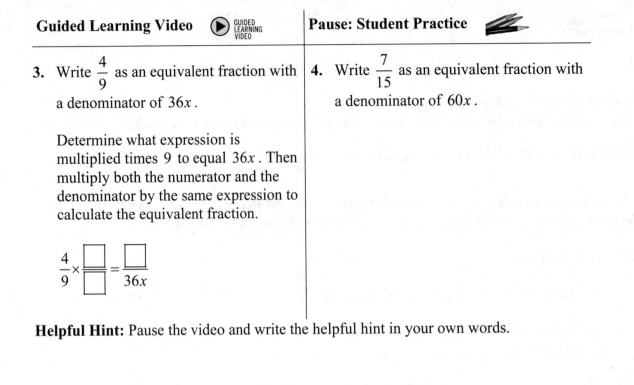	Pause: Student Practice
3. Write $\dfrac{4}{9}$ as an equivalent fraction with a denominator of $36x$. Determine what expression is multiplied times 9 to equal $36x$. Then multiply both the numerator and the denominator by the same expression to calculate the equivalent fraction. $$\frac{4}{9} \times \frac{\square}{\square} = \frac{\square}{36x}$$	**4.** Write $\dfrac{7}{15}$ as an equivalent fraction with a denominator of $60x$.

Helpful Hint: Pause the video and write the helpful hint in your own words.

For each Active Video Lesson, when the pencil icon appears, pause and work the problem. Then press play to check your work.

Active Video Lesson 1	Active Video Lesson 2
5. Write $\dfrac{3}{8}$ as an equivalent fraction with a denominator of $48x$.	**6.** Write $\dfrac{9}{11}$ as an equivalent fraction with a denominator of 77.

Name: _____   Date: _____
Instructor: _____   Section: _____

Guided Learning Video Worksheet: Changing Mixed Numbers to Improper Fractions

Text: *Basic College Mathematics*
Student Learning Objective 2.3.1: Change a mixed number to an improper fraction.

Follow along with *Guided Learning Video 2.3.1, Changing Mixed Numbers to Improper Fractions.*

Understanding the Big Picture: As you listen to the first part of the video, when the pencil icon appears, pause and fill in the blanks, choosing from the words listed.

improper • mixed • proper • one • zero

1. The value of a(n) _____ fraction is less than one.

2. When the value of a fraction's numerator is greater than or equal to its denominator, the fraction is called _____.

3. The sum of a whole number greater than _____ and a proper fraction is a _____ number.

Follow along with the two Guided Learning Video examples, and fill in the blanks on the left as you learn with the instructor. When the pencil icon appears, pause and try the Student Practice on your own.

Guided Learning Video ▶ GUIDED LEARNING VIDEO	Pause: Student Practice
1. Change $9\frac{5}{6}$ to an improper fraction.	2. Change $4\frac{3}{5}$ to an improper fraction.

Use the procedure for changing a mixed number to an improper fraction to write the improper fraction.

$$9\frac{5}{6} = \frac{\boxed{} \times \boxed{} + \boxed{}}{\boxed{}} = \frac{\boxed{}}{\boxed{}}$$

Guided Learning Video	Pause: Student Practice

3. Change $14\frac{7}{8}$ to an improper fraction.

Use the procedure for changing a mixed number to an improper fraction to write the improper fraction.

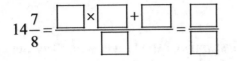

4. Change $17\frac{2}{9}$ to an improper fraction.

Helpful Hint: Pause the video and write the helpful hint in your own words.

For each **Active Video Lesson**, when the pencil icon appears, pause and work the problem. Then press play to check your work.

Active Video Lesson 1	Active Video Lesson 2

5. Change $19\frac{4}{7}$ to an improper fraction.

6. Change $8\frac{2}{7}$ to an improper fraction.

Guided Learning Video Worksheet: Change Improper Fractions to Mixed Numbers

Text: *Basic College Mathematics*
Student Learning Objective 2.3.2: Change an improper fraction to a mixed number.

Follow along with *Guided Learning Video 2.3.2, Change Improper Fractions to Mixed Numbers.*

Understanding the Big Picture: As you listen to the first part of the video, when the pencil icon appears, pause and fill in the blanks, choosing from the words listed.

quotient • denominator • divide • remainder • numerator

1. When changing an improper fraction to a mixed number, _____ the _____ by the _____.

2. The _____ is the whole number part of the mixed number.

3. The numerator of the fraction part of the mixed number is a _____.

Follow along with the two Guided Learning Video examples, and fill in the blanks on the left as you learn with the instructor. When the pencil icon appears, pause and try the Student Practice on your own.

Guided Learning Video ▶ GUIDED LEARNING VIDEO	**Pause: Student Practice**
1. Write $\dfrac{15}{4}$ as a mixed number.	2. Write $\dfrac{23}{6}$ as a mixed number.

Divide the numerator by the denominator and identify the remainder to write the mixed number.

$$\dfrac{15}{4} = \boxed{}\dfrac{\boxed{}}{\boxed{}}$$

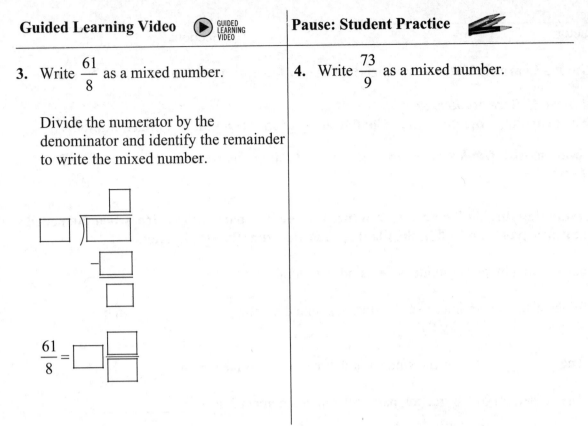
3. Write $\dfrac{61}{8}$ as a mixed number.

Divide the numerator by the denominator and identify the remainder to write the mixed number.

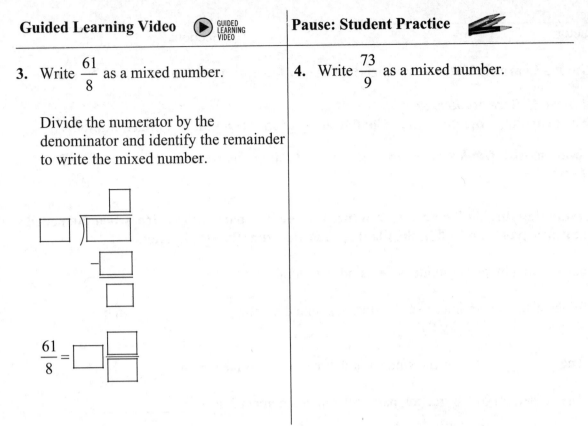

$$\dfrac{61}{8} = \boxed{}\ \dfrac{\boxed{}}{\boxed{}}$$

4. Write $\dfrac{73}{9}$ as a mixed number.

Helpful Hint: Pause the video and write the helpful hint in your own words.

For each Active Video Lesson, when the pencil icon appears, pause and work the problem. Then press play to check your work.

Active Video Lesson 1 | **Active Video Lesson 2**

5. Write $\dfrac{58}{9}$ as a mixed number.

6. Write $\dfrac{26}{3}$ as a mixed number.

Guided Learning Video Worksheet: Multiply Fractions

Text: *Basic College Mathematics*
Student Learning Objective 2.4.1: Multiply two fractions that are proper or improper.

Follow along with *Guided Learning Video 2.4.1, Multiply Fractions.*

Understanding the Big Picture: As you listen to the first part of the video, when the pencil icon appears, pause and fill in the blanks, choosing from the words listed.

division • denominators • numerators • multiplication • terms

1. The word "of" indicates the operation of _____.

2. To multiply fractions, multiply numerators times _____.

3. Multiplication of fractions also requires _____ to be multiplied.

Follow along with the two Guided Learning Video examples, and fill in the blanks on the left as you learn with the instructor. When the pencil icon appears, pause and try the Student Practice on your own.

Guided Learning Video	Pause: Student Practice
1. Multiply $\dfrac{4}{7} \times \dfrac{2}{9}$.	2. Multiply $\dfrac{5}{9} \times \dfrac{2}{11}$.

Multiply the numerators, and then multiply the denominators.

$$\frac{4}{7} \times \frac{2}{9} = \frac{\boxed{} \times \boxed{}}{\boxed{} \times \boxed{}} = \frac{\boxed{}}{\boxed{}}$$

3. Simplify first, then multiply $\dfrac{7}{15} \times \dfrac{9}{14}$.

4. Simplify first, then multiply $\dfrac{5}{28} \times \dfrac{21}{25}$.

Write the product as a single fraction. Find the prime factors of the values in the numerator and denominator, and factor out the common factors. Then, multiply the remaining factors.

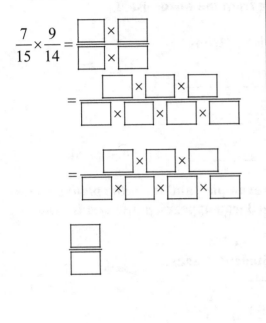

Helpful Hint: Pause the video and write the helpful hint in your own words.

For each Active Video Lesson, when the pencil icon appears, pause and work the problem. Then press play to check your work.

Active Video Lesson 1 | **Active Video Lesson 2**

5. Simplify first, and then multiply $\dfrac{15}{33} \times \dfrac{22}{35}$.

6. Simplify first, and then multiply $\dfrac{9}{14} \times \dfrac{7}{12}$.

Name: _____   Date: _____

Instructor: _____   Section: _____

Guided Learning Video Worksheet: Multiplying Mixed Numbers

Text: *Basic College Mathematics*
Student Learning Objective 2.4.3: Multiply mixed numbers.

Follow along with *Guided Learning Video 2.4.3, Multiplying Mixed Numbers.*

Understanding the Big Picture: As you listen to the first part of the video, when the pencil icon appears, pause and fill in the blanks, choosing from the words listed.

mixed number • improper • divide

1. To multiply two mixed numbers, first transform the mixed numbers into _____ fractions.

2. Whenever it's possible, _____ out common factors before multiplying.

3. When multiplying mixed numbers, a final step is to change the answer to a _____ if possible.

Follow along with the two Guided Learning Video examples, and fill in the blanks on the left as you learn with the instructor. When the pencil icon appears, pause and try the Student Practice on your own.

Guided Learning Video	Pause: Student Practice
1. Multiply $2\frac{3}{4} \times 5\frac{1}{3}$.	2. Multiply $4\frac{2}{3} \times 3\frac{3}{5}$.

Change the mixed numbers to improper fractions and multiply. Be sure to divide out common factors before multiplying.

$$2\frac{3}{4} \times 5\frac{1}{3} = \frac{\square}{\square} \times \frac{\square}{\square}$$

$$= \frac{\square}{\square} \times \frac{\square}{\square}$$

$$= \frac{\square}{\square} = \square \frac{\square}{\square}$$

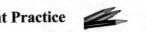
3. Multiply $3\frac{5}{12} \times 8$.

Change the mixed number to an improper fraction and the whole number to a fraction with a denominator of one. Then, multiply. Be sure to divide out common factors before multiplying.

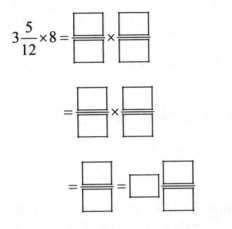

4. Multiply $5\frac{2}{9} \times 6$.

Helpful Hint: Pause the video and write the helpful hint in your own words.

For each Active Video Lesson, when the pencil icon appears, pause and work the problem. Then press play to check your work.

Active Video Lesson 1	Active Video Lesson 2
5. Find the area, in square inches, of the rectangle with a length of $4\frac{2}{3}$ inches and a width of $2\frac{1}{2}$ inches. Remember that Area = Length × Width.	6. Multiply $2\frac{5}{6} \times 3$.

Guided Learning Video Worksheet: Dividing Fractions

Text: *Basic College Mathematics*
Student Learning Objective 2.5.1: Divide two proper or improper fractions.

Follow along with *Guided Learning Video 2.5.1, Dividing Fractions.*

Understanding the Big Picture: As you listen to the first part of the video, when the pencil icon appears, pause and fill in the blanks, choosing from the words listed.

second • inverted • reciprocal • first

1. When a fraction is multiplied times its _____, the result is always one.

2. When a fraction is _____, its numerator and denominator are interchanged.

3. To divide two fractions, invert the _____ fraction and multiply.

Follow along with the two Guided Learning Video examples, and fill in the blanks on the left as you learn with the instructor. When the pencil icon appears, pause and try the Student Practice on your own.

Guided Learning Video ⏵ GUIDED LEARNING VIDEO	Pause: Student Practice
1. Divide the following. $\dfrac{7}{10} \div \dfrac{8}{9}$	**2.** Divide the following. $\dfrac{4}{11} \div \dfrac{5}{7}$
Rewrite the expression as multiplication by the reciprocal. Then, calculate the answer. $$\frac{7}{10} \div \frac{8}{9} = \frac{\Box}{\Box} \cdot \frac{\Box}{\Box}$$ $$= \frac{\Box}{\Box}$$	

3. Evaluate the following. $\dfrac{5}{9} \div \dfrac{2}{3}$

Rewrite the expression as multiplication by the reciprocal. Then, calculate the answer. Be sure to divide out common factors before multiplying.

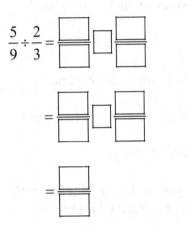

4. Evaluate the following. $\dfrac{8}{13} \div \dfrac{4}{5}$

Helpful Hint: Pause the video and write the helpful hint in your own words.

For each Active Video Lesson, when the pencil icon appears, pause and work the problem. Then press play to check your work.

Active Video Lesson 1

5. Simplify. $\dfrac{3}{7} \div \dfrac{12}{21}$

Active Video Lesson 2

6. Divide the following fractions. $\dfrac{5}{8} \div \dfrac{25}{36}$

Name: _____   Date: _____

Instructor: _____   Section: _____

Guided Learning Video Worksheet: Divide Mixed Numbers

Text: *Basic College Mathematics*
Student Learning Objective 2.5.3: Divide mixed numbers.

Follow along with *Guided Learning Video 2.5.3, Divide Mixed Numbers.*

Understanding the Big Picture: As you listen to the first part of the video, when the pencil icon appears, pause and fill in the blanks, choosing from the words listed.

addition • multiplication • improper

1. If mixed numbers are being divided, they should be converted to _____ fractions before dividing.

2. After inverting the second fraction, change the division symbol to a _____ symbol.

Follow along with the two Guided Learning Video examples, and fill in the blanks on the left as you learn with the instructor. When the pencil icon appears, pause and try the Student Practice on your own.

Guided Learning Video	Pause: Student Practice
1. Divide: $4\frac{1}{2} \div 3\frac{3}{8}$	2. Divide: $3\frac{3}{4} \div 5\frac{1}{2}$

Convert the mixed numbers to improper fractions. Then follow the rules for dividing fractions to calculate the answer. If necessary write the final answer as a mixed number.

$$4\frac{1}{20} \div 3\frac{3}{8} = \frac{\square}{\square} \div \frac{\square}{\square}$$

$$= \frac{\square}{\square} \cdot \frac{\square}{\square}$$

$$= \frac{\square}{\square} \cdot \frac{\square}{\square} = \frac{\square}{\square}$$

$$= \square\frac{\square}{\square}$$

3. Divide: $5\dfrac{1}{3} \div 8$

    Convert the mixed number to an improper fraction and the whole number to a fraction. Then follow the rules for dividing fractions to calculate the answer. Be sure to divide out common factors before multiplying.

    $$5\dfrac{1}{3} \div 8 = \dfrac{\boxed{}}{\boxed{}} \div \dfrac{\boxed{}}{\boxed{}}$$

    $$= \dfrac{\boxed{}}{\boxed{}} \cdot \boxed{} \cdot \dfrac{\boxed{}}{\boxed{}}$$

    $$= \dfrac{\boxed{}}{\boxed{}} \cdot \boxed{} \cdot \dfrac{\boxed{}}{\boxed{}} = \dfrac{\boxed{}}{\boxed{}}$$

4. Divide: $7\dfrac{3}{5} \div 19$

Helpful Hint: Pause the video and write the helpful hint in your own words.

For each Active Video Lesson, when the pencil icon appears, pause and work the problem. Then press play to check your work.

Active Video Lesson 1	**Active Video Lesson 2**
5. Divide: $7\dfrac{1}{2} \div 5$	6. Divide: $5\dfrac{1}{2} \div 6\dfrac{1}{4}$

Name: _____ Date: _____
Instructor: _____ Section: _____

Guided Learning Video Worksheet: Least Common Denominator

Text: *Basic College Mathematics*
Student Learning Objective 2.6.2: Find the least common denominator (LCD) given two or three fractions.

Follow along with *Guided Learning Video 2.6.2, Least Common Denominator*.

Understanding the Big Picture: As you listen to the first part of the video, when the pencil icon appears, pause and fill in the blanks, choosing from the words listed.

one • prime • compare • greatest

1. A least common denominator is the smallest denominator that allows us to _____ fractions directly.

2. To form an LCD, write each denominator as the product of _____ factors.

3. The LCD is composed of each factor the _____ number of times it appears in any _____ denominator.

Follow along with the two Guided Learning Video examples, and fill in the blanks on the left as you learn with the instructor. When the pencil icon appears, pause and try the Student Practice on your own.

Guided Learning Video ▶ GUIDED LEARNING VIDEO	**Pause: Student Practice**
1. Find the LCD of $\dfrac{1}{12}$ and $\dfrac{5}{9}$.	**2.** Find the LCD of $\dfrac{12}{25}$ and $\dfrac{7}{15}$.
Write each denominator as the product of primes, then form the LCD.	
$9 = \boxed{} \times \boxed{}$	
$12 = \boxed{} \times \boxed{} \times \boxed{}$	
$\text{LCD} = \boxed{} \times \boxed{} \times \boxed{} \times \boxed{} = \boxed{}$	

Guided Learning Video 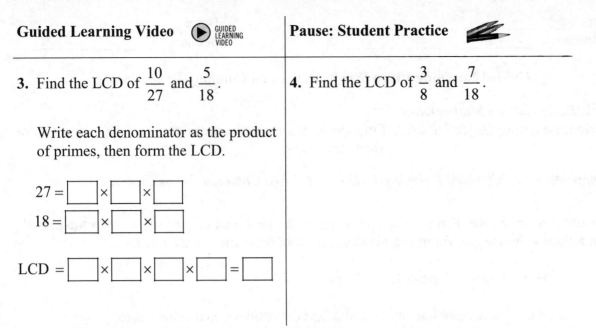 GUIDED LEARNING VIDEO	**Pause: Student Practice**
3. Find the LCD of $\frac{10}{27}$ and $\frac{5}{18}$.	**4.** Find the LCD of $\frac{3}{8}$ and $\frac{7}{18}$.

Write each denominator as the product of primes, then form the LCD.

$27 = \boxed{} \times \boxed{} \times \boxed{}$

$18 = \boxed{} \times \boxed{} \times \boxed{}$

$LCD = \boxed{} \times \boxed{} \times \boxed{} \times \boxed{} = \boxed{}$

Helpful Hint: Pause the video and write the helpful hint in your own words.

For each Active Video Lesson, when the pencil icon appears, pause and work the problem. Then press play to check your work.

Active Video Lesson 1	**Active Video Lesson 2**
5. Find the LCD of $\frac{9}{10}$ and $\frac{7}{15}$.	**6.** Find the LCD of $\frac{11}{24}$ and $\frac{7}{9}$.

Name: _____   Date: _____

Instructor: _____   Section: _____

Guided Learning Video Worksheet: Equivalent Fractions

Text: *Basic College Mathematics*
Student Learning Objective 2.6.3: Create equivalent fractions with a least common denominator.

Follow along with *Guided Learning Video 2.6.3, Equivalent Fractions*.

Understanding the Big Picture: As you listen to the first part of the video, when the pencil icon appears, pause and fill in the blanks, choosing from the words listed.

value • building • comparing • equivalent

1. Fractions that look different but have the same value are _____ fractions.

2. Whenever a fraction is multiplied by a fraction that is a form of one, the _____ of the original fraction does not change.

3. The process of multiplying a fraction by a form of one to create a new, equivalent fraction is called _____ a fraction.

Follow along with the two Guided Learning Video examples, and fill in the blanks on the left as you learn with the instructor. When the pencil icon appears, pause and try the Student Practice on your own.

Guided Learning Video ▶ GUIDED LEARNING VIDEO	Pause: Student Practice
1. Create a fraction equivalent to $\frac{5}{9}$ with a denominator of 36.	2. Create a fraction equivalent to $\frac{3}{5}$ with a denominator of 40.
Determine the number that is multiplied times the old denominator to give the new denominator. Then, build the equivalent fraction.	
$\frac{5}{9} \times \dfrac{\square}{\square} = \dfrac{\square}{36}$	

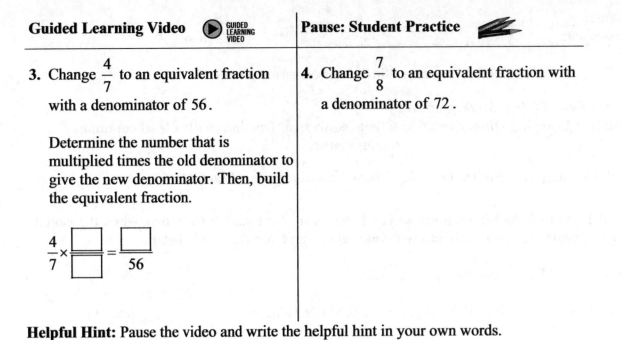
Pause: Student Practice

3. Change $\dfrac{4}{7}$ to an equivalent fraction with a denominator of 56.

Determine the number that is multiplied times the old denominator to give the new denominator. Then, build the equivalent fraction.

$$\dfrac{4}{7} \times \dfrac{\boxed{}}{\boxed{}} = \dfrac{\boxed{}}{56}$$

4. Change $\dfrac{7}{8}$ to an equivalent fraction with a denominator of 72.

Helpful Hint: Pause the video and write the helpful hint in your own words.

For each Active Video Lesson, when the pencil icon appears, pause and work the problem. Then press play to check your work.

Active Video Lesson 1

Active Video Lesson 2

5. Build an equivalent fraction with a denominator of 45 for the fraction $\dfrac{7}{15}$.

6. Build an equivalent fraction with a denominator of 60 for the fraction $\dfrac{11}{12}$.

Guided Learning Video Worksheet: Adding and Subtracting Fractions

Text: *Basic College Mathematics*
Student Learning Objective 2.7.2: Add and subtract fractions with different denominators.

Follow along with *Guided Learning Video 2.7.2, Adding and Subtracting Fractions.*

Understanding the Big Picture: As you listen to the first part of the video, when the pencil icon appears, pause and fill in the blanks, choosing from the words listed.

denominators • simplify • numerators • common

1. In order to add or subtract fractions, _____ denominators are needed.

2. To add or subtract fractions, add or subtract only the _____.

3. The last step in the process for adding or subtracting fractions is to _____ the answer if possible.

Follow along with the two Guided Learning Video examples, and fill in the blanks on the left as you learn with the instructor. When the pencil icon appears, pause and try the Student Practice on your own.

Guided Learning Video ▶ GUIDED LEARNING VIDEO	**Pause: Student Practice**
1. Subtract. $\dfrac{8}{14} - \dfrac{2}{21}$	2. Subtract. $\dfrac{15}{25} - \dfrac{3}{10}$

Use the process for adding/subtracting fractions with different denominators to calculate the answer.

LCD = ☐ × ☐ × ☐ = ☐

14 = ☐ × ☐

21 = ☐ × ☐

$\dfrac{☐}{☐} \times \dfrac{☐}{☐} = \dfrac{☐}{☐}$

$\dfrac{☐}{☐} \times \dfrac{☐}{☐} = \dfrac{☐}{☐}$

$\dfrac{☐}{☐} - \dfrac{☐}{☐} = \dfrac{☐}{☐} = \dfrac{☐}{☐}$

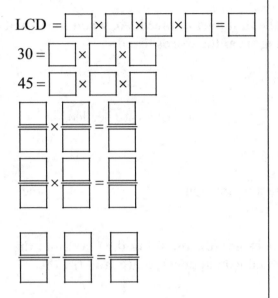
3. Add. $\dfrac{11}{30} + \dfrac{7}{45}$

Use the process for adding/subtracting fractions with different denominators to calculate the answer.

LCD = □ × □ × □ × □ = □

30 = □ × □ × □

45 = □ × □ × □

$\dfrac{□}{□} \times \dfrac{□}{□} = \dfrac{□}{□}$

$\dfrac{□}{□} \times \dfrac{□}{□} = \dfrac{□}{□}$

$\dfrac{□}{□} - \dfrac{□}{□} = \dfrac{□}{□}$

4. Add. $\dfrac{7}{20} + \dfrac{9}{70}$

Helpful Hint: Pause the video and write the helpful hint in your own words.

For each Active Video Lesson, when the pencil icon appears, pause and work the problem. Then press play to check your work.

Active Video Lesson 1

5. Add. $\dfrac{7}{25} + \dfrac{12}{45}$

Active Video Lesson 2

6. Subtract. $\dfrac{5}{12} - \dfrac{15}{42}$

Guided Learning Video Worksheet: Subtract Mixed Numbers

Text: *Basic College Mathematics*
Student Learning Objective 2.8.2: Subtract mixed numbers.

Follow along with *Guided Learning Video 2.8.2, Subtract Mixed Numbers*.

Understanding the Big Picture: As you listen to the first part of the video, when the pencil icon appears, pause and fill in the blanks, choosing from the words listed.

whole numbers • fractions • least • equivalent

1. When adding or subtracting fractions that do not have a common denominator, using the _____ common denominator makes the work easier.

2. When adding or subtracting mixed numbers, add or subtract the _____ first, then the _____.

Follow along with the two Guided Learning Video examples, and fill in the blanks on the left as you learn with the instructor. When the pencil icon appears, pause and try the Student Practice on your own.

Guided Learning Video ▶ GUIDED LEARNING VIDEO	**Pause: Student Practice**
1. Subtract. $8\frac{3}{5} - 3\frac{2}{3}$	2. Subtract. $6\frac{1}{5} - 3\frac{3}{4}$

Find the least common denominator and convert each fraction to an equivalent fraction. Subtract, borrowing if necessary, and express the final answer as a mixed number.

$\text{LCD} = \boxed{}$

$8\dfrac{3\times\boxed{}}{5\times\boxed{}} = 8\dfrac{\boxed{}}{\boxed{}} = \boxed{}\dfrac{\boxed{}}{\boxed{}}$

$-3\dfrac{2\times\boxed{}}{3\times\boxed{}} = -3\dfrac{\boxed{}}{\boxed{}} = -3\dfrac{\boxed{}}{\boxed{}}$

$\boxed{}\dfrac{\boxed{}}{\boxed{}} - \boxed{}\dfrac{\boxed{}}{\boxed{}} = \boxed{}\dfrac{\boxed{}}{\boxed{}}$

3. Subtract. $4\dfrac{1}{2} - 2\dfrac{2}{3}$

4. Subtract. $7\dfrac{1}{3} - 1\dfrac{3}{5}$

Change the mixed numbers to improper fractions, and then find the least common denominator. Convert each fraction to an equivalent fraction and subtract. Express the final answer as a mixed number.

LCD = □

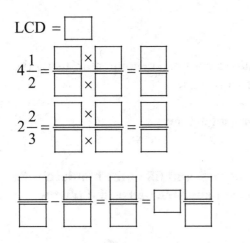

$4\dfrac{1}{2} = \dfrac{\square \times \square}{\square \times \square} = \dfrac{\square}{\square}$

$2\dfrac{2}{3} = \dfrac{\square \times \square}{\square \times \square} = \dfrac{\square}{\square}$

$\dfrac{\square}{\square} - \dfrac{\square}{\square} = \dfrac{\square}{\square} = \square\dfrac{\square}{\square}$

Helpful Hint: Pause the video and write the helpful hint in your own words.

For each Active Video Lesson, when the pencil icon appears, pause and work the problem. Then press play to check your work.

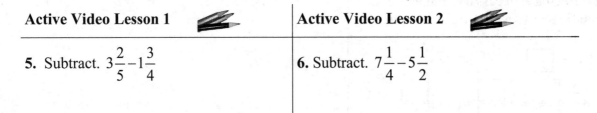

5. Subtract. $3\dfrac{2}{5} - 1\dfrac{3}{4}$

6. Subtract. $7\dfrac{1}{4} - 5\dfrac{1}{2}$

Name: _____   Date: _____

Instructor: _____   Section: _____

Guided Learning Video Worksheet: Order of Operations with Fractions

Text: *Basic College Mathematics*
Student Learning Objective 2.8.3: Evaluate fractional expressions using the order of operations.

Follow along with *Guided Learning Video 2.8.3, Order of Operations with Fractions*.

Understanding the Big Picture: As you listen to the first part of the video, when the pencil icon appears, pause and fill in the blanks, choosing from the words listed.

add • operations • denominator • subtract • dividing • multiplying • numerator

1. An order of _____ is needed in mathematics so the result of our calculations will be correct.

2. If a fraction has operations written in the _____ or in the _____, these operations must be done first.

3. When performing operations from left to right, be careful not to _____ or _____ before _____ or _____.

Follow along with the two Guided Learning Video examples, and fill in the blanks on the left as you learn with the instructor. When the pencil icon appears, pause and try the Student Practice on your own.

Guided Learning Video (▶) GUIDED LEARNING VIDEO	**Pause: Student Practice** ✐
1. Simplify the expression. $\dfrac{1}{3} + \dfrac{5}{12} \times \dfrac{4}{9}$	2. Simplify the expression. $\dfrac{1}{4} + \dfrac{3}{8} \times \dfrac{5}{6}$

Use the order of operations to calculate the answer.

$$\frac{1}{3} + \frac{5}{12} \times \frac{4}{9}$$

$$= \frac{\Box}{\Box} + \frac{\Box}{\Box} \times \frac{\Box}{\Box}$$

$$= \frac{\Box}{\Box} + \frac{\Box}{\Box}$$

$$= \frac{\Box}{\Box}$$

Guided Learning Video	**Pause: Student Practice**

3. Simplify the expression.

$$\left(\frac{1}{2}\right)^2 + \frac{1}{4} \div \frac{3}{4} - \left(\frac{1}{2} \times \frac{1}{3}\right)$$

Use the order of operations to calculate the answer.

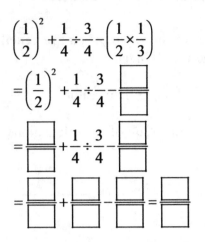

4. Simplify the expression.

$$\left(\frac{1}{3}\right)^2 + \frac{2}{5} \div \frac{4}{5} - \left(\frac{1}{8} \times \frac{2}{3}\right)$$

Helpful Hint: Pause the video and write the helpful hint in your own words.

For each Active Video Lesson, when the pencil icon appears, pause and work the problem. Then press play to check your work.

Active Video Lesson 1	**Active Video Lesson 2**

5. Evaluate the expression.

$$\frac{3}{8} \div \frac{1}{2} - \frac{2}{3} \times \frac{5}{6} + \left(\frac{2}{6}\right)^2$$

6. Evaluate the expression.

$$\left(\frac{2}{3}\right)^2 + \frac{5}{14} \times \frac{7}{15} - \frac{1}{9}$$

Name: _____ Date: _____

Instructor: _____ Section: _____

Guided Learning Video Worksheet: Solving Real-Life Problems Involving Fractions

Text: *Basic College Mathematics*
Student Learning Objective 2.9.1: Solve real-life problems with fractions.

Follow along with *Guided Learning Video 2.9.1, Solving Real-Life Problems Involving Fractions*.

Understanding the Big Picture: As you listen to the first part of the video, when the pencil icon appears, pause and fill in the blanks, choosing from the words listed.

facts • picture • reading • fractions

1. Often, the problems encountered in daily life involve _____.

2. Understanding a problem can be done by _____ carefully, drawing a _____, or using a mathematics blueprint.

3. The first step in completing a mathematics blueprint is to gather _____.

Follow along with the two Guided Learning Video examples, and fill in the blanks on the left as you learn with the instructor. When the pencil icon appears, pause and try the Student Practice on your own.

Guided Learning Video ▶ GUIDED LEARNING VIDEO	**Pause: Student Practice**
1. On Monday, the recycling center collected $5\frac{3}{10}$ tons of recyclable trash. On Tuesday, the center collected $4\frac{2}{5}$ tons of trash and on Wednesday, they collected $6\frac{1}{2}$ tons. What is the total weight of trash collected on those three days? After completing the mathematics blueprint, perform the necessary calculations to solve the problem. $\boxed{} + \boxed{} + \boxed{} + \dfrac{\boxed{}}{\boxed{}} + \dfrac{\boxed{}}{\boxed{}} + \dfrac{\boxed{}}{\boxed{}}$ $= \boxed{} + \dfrac{\boxed{}}{\boxed{}} + \dfrac{\boxed{}}{\boxed{}} + \dfrac{\boxed{}}{\boxed{}} = \boxed{}\dfrac{\boxed{}}{\boxed{}}$	2. On a 3-day trip, Nancy spends $7\frac{1}{3}$ hours, $6\frac{1}{2}$ hours, and $8\frac{1}{4}$ hours driving each day. What is the total number of hours she spent driving to her destination?

3. What is the inside diameter or "distance across" of a concrete storm drain pipe that has an outside diameter of $7\frac{3}{4}$ feet and is $\frac{5}{8}$ of a foot thick?

   After completing the mathematics blueprint, perform the necessary calculations to solve the problem.

   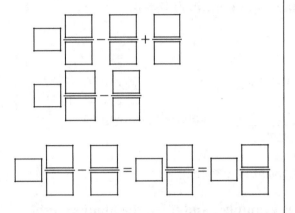

4. What is the inside diameter of a drain pipe that has an outside diameter of $3\frac{1}{2}$ feet and is $\frac{7}{8}$ of a foot thick?

Helpful Hint: Pause the video and write the helpful hint in your own words.

For each Active Video Lesson, when the pencil icon appears, pause and work the problem. Then press play to check your work.

Active Video Lesson 1

Active Video Lesson 2

5. Sheldon has a flag-making business. Large flags are $4\frac{1}{2}$ feet by $2\frac{4}{5}$ feet. Small flags are $2\frac{2}{5}$ feet by $1\frac{1}{2}$ feet. How much material does Sheldon need to fill an order of 2 large flags and 3 small flags?

6. A butcher has $18\frac{2}{3}$ pounds of steak that he wishes to place into packages that average $2\frac{1}{3}$ pounds each. How many packages can he make?

Guided Learning Video Worksheet: Change a Fraction with a Denominator That Is a Power of 10 to a Decimal

Text: *Basic College Mathematics*

Student Learning Objective 3.1.2: Change from fractional notation to decimal notation.

Follow along with *Guided Learning Video 3.1.2, Change a Fraction with a Denominator that is a Power of 10 to a Decimal.*

Understanding the Big Picture: As you listen to the first part of the video, when the pencil icon appears, pause and fill in the blanks, choosing from the words listed.

delete • equal • decimal • numerator • zeros

1. Fractions with denominators that are powers of ten are called _____ fractions.

2. When changing fractions with denominators that are powers of ten to decimals, move the decimal point in the _____ a number of places _____ to the number of _____ in the denominator.

3. After moving the decimal point in the numerator, _____ the denominator.

Follow along with the two Guided Learning Video examples, and fill in the blanks on the left as you learn with the instructor. When the pencil icon appears, pause and try the Student Practice on your own.

Guided Learning Video ⏵ GUIDED LEARNING VIDEO	Pause: Student Practice
1. Change $6\frac{59}{100}$ to a decimal.	2. Change $4\frac{6}{100}$ to a decimal.
Write the whole number part, and count the number of zeros in the denominator. Then, shift the decimal point in the numerator the necessary number of places.	
$6\frac{59}{100} = \boxed{}$	

3. Change $\dfrac{78}{1000}$ to a decimal.

Count the number of zeros in the denominator. Then, shift the decimal point in the numerator the necessary number of places.

$$\dfrac{78}{1000} = \boxed{}$$

4. Change $\dfrac{45}{1000}$ to a decimal.

Helpful Hint: Pause the video and write the helpful hint in your own words.

For each Active Video Lesson, when the pencil icon appears, pause and work the problem. Then press play to check your work.

Active Video Lesson 1 ✏️ | **Active Video Lesson 2** ✏️

5. Change $8\dfrac{3}{1000}$ to a decimal.

6. Change $\dfrac{7}{100}$ to a decimal.

Guided Learning Video Worksheet: Changing a Decimal to a Fraction

Text: *Basic College Mathematics*
Student Learning Objective 3.1.3: Change from decimal notation to fractional notation.

Follow along with *Guided Learning Video 3.1.3, Changing a Decimal to a Fraction.*

Understanding the Big Picture: As you listen to the first part of the video, when the pencil icon appears, pause and fill in the blanks, choosing from the words listed.

denominator • last • point • numerator • and • first • whole

1. Numbers written in decimal notation are composed of three parts; the _____ number part, the decimal part, and the decimal _____.

2. The word _____ represents the decimal _____.

3. When changing a decimal number to a fraction, the fraction's denominator is the same as the _____ place value used in the decimal part of the decimal number.

4. The decimal part of the decimal number becomes the _____ of the fraction to which the decimal number is changed.

Follow along with the two Guided Learning Video examples, and fill in the blanks on the left as you learn with the instructor. When the pencil icon appears, pause and try the Student Practice on your own.

Guided Learning Video ▶ GUIDED LEARNING VIDEO	**Pause: Student Practice** ✏
1. Write 6.317 as a fraction.	**2.** Write 4.629 as a fraction.
To write the fraction, identify the whole number part, the decimal part, and the last place value used in the decimal part.	
$6.317 = \boxed{}\dfrac{\boxed{}}{\boxed{}}$	

3. Write 0.75 as a fraction.

To write the fraction, identify the whole number part, the decimal part, and the last place value used in the decimal part. Remember to reduce the fraction if possible.

$0.75 = \dfrac{\boxed{}}{\boxed{}}$

4. Write 0.32 as a fraction.

Helpful Hint: Pause the video and write the helpful hint in your own words.

For each Active Video Lesson, when the pencil icon appears, pause and work the problem. Then press play to check your work.

| **Active Video Lesson 1** ✏️ | **Active Video Lesson 2** ✏️ |

5. Convert 0.135 to a fraction. Remember to reduce the fraction if possible.

6. Change 4.29 to a fraction.

Name: _____   Date: _____

Instructor: _____   Section: _____

Guided Learning Video Worksheet: Comparing Decimals

Text: *Basic College Mathematics*
Student Learning Objective 3.2.1: Compare decimals.

Follow along with *Guided Learning Video 3.2.1, Comparing Decimals.*

Understanding the Big Picture: As you listen to the first part of the video, when the pencil icon appears, pause and fill in the blanks, choosing from the words listed.

decrease • inequality • increase • greater • less

1. Moving to the right on a number line, the numbers _____ in value.

2. On a number line, if a first number is to the left of a second number, its value is _____ than the second number.

3. The symbols used for expressing the relationships "less than" and "greater than" are called _____ symbols.

Follow along with the two Guided Learning Video examples, and fill in the blanks on the left as you learn with the instructor. When the pencil icon appears, pause and try the Student Practice on your own.

Guided Learning Video ▶ GUIDED LEARNING VIDEO	Pause: Student Practice
1. Write an inequality statement for 5.78 and 5.79 using the symbols <, =, or >. Use the 2-step procedure for comparing two numbers in decimal notation. 5.78 5.79 5.78 ☐ 5.79	2. Write an inequality statement for 9.35 and 9.33 using the symbols <, =, or >.

3. Write an inequality comparing 0.666 and 0.66 using the symbols $<$, $=$, or $>$.

Be sure each decimal has the same number of digits to the right of each decimal point. Then use the 2-step procedure for comparing two numbers in decimal notation.

0.666
0.66

0.666 ☐ 0.66

4. Write an inequality comparing 0.88 and 0.888 using the symbols $<$, $=$, or $>$.

Helpful Hint: Pause the video and write the helpful hint in your own words.

For each Active Video Lesson, when the pencil icon appears, pause and work the problem. Then press play to check your work.

Active Video Lesson 1

Active Video Lesson 2

5. Write an inequality statement with 0.57 and 0.571.

6. Compare 2.032 and 2.031 using one of the symbols $<$, $=$, or $>$.

Name: _____   Date: _____

Instructor: _____   Section: _____

Guided Learning Video Worksheet: Rounding Decimals

Text: *Basic College Mathematics*
Student Learning Objective 3.2.3: Round decimals to a specified decimal place.

Follow along with *Guided Learning Video 3.2.3, Rounding Decimals*.

Understanding the Big Picture: As you listen to the first part of the video, when the pencil icon appears, pause and fill in the blanks, choosing from the words listed.

decrease • less • increase • place • one • number • greater

1. To round a decimal, first identify the given _____ value to which rounding is required.

2. When the digit to the right of the given place value is _____ than five, drop it and all the digits to its right.

3. If the first digit to the right of the given place value is five or more, _____ the number in the given place value by _____.

Follow along with the two Guided Learning Video examples, and fill in the blanks on the left as you learn with the instructor. When the pencil icon appears, pause and try the Student Practice on your own.

Guided Learning Video	Pause: Student Practice
1. Round 14.259 to the nearest hundredth.	2. Round 63.536 to the nearest hundredth.
Follow the rules for rounding decimals to round this number to the nearest hundredth.	
The number in the given place value is ☐ , and the first number to its right is ☐ .	
14.259 rounded to the nearest hundredth is ☐ .	

3. Round 0.70328 to the nearest thousandth.

   Follow the rules for rounding decimals to round this number to the nearest thousandth.

   The number in the given place value is ☐, and the first number to its right is ☐.

   0.70328 rounded to the nearest thousandth is ☐.

4. Round 0.09545 to the nearest thousandth.

Helpful Hint: Pause the video and write the helpful hint in your own words.

For each Active Video Lesson, when the pencil icon appears, pause and work the problem. Then press play to check your work.

Active Video Lesson 1 | **Active Video Lesson 2**

5. A student purchases a single cup coffee brewer. The sales tax is calculated to be $9.8736. Round this sales tax amount to the nearest cent.

6. Round 3.195 to the nearest tenth.

Guided Learning Video Worksheet: Adding Decimals

Text: *Basic College Mathematics*
Student Learning Objective 3.3.1: Add decimals.

Follow along with *Guided Learning Video 3.3.1, Adding Decimals.*

Understanding the Big Picture: As you listen to the first part of the video, when the pencil icon appears, pause and fill in the blanks, choosing from the words listed.

difference • sum • lining • fractions • same

1. The addition of decimals can be related to the addition of _____.

2. The procedure for adding decimals begins with writing the numbers vertically and _____ up the decimal points.

3. Add all digits with the _____ place value moving from the right column to the left, then place the decimal point in the _____ in line with the decimal points of the numbers added.

Follow along with the two Guided Learning Video examples, and fill in the blanks on the left as you learn with the instructor. When the pencil icon appears, pause and try the Student Practice on your own.

Guided Learning Video (▶) GUIDED LEARNING VIDEO	**Pause: Student Practice** ✏️
1. Add $1.2 + 3.7 + 6.9$.	**2.** Add $4.3 + 1.6 + 7.4$.
Align the numbers vertically, lining up the decimal points. Then, add the numbers.	

3. Add $8.73 + 25.1 + 0.679$.

Align the numbers vertically, lining up the decimal points. Fill in any empty place values with zeros. Then, add the numbers.

```
┌──────────────┐
│              │
└──────────────┘
┌──────────────┐
│              │
└──────────────┘
┌──────────────┐
+│              │
└──────────────┘
┌──────────────┐
│              │
└──────────────┘
```

4. Add $3.25 + 46.8 + 0.356$.

Helpful Hint: Pause the video and write the helpful hint in your own words.

For each Active Video Lesson, when the pencil icon appears, pause and work the problem. Then press play to check your work.

Active Video Lesson 1

Active Video Lesson 2

5. Add $0.528 + 36.9 + 7.41$.

6. Victoria has $739.52 in her savings account. If she deposits $214.86 into the account, how much does she have in savings?

Name: _____   Date: _____

Instructor: _____   Section: _____

Guided Learning Video Worksheet: Subtracting Decimals

Text: *Basic College Mathematics*
Student Learning Objective 3.3.2: Subtract decimals.

Follow along with *Guided Learning Video 3.3.2, Subtracting Decimals*.

Understanding the Big Picture: As you listen to the first part of the video, when the pencil icon appears, pause and fill in the blanks, choosing from the words listed.

right • left • difference • borrow • zeros • same

1. Just like subtraction of mixed numbers with the same denominators, when subtracting decimals, we must sometimes _____ from the whole number.

2. When subtracting decimals, additional _____ may be placed to the _____ of the decimal point if not all numbers have the _____ number of decimal places.

3. Place the decimal point in the _____ so it is in line with the decimal points of the two numbers being subtracted.

Follow along with the two Guided Learning Video examples, and fill in the blanks on the left as you learn with the instructor. When the pencil icon appears, pause and try the Student Practice on your own.

Guided Learning Video ▶ GUIDED LEARNING VIDEO	**Pause: Student Practice** ✐
1. a. Subtract $26.9 - 15.3$. **b.** Subtract $24.39 - 5.8$. For both examples, align the numbers vertically, lining up the decimal points and adding zeros where needed. Then, subtract the numbers, borrowing when necessary.	**2. a.** Subtract $47.5 - 26.3$. **b.** Subtract $61.27 - 9.3$.

a. ☐ − ☐ = ☐   **b.** ☐ − ☐ = ☐

3. Subtract $17 - 8.769$.

Align the numbers vertically, lining up the decimal points and adding zeros where needed. Then, subtract the numbers, borrowing when necessary.

```
┌─────────────┐
│             │
├─────────────┤
─│             │
├─────────────┤
│             │
└─────────────┘
```

4. Subtract $23 - 6.592$.

Helpful Hint: Pause the video and write the helpful hint in your own words.

For each Active Video Lesson, when the pencil icon appears, pause and work the problem. Then press play to check your work.

Active Video Lesson 1 | **Active Video Lesson 2**

5. Subtract $26 - 11.045$.

6. Subtract $96.35 - 48.29$.

Name: _____   Date: _____

Instructor: _____   Section: _____

Guided Learning Video Worksheet: Multiplying Decimals

Text: *Basic College Mathematics*
Student Learning Objective 3.4.1: Multiply a decimal by a decimal or a whole number.

Follow along with *Guided Learning Video 3.4.1, Multiplying Decimals*.

Understanding the Big Picture: As you listen to the first part of the video, when the pencil icon appears, pause and fill in the blanks, choosing from the words listed.

left • total • equal • zeros • whole

1. To multiply decimal numbers, multiply the numbers as if they were _____ numbers.

2. To position the decimal point in the product of decimal numbers, the _____ number of decimal places in the factors must be determined.

3. The number of decimal places in the product of decimal numbers must _____ the total number of decimal places in the factors.

4. Sometimes, _____ need to be inserted to the _____ of the product of decimal numbers.

Follow along with the two Guided Learning Video examples, and fill in the blanks on the left as you learn with the instructor. When the pencil icon appears, pause and try the Student Practice on your own.

Guided Learning Video	**Pause: Student Practice**
1. Multiply 0.07×0.8. Write the factors in a vertical format. Then, multiply and position the decimal point in the answer.	2. Multiply 0.06×0.4.

| **Guided Learning Video** ▶ GUIDED LEARNING VIDEO | **Pause: Student Practice** |

3. Multiply 4.679×53.

Write the factors in a vertical format. Then, multiply and position the decimal point in the answer.

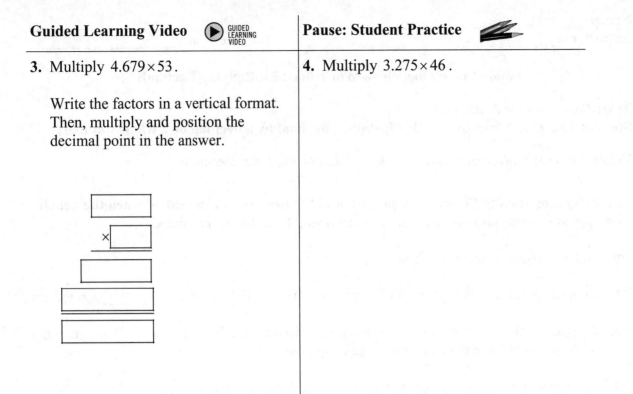

4. Multiply 3.275×46.

Helpful Hint: Pause the video and write the helpful hint in your own words.

For each Active Video Lesson, when the pencil icon appears, pause and work the problem. Then press play to check your work.

Active Video Lesson 1	**Active Video Lesson 2**
5. Multiply 0.9×0.04.	**6.** Multiply 0.758×96.

Name: _____   Date: _____

Instructor: _____   Section: _____

Guided Learning Video Worksheet: Multiplying Decimals by a Power of Ten

Text: *Basic College Mathematics*
Student Learning Objective 3.4.2: Multiply a decimal by a power of ten.

Follow along with *Guided Learning Video 3.4.2, Multiplying Decimals by a Power of Ten.*

Understanding the Big Picture: As you listen to the first part of the video, when the pencil icon appears, pause and fill in the blanks, choosing from the words listed.

left • right • multiplication • zeros • places • less

1. The operation of _____ is used to write a number with digits only when its value is expressed in word form.

2. When multiplying by a power of ten, the number of _____ in the power of ten is the number of _____ the decimal point is moved to the _____.

3. Zeros are sometimes needed when the number of digits to the right of the decimal point is _____ than the number of zeros in the power of ten.

Follow along with the two Guided Learning Video examples, and fill in the blanks on the left as you learn with the instructor. When the pencil icon appears, pause and try the Student Practice on your own.

Guided Learning Video (▶) GUIDED LEARNING VIDEO	**Pause: Student Practice** ✏
1. Multiply 81.74×10^5. To multiply, move the decimal point the same number of places as there are zeros in the power of ten. If necessary, add zeros. $81.74 \times 10^5 =$ ☐	2. Multiply 66.73×10^4.

Guided Learning Video	Pause: Student Practice
3. Multiply $9.16 \times 10,000$. To multiply, move the decimal point the same number of places as there are zeros in the power of ten. If necessary, add zeros. $9.16 \times 10,000 =$ [ ]	**4.** Multiply 4.95×1000.

Helpful Hint: Pause the video and write the helpful hint in your own words.

For each Active Video Lesson, when the pencil icon appears, pause and work the problem. Then press play to check your work.

Active Video Lesson 1	Active Video Lesson 2
5. Multiply 1.7×10^3.	**6.** Change $62.8 \, \text{km}$ to m. $1 \, \text{km} = 1000 \, \text{m}$.

Name: _____   Date: _____
Instructor: _____   Section: _____

Guided Learning Video Worksheet: Dividing Decimals

Text: *Basic College Mathematics*
Student Learning Objective 3.5.2: Divide a decimal by a decimal.

Follow along with *Guided Learning Video 3.5.2, Dividing Decimals*.

Understanding the Big Picture: As you listen to the first part of the video, when the pencil icon appears, pause and fill in the blanks, choosing from the words listed.

left • quotient • right • divisor • dividend

1. When dividing a decimal by a decimal, make the _____ a whole number by moving the decimal point to the _____.

2. Move the decimal point in the _____ the same number of places it was moved in the _____.

3. Position the decimal point in the _____ directly above its position in the _____.

Follow along with the two Guided Learning Video examples, and fill in the blanks on the left as you learn with the instructor. When the pencil icon appears, pause and try the Student Practice on your own.

Guided Learning Video	Pause: Student Practice
1. Divide: $0.537 \div 0.3$	**2.** Divide: $0.918 \div 0.6$
Write the expression in long division form. Then, follow the procedure for dividing a decimal by a decimal to calculate the answer.	

3. Divide: $0.756 \div 0.24$

Write the expression in long division form. Then, follow the procedure for dividing a decimal by a decimal to calculate the answer.

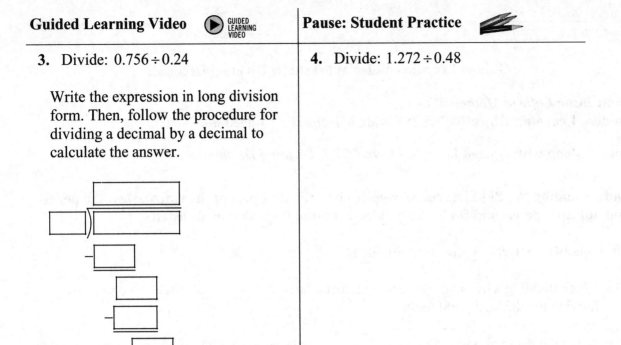

4. Divide: $1.272 \div 0.48$

Helpful Hint: Pause the video and write the helpful hint in your own words.

For each Active Video Lesson, when the pencil icon appears, pause and work the problem. Then press play to check your work.

Active Video Lesson 1 🖎 | **Active Video Lesson 2** 🖎

5. Divide: $8.93 \div 3.8$

6. Divide: $15.75 \div 3.5$

Name: _____   Date: _____

Instructor: _____   Section: _____

Guided Learning Video Worksheet: Convert a Fraction to a Decimal

Text: *Basic College Mathematics*
Student Learning Objective 3.6.1: Convert a fraction to a decimal.

Follow along with *Guided Learning Video 3.6.1, Convert a Fraction to a Decimal.*

Understanding the Big Picture: As you listen to the first part of the video, when the pencil icon appears, pause and fill in the blanks, choosing from the words listed.

denominator • terminating • repeating • numerator • repeat

1. To convert a fraction to a decimal, divide the _____ by the _____.

2. When converting fractions to decimals, if the division yields a remainder of zero, the decimal is called a _____ decimal.

3. Decimals having a digit or group of digits that _____ are called _____ decimals.

Follow along with the two Guided Learning Video examples, and fill in the blanks on the left as you learn with the instructor. When the pencil icon appears, pause and try the Student Practice on your own.

Guided Learning Video	Pause: Student Practice
1. Write $\frac{7}{8}$ as an equivalent decimal.	2. Write $\frac{3}{8}$ as an equivalent decimal.
Perform the division operation until a remainder of zero occurs, or the remainder repeats itself.	

3. Write $\frac{4}{9}$ as an equivalent decimal.

Write the expression in long division form. Then, follow the procedure for dividing a decimal by a decimal to calculate the answer.

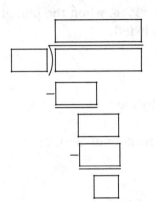

4. Write $\frac{5}{6}$ as an equivalent decimal.

Helpful Hint: Pause the video and write the helpful hint in your own words.

For each Active Video Lesson, when the pencil icon appears, pause and work the problem. Then press play to check your work.

Active Video Lesson 1 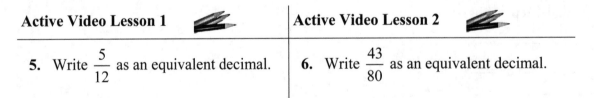 | **Active Video Lesson 2**

5. Write $\frac{5}{12}$ as an equivalent decimal.

6. Write $\frac{43}{80}$ as an equivalent decimal.

Name: _____   Date: _____

Instructor: _____   Section: _____

Guided Learning Video Worksheet: Order of Operations with Decimals

Text: *Basic College Mathematics*
Student Learning Objective 3.6.2: Use the order of operations with decimals.

Follow along with *Guided Learning Video 3.6.2, Order of Operations with Decimals.*

Understanding the Big Picture: As you listen to the first part of the video, when the pencil icon appears, pause and fill in the blanks, choosing from the words listed.

multiplication • decimals • simplify • priorities • exponents

1. An order of operations is a list of _____ for working with numbers in a computation.

2. The second step in the order of operations is to _____ expressions with _____.

3. To evaluate expressions with exponents, use repeated _____.

Follow along with the two Guided Learning Video examples, and fill in the blanks on the left as you learn with the instructor. When the pencil icon appears, pause and try the Student Practice on your own.

Guided Learning Video ▶ GUIDED LEARNING VIDEO	Pause: Student Practice
1. Evaluate the expression $12.1 - 3.2^2 + 5 \times 3.6$. Use the order of operations to calculate the answer. $12.1 - 3.2^2 + 5 \times 3.6$ = [ ] = [ ] = [ ] = [ ]	2. Evaluate the expression $14.7 - 2.6^2 + 7 \times 6.8$.

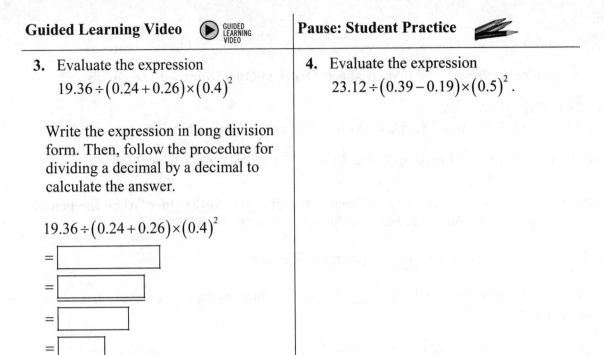

3. Evaluate the expression
$$19.36 \div (0.24 + 0.26) \times (0.4)^2$$

Write the expression in long division form. Then, follow the procedure for dividing a decimal by a decimal to calculate the answer.

$$19.36 \div (0.24 + 0.26) \times (0.4)^2$$

= ☐

= ☐

= ☐

= ☐

4. Evaluate the expression
$$23.12 \div (0.39 - 0.19) \times (0.5)^2.$$

Helpful Hint: Pause the video and write the helpful hint in your own words.

For each Active Video Lesson, when the pencil icon appears, pause and work the problem. Then press play to check your work.

Active Video Lesson 1	**Active Video Lesson 2**

5. Evaluate the expression
$$0.6 \times 0.3 + 0.1^3 - 0.07.$$

6. Evaluate the expression
$$0.25 \div 5 - 0.2^2 + (4 - 0.85).$$

Guided Learning Video Worksheet: Solving Real-Life Problems Involving Decimals

Text: *Basic College Mathematics*
Student Learning Objective 3.7.2: Solve applied problems using operations with decimals.

Follow along with *Guided Learning Video 3.7.2, Solving Real-Life Problems Involving Decimals.*

Understanding the Big Picture: As you listen to the first part of the video, when the pencil icon appears, pause and fill in the blanks, choosing from the words listed.

solution • decimals • numerical • process

1. When gathering the facts, write down all the _____ information given in the problem.

2. To determine what to do, start by identifying what type of _____ the problem is asking for.

3. To solve the problem, identify the _____ for calculating the problem's answer.

Follow along with the two Guided Learning Video examples, and fill in the blanks on the left as you learn with the instructor. When the pencil icon appears, pause and try the Student Practice on your own.

Guided Learning Video	Pause: Student Practice
1. Ashley earns $12.49/hr as a manager at a sandwich shop. She works an average of 53 hours per week. 40 hours is the regular full-time work week and any hours over that is considered overtime. The overtime rate is 1.5 times the employee's hourly pay. How much does Ashley make in an average week?	2. Dave typically works 55 hours a week as dispatcher. His hourly pay rate for a 40 hour work week is $17.50/hr. Overtime is paid at the rate of 1.5 times the number of hours over 40. What is Dave's pay for a typical work week?

Use a math blueprint to organize the problem. Then, calculate the answer.

$12.49 \times 40 =$ ☐

$12.49 \times 1.5 =$ ☐

☐ ☐ ☐ = ☐

☐ ☐ ☐ = ☐

3. The winning times for the Olympic 100-meter dash are listed below.

   London 2012: 10.70
   Beijing 2008: 10.78
   Greece 2004: 10.93
   Sydney 2000: 11.12

   What is the average winning time for years 2000 to 2012?

   Use a math blueprint to organize the problem. Then, calculate the answer.

   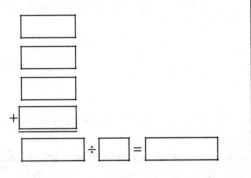

4. An accountant's record of hours worked per week over a five-week period is 47 hours, 58 hours, 45 hours, 60 hours, and 45 hours. What is the average number of hours worked per week over this five-week period?

Helpful Hint: Pause the video and write the helpful hint in your own words.

For each Active Video Lesson, when the pencil icon appears, pause and work the problem. Then press play to check your work.

Active Video Lesson 1 | **Active Video Lesson 2**

5. You are working in a chemistry lab testing 25.5 liters of cleaning fluid. You need to pour it into several smaller containers that each holds 1.5 liters of fluid. How many containers will you need?

6. Iesha and Angelo borrowed $147,000 to buy their new home. They make monthly payments to the bank of $773.29 to repay the loan. They will be making these payments for the next 30 years. How much money will they pay to the bank in the next 30 years?

Guided Learning Video Worksheet: Ratios

Text: *Basic College Mathematics*
Student Learning Objective 4.1.1: Use a ratio to compare two quantities with the same units.

Follow along with *Guided Learning Video 4.1.1, Ratios*.

Understanding the Big Picture: As you listen to the first part of the video, when the pencil icon appears, pause and fill in the blanks, choosing from the words listed.

comma • simplified • colon • ratio • fractions • comparison

1. A(n) _____ is a _____ of two quantities having the same units of measurement.

2. Ratios can be written as _____ but always in _____ form.

3. Ratios can also be expressed using a(n) _____ positioned between the two quantities.

Follow along with the two Guided Learning Video examples, and fill in the blanks on the left as you learn with the instructor. When the pencil icon appears, pause and try the Student Practice on your own.

Guided Learning Video	Pause: Student Practice
1. Express the ratio as a fraction in simplest form: $350:400$	2. Write the ratio as a fraction in simplest form: $720:1080$
Write the ratio in fraction form, reduced to lowest terms.	
$350:400 = \dfrac{\boxed{}}{\boxed{}} = \dfrac{\boxed{}}{\boxed{}} = \dfrac{\boxed{}}{\boxed{}}$	

3. A mixture consists of 56 milliliters of water and 24 milliliters of alcohol. What is the ratio of alcohol to water?

   Write the ratio in fraction form, reduced to lowest terms.

   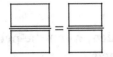

4. A mixture consists of 64 ounces of water and 12 ounces of saline. What is the ratio of saline to water?

Helpful Hint: Pause the video and write the helpful hint in your own words.

For each Active Video Lesson, when the pencil icon appears, pause and work the problem. Then press play to check your work.

Active Video Lesson 1 | **Active Video Lesson 2**

5. There are 1450 freshmen at Wyatt Community College this year. 500 of the freshman are in the Liberal Arts program, 350 students are in the Business program, and the other students are enrolled in other programs. Write the ratio of freshmen in the Business program to the number of freshmen students.

6. Write the ratio of 14 pounds to 21 pounds as a fraction in simplest form.

Name: _____   Date: _____

Instructor: _____   Section: _____

Guided Learning Video Worksheet: Rates

Text: *Basic College Mathematics*
Student Learning Objective 4.1.2: Use a rate to compare two quantities with different units.

Follow along with *Guided Learning Video 4.1.2, Rates.*

Understanding the Big Picture: As you listen to the first part of the video, when the pencil icon appears, pause and fill in the blanks, choosing from the words listed.

multiplication • different • division • unit • one • comparison

1. A comparison of two quantities with _____ units is called a _____.

2. A(n) _____ rate is a ratio with a denominator of _____.

3. The operation of _____ is used for finding a unit rate.

Follow along with the two Guided Learning Video examples, and fill in the blanks on the left as you learn with the instructor. When the pencil icon appears, pause and try the Student Practice on your own.

Guided Learning Video ▶ GUIDED LEARNING VIDEO	Pause: Student Practice ✏
1. An average avocado contains 242 calories and 22 grams of fat. How many calories are there per gram of fat? Set up the rate and divide to calculate the number of calories per gram of fat. $$\frac{\rule{3cm}{0.4pt}}{\rule{3cm}{0.4pt}} = \frac{\rule{3cm}{0.4pt}}{\rule{3cm}{0.4pt}}$$	2. A 2-tablespoon serving of peanut butter contains 16 grams of fat. How many grams of fat are there per tablespoon?

3. A coffee shop bought 215 pounds of coffee for $688. The coffee was sold to customers for a total of $1,118. How much profit did the shop make per pound of coffee?

Calculate the profit and set up the rate. Then divide to calculate the profit per pound of coffee.

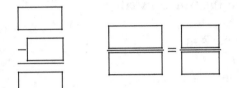

4. A grocery store purchases 88 gallons of ice cream for $220, and sells all of it to customers for $836. How much profit did the grocery store make per gallon of ice cream?

Helpful Hint: Pause the video and write the helpful hint in your own words.

For each Active Video Lesson, when the pencil icon appears, pause and work the problem. Then press play to check your work.

Active Video Lesson 1 | **Active Video Lesson 2**

5. Zack drove 414 miles from Minneapolis, Minnesota, to Kansas City, Missouri. It took him 6 hours. What was his average speed?

6. Last year, the Fischer family used a total of 96 gigabytes of data on their phone plan. What was the average amount of data used per month?

Guided Learning Video Worksheet: Writing Proportions

Text: *Basic College Mathematics*
Student Learning Objective 4.2.1: Write a proportion.

Follow along with *Guided Learning Video 4.2.1, Writing Proportions*.

Understanding the Big Picture: As you listen to the first part of the video, when the pencil icon appears, pause and fill in the blanks, choosing from the words listed.

denominators • equal • proportional • rates • numerators • ratios

1. Equivalent fractions representing rates are said to be _____.

2. A proportion states that two _____ or _____ are _____.

3. To write a proportion correctly, be sure the same unit is used in both _____ and the same unit is used in both _____.

Follow along with the two Guided Learning Video examples, and fill in the blanks on the left as you learn with the instructor. When the pencil icon appears, pause and try the Student Practice on your own.

Guided Learning Video	Pause: Student Practice
1. Write the proportion 6 is to 7 as 18 is to 21. Write the proportion in fractional form. $\frac{}{} = \frac{}{}$	2. Write the proportion 9 is to 11 as 45 is to 55.

3. Write the proportion to express the following: "If it takes 5 hours to drive 350 miles, then it will take 8 hours to drive 560 miles."

   Write the first rate and the second rate, making sure the same unit appears in both numerators, and the same unit appears in both denominators.

   $$\frac{\boxed{}}{\boxed{}} = \frac{\boxed{}}{\boxed{}}$$

4. Write the proportion to express the following: "If 330 miles can be traveled on 15 gallons of gas, then 1320 miles can be traveled on 60 gallons of gas."

Helpful Hint: Pause the video and write the helpful hint in your own words.

For each Active Video Lesson, when the pencil icon appears, pause and work the problem. Then press play to check your work.

Active Video Lesson 1 | **Active Video Lesson 2**

5. If a 3-credit-hour class at the local community college costs $669, then a 5-credit hour class should cost $1115.

6. If a pulley can complete 5 rotations in 2 minutes, then it should complete 30 rotations in 12 minutes.

Guided Learning Video Worksheet: Identifying Proportions

Text: *Basic College Mathematics*
Student Learning Objective 4.2.2: Determine whether a statement is a proportion.

Follow along with *Guided Learning Video 4.2.2, Identifying Proportions.*

Understanding the Big Picture: As you listen to the first part of the video, when the pencil icon appears, pause and fill in the blanks, choosing from the words listed.

denominators • proportion • cross • numerators

1. Two fractions are considered equal if their _____ products are equal.

2. If two fractions are equal, then the statement is considered a _____.

3. Cross products are the result of multiplying _____ times _____.

Follow along with the two Guided Learning Video examples, and fill in the blanks on the left as you learn with the instructor. When the pencil icon appears, pause and try the Student Practice on your own.

Guided Learning Video	Pause: Student Practice
1. Determine if the statement is a proportion: four is to five as twelve is to fifteen. Write the statement using fractions and calculate the cross products. $$\frac{\boxed{}}{\boxed{}} = \frac{\boxed{}}{\boxed{}}$$ $$\boxed{} = \boxed{}$$	2. Determine if the statement is a proportion: twelve is to nineteen as thirty-six is to fifty-seven.

3. Is the rate ninety-five instructors per two thousand students equal to the rate one hundred fifty instructors per three thousand students?

   Write the rates as fractions and determine if the statement is a proportion.

   $$\frac{\boxed{}}{\boxed{}} = \frac{\boxed{}}{\boxed{}}$$

   $$\boxed{} = \boxed{}$$

4. Is the rate six party platters per seventy guests equal to the rate two party platters per twenty-five guests?

Helpful Hint: Pause the video and write the helpful hint in your own words.

For each Active Video Lesson, when the pencil icon appears, pause and work the problem. Then press play to check your work.

Active Video Lesson 1	**Active Video Lesson 2**

5. Determine whether the statement is a proportion: the rate two dollars and fifty cents for three pounds equal to the rate ten dollars for twelve pounds.

6. Determine whether the statement is a proportion: fourteen is to sixteen as twenty-one is to twenty-seven.

Guided Learning Video Worksheet: Solving Proportions

Text: *Basic College Mathematics*
Student Learning Objective 4.3.2: Find the missing number in a proportion.

Follow along with *Guided Learning Video 4.3.2, Solving Proportions*.

Understanding the Big Picture: As you listen to the first part of the video, when the pencil icon appears, pause and fill in the blanks, choosing from the words listed.

equation • unknown • positions • cross • denominators

1. To solve a proportion, begin by _____ multiplying numerators times _____.

2. Create a(n) _____ by setting the cross products equal to each other.

3. The letter n is often used to represent the _____ quantity.

4. When setting up the proportion, the same units should be in the same _____ in each of the fractions.

Follow along with the two Guided Learning Video examples, and fill in the blanks on the left as you learn with the instructor. When the pencil icon appears, pause and try the Student Practice on your own.

Guided Learning Video ▶ GUIDED LEARNING VIDEO	**Pause: Student Practice** ✎
1. Find the value of n: $\dfrac{13}{3} = \dfrac{26}{n}$	2. Find the value of n: $\dfrac{17}{4} = \dfrac{51}{n}$

Find the cross products, then divide to solve for n.

$$\frac{13}{3} = \frac{26}{n}$$

$\boxed{} = \boxed{}$ Check

$\boxed{} = \boxed{}$

$\dfrac{\boxed{}}{\boxed{}} = \dfrac{\boxed{}}{\boxed{}}$

$\boxed{} = \boxed{}$

$\boxed{} = \boxed{}$

3. If the current dose of an antibiotic for a 180-pound man is 15 milligrams, what is the correct dose for a 120-pound woman?

   Set up the proportion, find the cross products, then divide to solve for *n*.

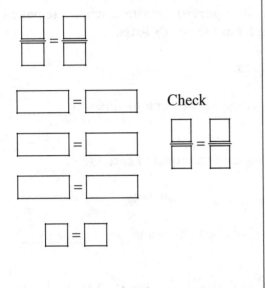

   Check

4. If 3 grams of saline need to be added to a 15 liter solution, how many grams need to be added to a 20 liter solution?

Helpful Hint: Pause the video and write the helpful hint in your own words.

For each Active Video Lesson, when the pencil icon appears, pause and work the problem. Then press play to check your work.

Active Video Lesson 1	**Active Video Lesson 2**
5. Find the value of *n*: $\dfrac{0.5}{n} = \dfrac{3}{4.2}$	6. Find the value of the unknown amount, in dollars, of: 54 dollars per 18 square feet is equal to how much per 15 square feet?

Guided Learning Video Worksheet: Solving Applied Problems Involving Proportions

Text: *Basic College Mathematics*
Student Learning Objective 4.4.1: Solve applied problems using proportions.

Follow along with *Guided Learning Video 4.4.1, Solving Applied Problems Involving Proportions.*

Understanding the Big Picture: As you listen to the first part of the video, when the pencil icon appears, pause and fill in the blanks, choosing from the words listed.

requested • unknown • sample • facts • proportion

1. The first step in the math blueprint is to gather the _____.

2. For the problems in this video, the answer to the question of how to proceed is to set up a(n) _____.

3. To begin solving the compact car problem, first set up the proportion comparing the _____ rate to the _____ rate.

Follow along with the two Guided Learning Video examples, and fill in the blanks on the left as you learn with the instructor. When the pencil icon appears, pause and try the Student Practice on your own.

Guided Learning Video GUIDED LEARNING VIDEO	**Pause: Student Practice**
1. Students in a math class counted that 26 out of 234 cars in one of the school's parking lots were compact cars. If that rate is true for all of the parking lots, how many of the total of 1080 cars in those lots are compact? Use the math blueprint to help organize, set up, and solve the problem.	2. In one statistics class at North Central Community College, 17 students out of a total of 35 students are female. If this rate is true for all statistics classes at the college, how many of the total of 280 students enrolled in statistics classes are female?

$$\frac{\boxed{}}{\boxed{}} = \frac{\boxed{}}{\boxed{}}$$

$$\boxed{} = \boxed{}$$

$$\boxed{} = \boxed{}$$

$$\boxed{} = \boxed{}$$

3. A document copier can produce 125 double-sided copies in 1.2 minutes. How long will it take the copier to make 500 double-sided copies?

   Use the math blueprint to help organize, set up, and solve the problem.

   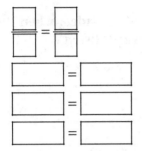

4. A six-page electric collator can process 280 sheets of standard sized sheets of paper per 1.5 minutes. How long will it take the collator to process 5600 sheets of standard sized paper?

Helpful Hint: Pause the video and write the helpful hint in your own words.

For each Active Video Lesson, when the pencil icon appears, pause and work the problem. Then press play to check your work.

Active Video Lesson 1 | **Active Video Lesson 2**

5. If 9 inches on a map represents 57 miles, what distance does 3 inches represent?

6. Park rangers captured, tagged, and released 48 deer. A month later, they captured deer. 6 of the 20 had tags from the earlier capture. Estimate the number of deer in this forest area.

Name: _____   Date: _____

Instructor: _____   Section: _____

Guided Learning Video Worksheet: Changing a Decimal to a Percent

Text: *Basic College Mathematics*
Student Learning Objective 5.1.3: Write a decimal as a percent.

Follow along with *Guided Learning Video 5.1.3, Changing a Decimal to a Percent.*

Understanding the Big Picture: As you listen to the first part of the video, when the pencil icon appears, pause and fill in the blanks, choosing from the words listed.

parts • ratios • numerators • percent • denominators

1. The word _____ means per hundred.

2. Percents can be described as _____ with _____ of one hundred.

3. The percent symbol means _____ per one hundred.

Follow along with the two Guided Learning Video examples, and fill in the blanks on the left as you learn with the instructor. When the pencil icon appears, pause and try the Student Practice on your own.

Guided Learning Video ▶ GUIDED LEARNING VIDEO	Pause: Student Practice
1. Change 0.726 to a percent.	2. Change 0.539 to a percent.
Move the decimal point the necessary number of places, and write the percent symbol at the end of the number.	
$0.726 \Rightarrow$ ☐	

Guided Learning Video	**Pause: Student Practice**
3. Change 5.8 to a percent. Move the decimal point the necessary number of places, and write the percent symbol at the end of the number. 5.8 ⇒ ☐	4. Change 3.1 to a percent.

Helpful Hint: Pause the video and write the helpful hint in your own words.

For each Active Video Lesson, when the pencil icon appears, pause and work the problem. Then press play to check your work.

Active Video Lesson 1 ✎	**Active Video Lesson 2** ✎
5. Convert 2.9 to a percent.	6. Convert 0.165 to a percent.

Name: _____   Date: _____

Instructor: _____   Section: _____

Guided Learning Video Worksheet: Changing a Fraction to a Percent

Text: *Basic College Mathematics*
Student Learning Objective 5.2.2: Change a fraction to a percent.

Follow along with *Guided Learning Video 5.2.2, Changing a Fraction to a Percent.*

Understanding the Big Picture: As you listen to the first part of the video, when the pencil icon appears, pause and fill in the blanks, choosing from the words listed.

multiplying • hundred • decimal • percent • two

1. To change a fraction to a percent, first write the fraction as a _____.

2. After changing the fraction to a _____, convert it to a percent.

3. Moving the decimal point _____ places to the right is the same as _____ the decimal times a _____.

Follow along with the two Guided Learning Video examples, and fill in the blanks on the left as you learn with the instructor. When the pencil icon appears, pause and try the Student Practice on your own.

Guided Learning Video ▶ GUIDED LEARNING VIDEO	Pause: Student Practice
1. Convert $\dfrac{11}{20}$ to a percent.	2. Convert $\dfrac{7}{25}$ to a percent.

Write the fraction in decimal form, and then convert the decimal to a percent.

$$\boxed{}$$
$$\boxed{})\boxed{}$$
$$-\boxed{}$$
$$\boxed{}$$
$$\boxed{}$$

$$\boxed{} \Rightarrow \boxed{}$$

3. Convert $\dfrac{9}{40}$ to a percent. | **4.** Convert $\dfrac{3}{8}$ to a percent.

Write the fraction in decimal form, and then convert the decimal to a percent.

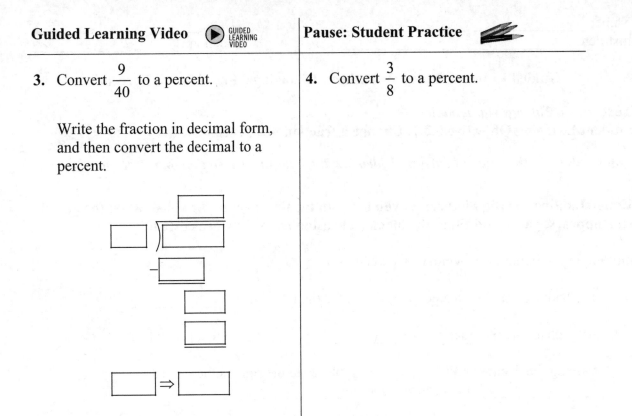

Helpful Hint: Pause the video and write the helpful hint in your own words.

For each Active Video Lesson, when the pencil icon appears, pause and work the problem. Then press play to check your work.

Active Video Lesson 1 | **Active Video Lesson 2**

5. Convert $\dfrac{16}{25}$ to a percent. | **6.** Convert $\dfrac{9}{16}$ to a percent.

Name: _____   Date: _____

Instructor: _____   Section: _____

Guided Learning Video Worksheet: Changing a Fraction to a Decimal and a Percent

Text: *Basic College Mathematics*
Student Learning Objective 5.2.3: Change a percent, a decimal, or a fraction to equivalent forms.

Follow along with *Guided Learning Video 5.2.3, Changing a Fraction to a Decimal and a Percent.*

Understanding the Big Picture: As you listen to the first part of the video, when the pencil icon appears, pause and fill in the blanks, choosing from the words listed.

multiplying • hundred • decimal • percent • two

1. To change a fraction to a percent, first write the fraction as a _____.

2. After changing the fraction to a _____, convert it to a percent.

3. Moving the decimal point _____ places to the right is the same as
   _____ the decimal times a _____.

Follow along with the two Guided Learning Video examples, and fill in the blanks on the left as you learn with the instructor. When the pencil icon appears, pause and try the Student Practice on your own.

Guided Learning Video $\blacktriangleright$ GUIDED LEARNING VIDEO	Pause: Student Practice
1. Convert $\dfrac{11}{20}$ to a percent.	2. Convert $\dfrac{13}{20}$ to a percent.
Write the fraction in decimal form, and then convert the decimal to a percent.	

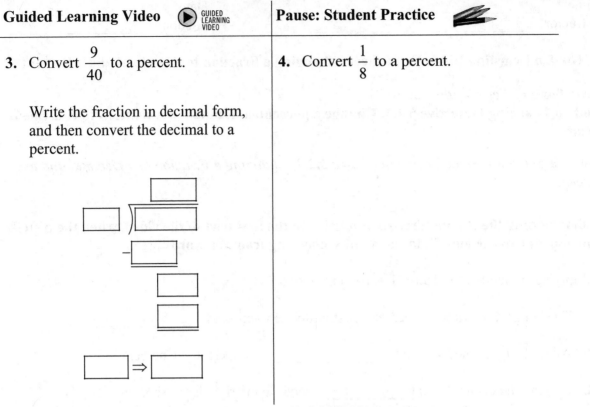
3. Convert $\dfrac{9}{40}$ to a percent.

Write the fraction in decimal form, and then convert the decimal to a percent.

4. Convert $\dfrac{1}{8}$ to a percent.

Helpful Hint: Pause the video and write the helpful hint in your own words.

For each Active Video Lesson, when the pencil icon appears, pause and work the problem. Then press play to check your work.

Active Video Lesson 1

5. Convert $\dfrac{16}{25}$ to a percent.

Active Video Lesson 2

6. Convert $\dfrac{9}{16}$ to a percent.

Guided Learning Video Worksheet: Solving Applied Percentage Problems

Text: *Basic College Mathematics*
Student Learning Objective 5.4.1: Solve general applied percent problems.

Follow along with *Guided Learning Video 5.4.1, Solving Applied Percentage Problems.*

Understanding the Big Picture: As you listen to the first part of the video, when the pencil icon appears, pause and fill in the blanks, choosing from the words listed.

base • two • equations • amount • proportions

1. Percent problems can be set up as _____ or _____.

2. The formula for the percent equation is _____ equals percent times the
    _____.

3. To solve a percent problem _____ out of the three parts of the percent equation
    must be known.

Follow along with the two Guided Learning Video examples, and fill in the blanks on the left as you learn with the instructor. When the pencil icon appears, pause and try the Student Practice on your own.

Guided Learning Video	Pause: Student Practice
1. Joshua spends 12% of his monthly income on his truck payment. If this payment is $324, what is Joshua's monthly income? Substitute the information from the problem into the percent equation. Solve the equation for *n*, the unknown. $\boxed{} = \boxed{} \times \boxed{}$ $\boxed{} = \boxed{}\ \boxed{}$ $\dfrac{\boxed{}}{\boxed{}} = \dfrac{\boxed{}}{\boxed{}}$ $\boxed{} = \boxed{}$	2. The Jones's family budget reveals that 22% of their total monthly income is spent on their mortgage payment. If the payment is $1650, what is their total monthly income?

3. Keisha's lunch at the local deli cost $12.50 before tax. If the sales tax for her city is 9%, how much did Keisha pay for her lunch?

   Substitute the information from the problem into the percent proportion. Solve the equation for n, the unknown.

   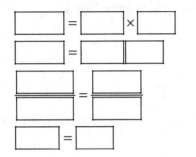

4. Don's total purchase before tax at an automotive supply store is $82.75. If the sales tax rate is 6%, what did Don pay for his purchase?

Helpful Hint: Pause the video and write the helpful hint in your own words.

For each Active Video Lesson, when the pencil icon appears, pause and work the problem. Then press play to check your work.

Active Video Lesson 1 | **Active Video Lesson 2**

5. Sumiko spends 20% of her monthly income on rent. If she spends $834 a month for her apartment, what is her monthly income?

6. Miguel purchased a new cordless power drill for $159 before tax. If sales tax in his city is 9.5%, how much did he pay for the drill?

Guided Learning Video Worksheet: Solving Discount Problems

Text: *Basic College Mathematics*
Student Learning Objective 5.4.3: Solve discount problems.

Follow along with *Guided Learning Video 5.4.3, Solving Discount Problems.*

Understanding the Big Picture: As you listen to the first part of the video, when the pencil icon appears, pause and fill in the blanks, choosing from the words listed.

percent • subtracted • rate • amount • proportion

1. The _____ of a discount is the product of the discount _____ and the list price.

2. The discount rate is expressed as a(n) _____.

3. The amount of a discount is _____ from the list price to determine how much will be paid for the discounted item(s).

Follow along with the two Guided Learning Video examples, and fill in the blanks on the left as you learn with the instructor. When the pencil icon appears, pause and try the Student Practice on your own.

Guided Learning Video 🔘 GUIDED LEARNING VIDEO	Pause: Student Practice ✏️
1. Garret purchased a flat-panel LCD TV on sale at a 15% discount. The list price was $390. What was the amount of the discount? Identify the list price and the discount rate. Then, determine the amount of the discount. discount = [ ] × [ ] discount = [ ] × [ ] discount = [ ]	2. A laptop computer with a list price of $1030 is being sold at a 20% discount. What is the amount of the discount?

3. Angie purchased a new blouse, jeans, and a sweater. All of the clothes were discounted 45%. Before the sale, the total purchase would have been $270 for these 3 items. How much did she pay for them with the discount?

   Identify the list price and the discount rate. Then, determine the amount of the discount to calculate how much was paid for the items.

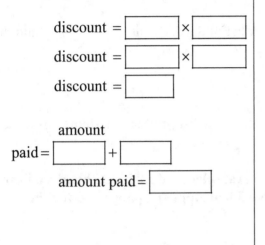

   discount = [   ] × [   ]

   discount = [   ] × [   ]

   discount = [   ]

   amount
   paid = [   ] + [   ]

   amount paid = [   ]

4. Mauricio purchases painting supplies totaling $450. He receives a contractor's discount of 35%. How much does Mauricio pay for the supplies?

Helpful Hint: Pause the video and write the helpful hint in your own words.

For each Active Video Lesson, when the pencil icon appears, pause and work the problem. Then press play to check your work.

Active Video Lesson 1 | **Active Video Lesson 2**

5. Craig is buying a bicycle to make getting across campus faster. The bike he is looking at has a list price of $470 but is discounted this week at 35% off. How much will Craig pay for the bicycle?

6. Mattress King is having a Labor Day blowout sale. A queen mattress is being sold at 25% off of the original price of $1340. How much is the set being sold for?

Guided Learning Video Worksheet: Solving Commission Problems

Text: *Basic College Mathematics*
Student Learning Objective 5.5.1: Solve commission problems.

Follow along with *Guided Learning Video 5.5.1, Solving Commission Problems.*

Understanding the Big Picture: As you listen to the first part of the video, when the pencil icon appears, pause and fill in the blanks, choosing from the words listed.

sales • commission • rate • amount • dividing

1. The _____ rate is the percent rate.

2. The _____ of commission is equal to the commission rate multiplied times the total _____.

3. When the amount of commission is known, the total sales can be found by
   _____ the amount of commission by the commission _____.

Follow along with the two Guided Learning Video examples, and fill in the blanks on the left as you learn with the instructor. When the pencil icon appears, pause and try the Student Practice on your own.

Guided Learning Video 🎦 GUIDED LEARNING VIDEO	**Pause: Student Practice** ✏️
1. Chris works in the sales department of a large company that does kitchen remodeling. He visits homeowners and gives them an estimate on the price of the remodeling job. His commission rate is 15%. If Chris earned $2868 last month in commission, what were his total sales for the month? Use the commission equation to determine the total sales for the month. Let n represent the unknown.	2. Each month, Mark receives a 2% commission on the total sales his sales team generates. His commission this month amounted to $2570. What was the total value of his team's sales for the month?

$$\boxed{} = \boxed{} \times \boxed{}$$

$$\boxed{} = \boxed{} \times \boxed{}$$

$$\boxed{} = \boxed{}$$

3. Mike works for Max's Auto Super Center as a car salesman. He has a commission rate of 2%. If he sold a car for $17,500, what was his commission?

   Use the commission equation to determine Mike's commission. Let n represent the unknown.

   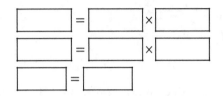

4. Paula is to receive a 5% commission on the sale of a newly constructed condominium. The condominium's final negotiated sale price is $395,000. What is Paula's commission?

Helpful Hint: Pause the video and write the helpful hint in your own words.

For each Active Video Lesson, when the pencil icon appears, pause and work the problem. Then press play to check your work.

Active Video Lesson 1 | **Active Video Lesson 2**

5. You must pay your real estate agent a 6% commission on the sale price of your home. If she earns an $8700 commission, what is the sales price of your home?

6. Will works at the local computer store. He earns an 11% commission on all of his sales. If Will sold $53,400 worth of computer equipment last month, what was his commission?

Guided Learning Video Worksheet: Solve Simple Interest Problems

Text: *Basic College Mathematics*
Student Learning Objective 5.5.3: Solve simple-interest problems.

Follow along with *Guided Learning Video 5.5.3, Solve Simple Interest Problems.*

Understanding the Big Picture: As you listen to the first part of the video, when the pencil icon appears, pause and fill in the blanks, choosing from the words listed.

yearly • borrowed • rate • interest • amount • principal

1. Money that is earned or paid for the use of money is called _____.

2. In simple interest problems, the _____ is the amount of money _____ or deposited.

3. An interest _____ is the percent used in computing an interest _____.

4. The simple interest formula is based on a(n) _____ interest rate.

Follow along with the two Guided Learning Video examples, and fill in the blanks on the left as you learn with the instructor. When the pencil icon appears, pause and try the Student Practice on your own.

Guided Learning Video	Pause: Student Practice
1. Lisbeth borrowed $5000 from the bank in order to purchase an ATV at a simple interest rate of 12% . Find the simple interest on her loan for two years. Identify the values of the variables and substitute them in the simple interest formula. Calculate the simple interest. $\boxed{} = \boxed{} \times \boxed{} \times \boxed{}$ $\boxed{} = \boxed{}$	2. How much interest is earned over a 5- year period if a deposit of $15,000 is made into an account earning 3% simple interest.

3. Uta borrowed $1800 from her bank in order to buy a new computer and printer at a simple interest rate of 9% for 3 months. Find the simple interest on her loan and determine how much she pays back to the bank at the end of the loan.

    Identify the values of the variables and substitute them in the simple interest formula. Calculate the simple interest and the amount to be paid to the bank.

    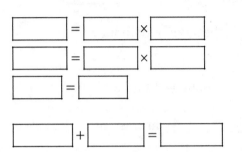

4. Ray borrows $2500 at a simple interest rate of 5% for 9 months. Calculate the simple interest on the loan and the total amount that must be paid to the bank at the end of 9 months.

Helpful Hint: Pause the video and write the helpful hint in your own words.

For each Active Video Lesson, when the pencil icon appears, pause and work the problem. Then press play to check your work.

Active Video Lesson 1 | **Active Video Lesson 2**

5. Steve borrowed $2000 for some much needed car repairs. If his loan is borrowed at a simple interest rate of 11% for 9 months, how much interest will he pay?

6. Daniella borrowed $6500 for a student loan to finish college this year. Next year, she will need to pay 7% simple interest on the amount she borrowed. How much interest will she need to pay next year?

Guided Learning Video Worksheet: Solve Percent Equations

Text: *Basic College Mathematics*
Student Learning Objective 5.3A.2: Solve a percent problem by solving an equation.

Follow along with *Guided Learning Video 5.3A.2, Solve Percent Equations.*

Understanding the Big Picture: As you listen to the first part of the video, when the pencil icon appears, pause and fill in the blanks, choosing from the words listed.

multiplication • base • decimal • percent • amount

1. The percent equation is _____ equals _____ times
   _____ .

2. When translating percent statements into equations, the word "of" means
   _____ .

3. Before performing any calculations with percents, change the _____ to an
   equivalent _____ .

Follow along with the two Guided Learning Video examples, and fill in the blanks on the left as you learn with the instructor. When the pencil icon appears, pause and try the Student Practice on your own.

Guided Learning Video ⏵ GUIDED LEARNING VIDEO	**Pause: Student Practice**
1. What is 32% of 260 ?	2. What is 45% of 730 ?
Use the procedure for solving applied percent problems to find the answer.	

$$\boxed{} = \boxed{} \times \boxed{}$$

$$\boxed{} = \boxed{} \times \boxed{}$$

$$\boxed{} = \boxed{}$$

3. 76% of college freshman receive financial aid. If 1862 freshman receive financial aid, how many freshmen are there in total?

Use the procedure for solving applied percent problems to find the answer.

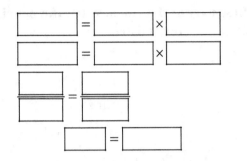

4. 62% of the total number of ballots mailed to voting members of a union are returned and considered valid. If 3565 ballots were returned and considered valid, what was the total number of ballots mailed?

Helpful Hint: Pause the video and write the helpful hint in your own words.

For each Active Video Lesson, when the pencil icon appears, pause and work the problem. Then press play to check your work.

Active Video Lesson 1 | **Active Video Lesson 2**

5. 15% of the students at West Lake College are currently taking math courses. There are 3400 students at West Lake College. How many students are currently taking math courses?

6. 12% of the residents of Clinton live in apartment buildings. 1524 of the residents of Clinton live in apartment buildings. How many people live in Clinton?

Name: _____   Date: _____

Instructor: _____   Section: _____

Guided Learning Video Worksheet: Solve for a Missing Percent

Text: *Basic College Mathematics*
Student Learning Objective 5.3A.3: Solve a percent problem by solving an equation.

Follow along with *Guided Learning Video 5.3A.3, Solve for a Missing Percent.*

Understanding the Big Picture: As you listen to the first part of the video, when the pencil icon appears, pause and fill in the blanks, choosing from the words listed.

right • base • divide • two • percent • amount

1. The percent equation is _____ equals _____ times _____.

2. When solving the percent equation for the percent quantity, _____ the _____ by the _____.

3. To convert a decimal to a percent, shift the decimal point _____ places to the _____.

Follow along with the two Guided Learning Video examples, and fill in the blanks on the left as you learn with the instructor. When the pencil icon appears, pause and try the Student Practice on your own.

Guided Learning Video ▶ GUIDED LEARNING VIDEO	Pause: Student Practice
1. 75 is what percent of 300? Use the procedure for solving applied percent problems to find the answer. ☐ = ☐ × ☐ ☐/☐ = ☐/☐ ☐ = ☐	2. 21 is what percent of 35?

3. Adelina purchases a new laptop for $950 and paid $76 in sales tax. What percent is the sales tax in her area? Use the procedure for solving applied percent problems to find the answer. 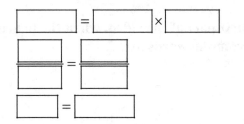	4. A new set of golf clubs costs $1250 plus $75 in sales tax. What percent is the sales tax rate?

Helpful Hint: Pause the video and write the helpful hint in your own words.

For each Active Video Lesson, when the pencil icon appears, pause and work the problem. Then press play to check your work.

Active Video Lesson 1	**Active Video Lesson 2**
5. Tammy correctly answered 28 problems out of 35 problems on her Chemistry quiz. What percent did she get correct?	6. What percent of 75 is 30?

Guided Learning Video Worksheet: Solving Percent Proportions

Text: *Basic College Mathematics*
Student Learning Objective 5.3B.2: Use the percent proportion to solve percent problems.

Follow along with *Guided Learning Video 5.3B.2, Solving Percent Proportions.*

Understanding the Big Picture: As you listen to the first part of the video, when the pencil icon appears, pause and fill in the blanks, choosing from the words listed.

right • base • divide • two • percent • amount

1. The percent proportion can be described as the amount is to the _____ as the
    _____ is to one hundred.

2. In the percent proportion, the letter "a" represents the _____, the letter "b"
    represents the _____, and the letter "p" represents the _____.

3. To solve percent problems using the percent proportion, _____ out of three variables
    must be known.

Follow along with the two Guided Learning Video examples, and fill in the blanks on the left as you learn with the instructor. When the pencil icon appears, pause and try the Student Practice on your own.

Guided Learning Video ▶ GUIDED LEARNING VIDEO	Pause: Student Practice ✏
1. Use a percent proportion to find 45% of 60. Substitute the values in the problem into the percent proportion and solve for the unknown. $\dfrac{\Box}{\Box} = \dfrac{\Box}{\Box}$ $\Box = \Box$ $\dfrac{\Box}{\Box} = \dfrac{\Box}{\Box}$ $\Box = \Box$	2. Use a percent proportion to find 35% of 120.

3. Use a percent proportion to find the following: 65% of what is 91?

Substitute the values in the problem into the percent proportion and solve for the unknown.

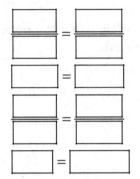

4. Use a percent proportion to find the following: 25% of what is 57 ?

Helpful Hint: Pause the video and write the helpful hint in your own words.

For each Active Video Lesson, when the pencil icon appears, pause and work the problem. Then press play to check your work.

Active Video Lesson 1 | **Active Video Lesson 2**

5. Use a percent proportion to find the following: 136 is 17% of what?

6. Use a percent proportion to find the following: What is 68% percent of 225 ?

Guided Learning Video Worksheet: Convert U.S. Units

Text: *Basic College Mathematics*
Student Learning Objective 6.1.2: Convert from one unit of measure to another.

Follow along with *Guided Learning Video 6.1.2, Convert U.S. Units*.

Understanding the Big Picture: As you listen to the first part of the video, when the pencil icon appears, pause and fill in the blanks, choosing from the words listed.

simplified • denominator • unit • numerator • one

1. A _____ fraction is a fraction showing the relationship between units, and is equal to _____.

2. In a unit fraction, the measurement in the _____ is equivalent to the measurement in the _____.

3. When multiplying by a unit fraction, the unit of measurement desired should be in the _____.

Follow along with the two Guided Learning Video examples, and fill in the blanks on the left as you learn with the instructor. When the pencil icon appears, pause and try the Student Practice on your own.

Guided Learning Video ▶ GUIDED LEARNING VIDEO	Pause: Student Practice
1. A flight from Boston to Pittsburgh takes 2 hours 35 minutes. How many minutes long is the flight? Identify the unit fraction to use for the conversion, then multiply by the unit fraction to solve the problem. ☐/☐ × ☐/☐ = ☐ ☐ + ☐ = ☐	2. A distance between point A and point B is measured as 2 miles 108 feet. Express the distance between point A and point B in feet.

3. Convert 5 cups to quarts. Express your answer as a decimal rounded to the nearest hundredth.

   Identify the unit fraction to use for the conversion, then multiply by the unit fraction to solve the problem.

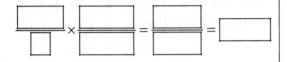

4. Convert 84 ounces to cups. If necessary, express the answer as a decimal rounded to the nearest tenth.

Helpful Hint: Pause the video and write the helpful hint in your own words.

For each Active Video Lesson, when the pencil icon appears, pause and work the problem. Then press play to check your work.

Active Video Lesson 1 | **Active Video Lesson 2**

5. A butcher has 56 ounces of beef that he sells in packages that weigh 1 pound each. How many packages can he make?

6. Juan is 6 feet tall. For his graduation robe, he has to write down his height in inches on the order form. How many inches tall is Juan?

Guided Learning Video Worksheet: Convert Metric Units: Length

Text: *Basic College Mathematics*
Student Learning Objective 6.2.2: Convert from one metric unit of length to another.

Follow along with *Guided Learning Video 6.2.2, Convert Metric Units: Length.*

Understanding the Big Picture: As you listen to the first part of the video, when the pencil icon appears, pause and fill in the blanks, choosing from the words listed.

longer • shorter • kilometer • foot • meter

1. In the metric system, a _____ is the basic measure of length.

2. A centimeter is _____ than a decimeter.

3. If a measurement is given in meters and the decimal point is moved three places to the left, the measurement would be converted to _____.

Follow along with the two Guided Learning Video examples, and fill in the blanks on the left as you learn with the instructor. When the pencil icon appears, pause and try the Student Practice on your own.

Guided Learning Video ▶ GUIDED LEARNING VIDEO	**Pause: Student Practice**
1. Change 8 cm to mm. Identify the location of the unit of measurement to be changed, and move the decimal point the necessary number of places in the proper direction to make the conversion. 8 *cm* = [ ]	2. Change 3.2 decimeters to millimeters.

3. Change 130 dekameters to kilometers.

    Identify the location of the unit of measurement to be changed, and move the decimal point the necessary number of places in the proper direction to make the conversion.

    130 dekameters = [   ]

4. Change 4.5 centimeters to kilometers.

Helpful Hint: Pause the video and write the helpful hint in your own words.

For each Active Video Lesson, when the pencil icon appears, pause and work the problem. Then press play to check your work.

Active Video Lesson 1 | **Active Video Lesson 2**

5. Convert 152 meters to kilometers.

6. Convert three point five six meters to centimeters.

Name: _____   Date: _____

Instructor: _____   Section: _____

Guided Learning Video Worksheet: Convert Metric Units: Volume

Text: *Basic College Mathematics*

Student Learning Objective 6.3.1: Convert between metric units of volume.

Follow along with *Guided Learning Video 6.3.1, Convert Metric Units: Volume.*

Understanding the Big Picture: As you listen to the first part of the video, when the pencil icon appears, pause and fill in the blanks, choosing from the words listed.

pint • centimeters • larger • liter • quart • smaller

1. The basic metric unit of a _____ is a measure of volume.

2. One liter is slightly larger than a _____

3. One thousand cubic _____ is equal to one liter.

4. When moving from left to right on the metric prefix chart, the units of measurement become _____.

Follow along with the two Guided Learning Video examples, and fill in the blanks on the left as you learn with the instructor. When the pencil icon appears, pause and try the Student Practice on your own.

Guided Learning Video	Pause: Student Practice
1. Change 28 liters to milliliters. Identify the location of the unit of measurement to be changed, and move the decimal point the necessary number of places in the proper direction to make the conversion. 28 liters = [____]	2. Convert 7 kiloliters to deciliters.

Guided Learning Video 🔘 GUIDED LEARNING VIDEO	Pause: Student Practice ✏
3. Convert 4000 liters to kiloliters. Identify the location of the unit of measurement to be changed, and move the decimal point the necessary number of places in the proper direction to make the conversion. 4000 liters = ☐	**4.** Convert 135 milliliters to liters.

Helpful Hint: Pause the video and write the helpful hint in your own words.

For each Active Video Lesson, when the pencil icon appears, pause and work the problem. Then press play to check your work.

Active Video Lesson 1 ✏	Active Video Lesson 2 ✏
5. Convert 382 centiliters to liters.	**6.** Convert 12 liters to deciliters.

Name: _____   Date: _____

Instructor: _____   Section: _____

Guided Learning Video Worksheet: Converting Metric Units: Weight

Text: *Basic College Mathematics*
Student Learning Objective 6.3.2: Convert between metric units of weight.

Follow along with *Guided Learning Video 6.3.2, Converting Metric Units: Weight.*

Understanding the Big Picture: As you listen to the first part of the video, when the pencil icon appears, pause and fill in the blanks, choosing from the words listed.

pound • gram • weight • mass • gravity

1. The amount of material in an object is called its _____.

2. The measure of the pull of _____ on an object is called _____.

3. A _____ is the basic measure of weight in the metric system.

Follow along with the two Guided Learning Video examples, and fill in the blanks on the left as you learn with the instructor. When the pencil icon appears, pause and try the Student Practice on your own.

Guided Learning Video	Pause: Student Practice
1. Convert 6.25 grams to milligrams. Identify the location of the unit of measurement to be changed, and move the decimal point the necessary number of places in the proper direction to make the conversion. 6.25 grams = ☐	2. Convert 3.2 kilograms to grams.

3. Convert 79 milligrams to centigrams.

Identify the location of the unit of measurement to be changed, and move the decimal point the necessary number of places in the proper direction to make the conversion.

4000 liters = [   ]

4. Convert 1750 decigrams to kilograms.

Helpful Hint: Pause the video and write the helpful hint in your own words.

For each Active Video Lesson, when the pencil icon appears, pause and work the problem. Then press play to check your work.

Active Video Lesson 1	Active Video Lesson 2
5. Convert 983 grams to kilograms.	**6.** Convert 14.8 hectograms to decigrams.

Name: _____   Date: _____

Instructor: _____   Section: _____

Guided Learning Video Worksheet: Convert Between U.S. and Metric Systems

Text: *Basic College Mathematics*
Student Learning Objective 6.4.1: Convert units of length, volume, or weight between the metric and American systems.

Follow along with *Guided Learning Video 6.4.1, Convert Between U.S. and Metric Systems*.

Understanding the Big Picture: As you listen to the first part of the video, when the pencil icon appears, pause and fill in the blanks, choosing from the words listed.

divide • one • approximations • multiply • numerator

1. The equivalent measures for the U.S. and metric systems are _____.

2. To convert from one system of measurement to another, _____ by a unit fraction equal to _____.

3. The unit in the _____ is the unit of measurement desired.

Follow along with the two Guided Learning Video examples, and fill in the blanks on the left as you learn with the instructor. When the pencil icon appears, pause and try the Student Practice on your own.

Guided Learning Video 🄿 GUIDED LEARNING VIDEO	Pause: Student Practice
1. Adele went to the store and purchased 4 quarts of oil for her car. How many liters of oil were purchased?	2. Charles purchases 8 liters of bottled beverages for a party. Approximately how many quarts did he purchase?
Identify the unit of measurement desired, and multiply by the appropriate unit fraction to make the conversion and approximate the solution.	

$$\boxed{} \times \boxed{} = \boxed{}$$

Guided Learning Video ▶ GUIDED LEARNING VIDEO	**Pause: Student Practice** ✏
3. Bruce purchased 3 kilograms of hamburger meat. How many pounds did he purchase? Identify the unit of measurement desired, and multiply by the appropriate unit fraction to make the conversion and approximate the solution.	4. A butcher buys a 400 gram package of sausage casing. Approximately how many ounces did he purchase?

$$\boxed{} \times \boxed{} = \boxed{}$$

Helpful Hint: Pause the video and write the helpful hint in your own words.

For each Active Video Lesson, when the pencil icon appears, pause and work the problem. Then press play to check your work.

Active Video Lesson 1 ✏	**Active Video Lesson 2** ✏
5. Catherine drove 86 kilometers from Red Wing to Bloomington to work with her colleague, Jen. How many miles did Catherine drive? Round your answer to the nearest tenth of a mile. Recall that 1 kilometer ≈ 0.62 mile.	6. Pepper purchased a new 10-inch tablet. What is the size of the tablet in centimeters?

Guided Learning Video Worksheet: Convert Between Fahrenheit and Celsius

Text: *Basic College Mathematics*
Student Learning Objective 6.4.2: Convert between Fahrenheit and Celsius degrees of temperature.

Follow along with *Guided Learning Video 6.4.2, Convert Between Fahrenheit and Celsius.*

Understanding the Big Picture: As you listen to the first part of the video, when the pencil icon appears, pause and fill in the blanks, choosing from the words listed.

thousand • Celsius • zero • Fahrenheit • hundred

1. The _____ scale is used to measure temperature in the metric system.

2. In the _____ system, water freezes at thirty-two degrees.

3. In the Celsius scale, water boils at one _____ degrees and freezes at _____ degrees.

Follow along with the two Guided Learning Video examples, and fill in the blanks on the left as you learn with the instructor. When the pencil icon appears, pause and try the Student Practice on your own.

Guided Learning Video	Pause: Student Practice
1. Anne wanted to try baking homemade bread. The recipe that she found on the internet told her to bake it at $218°C$ for 30 minutes. Anne's oven uses a Fahrenheit scale. What temperature should she set her oven to? Round the answer to the nearest whole number. Use the formula for converting Celsius degrees to Fahrenheit degrees to approximate the solution. $F = \boxed{} \times \boxed{} + \boxed{}$ $F = \boxed{} \times \boxed{} + \boxed{}$ $F = \boxed{} + \boxed{}$ $F = \boxed{}$ $218°C = \boxed{}$	2. What is the approximate Fahrenheit equivalent of $150°C$?

3. The average daily high temperature in July for Boston is 82°F. What is the Celsius equivalent for this temperature. Rounded to the nearest tenth of a degree.

Use the formula for converting Fahrenheit degrees to Celsius degrees to approximate the solution.

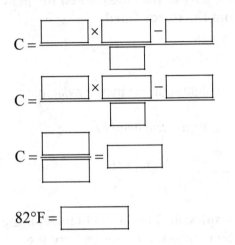

$$C = \frac{\boxed{} \times \boxed{} - \boxed{}}{\boxed{}}$$

$$C = \frac{\boxed{} \times \boxed{} - \boxed{}}{\boxed{}}$$

$$C = \frac{\boxed{}}{\boxed{}} = \boxed{}$$

82°F = \boxed{}

4. A meteorologist states that the average daily high temperature for the month of May is 64°F. What is the approximate Celsius equivalent rounded to the nearest tenth of a degree?

Helpful Hint: Pause the video and write the helpful hint in your own words.

For each Active Video Lesson, when the pencil icon appears, pause and work the problem. Then press play to check your work.

Active Video Lesson 1

5. The daily minimum temperature in Australia in 1997 averaged 22°C. What is the Fahrenheit equivalent of this temperature?

Active Video Lesson 2

6. Allen purchased a new refrigerator for his restaurant from a company in Germany. United States food regulations require that the refrigerator setting be 40°F for all cooling units. His new refrigerator has a Celsius thermostat. What Celsius temperature should Allen set the refrigerator on to meet U.S. guidelines? Round your answer to the nearest tenth of a degree.

Guided Learning Video Worksheet: Solve Applied Problems: U.S. Units

Text: *Basic College Mathematics*
Student Learning Objective 6.5.1: Solve applied problems involving metric and American units.

Follow along with *Guided Learning Video 6.5.1, Solve Applied Problems: U.S. Units*.

Understanding the Big Picture: As you listen to the first part of the video, when the pencil icon appears, pause and fill in the blanks, choosing from the words listed.

multiplied • denominator • relationship • numerator

1. To convert from one unit to another, first write the _____ between the units.

2. The unit desired must be in the _____ of the unit fraction.

3. The unit fraction is _____ by the unit being converted.

Follow along with the two Guided Learning Video examples, and fill in the blanks on the left as you learn with the instructor. When the pencil icon appears, pause and try the Student Practice on your own.

Guided Learning Video ▶ GUIDED LEARNING VIDEO	Pause: Student Practice ✏️
1. Aunt Martha is putting a fence around her triangular shaped vegetable garden to keep the animals out. One side is 26 feet long and a second side is 10 feet long. If there are 20 yards of fencing available, how much is left for the third side? Express your answer in feet. Use a math blueprint to identify the facts in the problem and organize the process for solving it. Then, solve the problem. ☐ + ☐ = ☐ ☐ × ☐/☐ = ☐ ☐ − ☐ = ☐	2. A landscaper plans to install 40 yards of edging around a 3-sided lawn. One side measures 29 feet and a second side measures 46 feet. How many feet of edging remain for the third side?

3. Brianna purchased an 8-ounce bag of chocolate covered almonds. If the price for this candy is $3 per pound, how much did she pay for her 8-ounce bag?

    Use a math blueprint to identify the facts in the problem and organize the process for solving it. Then, solve the problem.

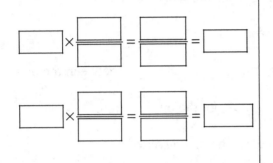

4. A coffee shop sells ground hazelnut coffee for $6 per pound. How much will a 24-ounce bag of ground hazelnut coffee cost?

Helpful Hint: Pause the video and write the helpful hint in your own words.

For each Active Video Lesson, when the pencil icon appears, pause and work the problem. Then press play to check your work.

Active Video Lesson 1 | **Active Video Lesson 2**

5. A rectangular doorway measures 35 inches by 80 inches. Weatherstripping is applied on the top and the two sides. The weatherstripping costs $2 per foot. What did it cost to weather strip the door?

6. A catering company purchases 18 cans of orange juice concentrate for a breakfast event. If each can makes 2 quarts of juice, how many gallons of juice will they provide for breakfast?

Guided Learning Video Worksheet: Solve Applied Problems Involving Metric Units

Text: *Basic College Mathematics*
**Student Learning Objective 6.5.1: Solve applied problems involving metric and
American units.**

Follow along with *Guided Learning Video 6.5.1, Solve Applied Problems Involving Metric
Units.*

**Understanding the Big Picture: As you listen to the first part of the video, when the pencil
icon appears, pause and fill in the blanks, choosing from the words listed.**

less • same • equal • greater • milliliter

1. When using the metric prefix chart to convert from one unit to another, the decimal point
   must be moved an _____ number of places in the _____ direction.

2. A kiloliter is _____ than a hectoliter.

3. The smallest unit for volume in the metric prefix chart is a _____.

**Follow along with the two Guided Learning Video examples, and fill in the blanks on the
left as you learn with the instructor. When the pencil icon appears, pause and try the
Student Practice on your own.**

Guided Learning Video 🔘 GUIDED LEARNING VIDEO	Pause: Student Practice ✏️
1. A seed company has 250 kilograms of flower seed that is to be packaged into 50-gram packets. How many packets can be made from this seed? Use a math blueprint to identify the facts in the problem and organize the process for solving it. Then, solve the problem. 250 kilograms = [＿＿＿] [＿＿＿]/[＿＿＿] = [＿＿＿]	2. Calculate how many 2-liter bottles of vegetable oil can be filled from total supply of 1.34 kiloliters of vegetable oil.

3. A special cleaning fluid is used to rinse test tubes in a chemistry lab. It costs $35 per liter. What is the cost per centiliter?

   Use a math blueprint to identify the facts in the problem and organize the process for solving it. Then, solve the problem.

   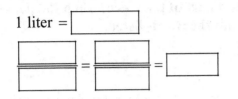

   1 liter = [____]

4. A cardiac medication costs $230 per liter. What is the cost of this medication per milliliter?

Helpful Hint: Pause the video and write the helpful hint in your own words.

For each Active Video Lesson, when the pencil icon appears, pause and work the problem. Then press play to check your work.

Active Video Lesson 1 | **Active Video Lesson 2**

5. A new anticancer drug costs $90 per gram. How much does it cost per milligram?

6. A stretch of road that is 1.245 kilometers long has 150 parking spaces of equal length painted in white on the pavement. How many meters long is each parking space?

Name: _____   Date: _____

Instructor: _____   Section: _____

Guided Learning Video Worksheet: Complementary and Supplementary Angles

Text: *Basic College Mathematics*
Student Learning Objective 7.1.1: Understand and use angles.

Follow along with *Guided Learning Video 7.1.1, Complementary and Supplementary Angles*.

Understanding the Big Picture: As you listen to the first part of the video, when the pencil icon appears, pause and fill in the blanks, choosing from the words listed.

side • angle • supplementary • ray • complementary

1. A(n) _____ starts at a point and extends indefinitely in one direction.

2. A(n) _____ is formed by two rays with the same endpoint called the
    _____.

3. Two angles with a sum of ninety degrees are called _____ angles.

4. Two angles with a sum of one hundred and eighty degrees are called
    _____ angles.

Follow along with the two Guided Learning Video examples, and fill in the blanks on the left as you learn with the instructor. When the pencil icon appears, pause and try the Student Practice on your own.

Guided Learning Video ▶ GUIDED LEARNING VIDEO	Pause: Student Practice
1. The measure of angle A is $142°$. Find the measure of angle S, the supplement of angle A. To calculate the answer, remember that the sum of supplementary angles is $180°$. Angle $S = \boxed{} - \boxed{}$ Angle $S = \boxed{}$	2. The measure of angle A is $99°$. Find the measure of angle S, the supplement of angle A.

3. The measure of angle y is $37°$. Find the complement of angle y. Let angle C represent the complement of angle y.

   To calculate the answer, remember that the sum of complementary angles is $90°$.

   Angle $C =$ ☐ $-$ ☐

   Angle $C =$ ☐

4. The measure of angle A is $17°$. Find the measure of angle C, the complement of angle A.

Helpful Hint: Pause the video and write the helpful hint in your own words.

For each Active Video Lesson, when the pencil icon appears, pause and work the problem. Then press play to check your work.

Active Video Lesson 1 🖉 | **Active Video Lesson 2** 🖉

5. The measure of angle A is $49°$. Find the complement of angle A, and then find the supplement of angle A.

6. The measure of angle B is $63°$. Find the complement of angle B, and then find the supplement of angle B.

Guided Learning Video Worksheet: Perimeter of Shapes

Text: *Basic College Mathematics*
Student Learning Objective 7.2.2: Find the perimeters of shapes made up of rectangles and squares.

Follow along with *Guided Learning Video 7.2.2, Perimeter of Shapes.*

Understanding the Big Picture: As you listen to the first part of the video, when the pencil icon appears, pause and fill in the blanks, choosing from the words listed.

square • perpendicular • perimeter • opposite

1. The distance around an object is called its _____.

2. In a rectangle, _____ sides are equal.

3. A(n) _____ is a rectangle whose sides are all equal.

4. Two sides of a rectangle forming a right angle are _____.

Follow along with the two Guided Learning Video examples, and fill in the blanks on the left as you learn with the instructor. When the pencil icon appears, pause and try the Student Practice on your own.

Guided Learning Video ▶ GUIDED LEARNING VIDEO	**Pause: Student Practice**

1. Find the perimeter of the shape consisting of a rectangle and a square.

2. Find the perimeter of the shape consisting of a rectangle and a square.

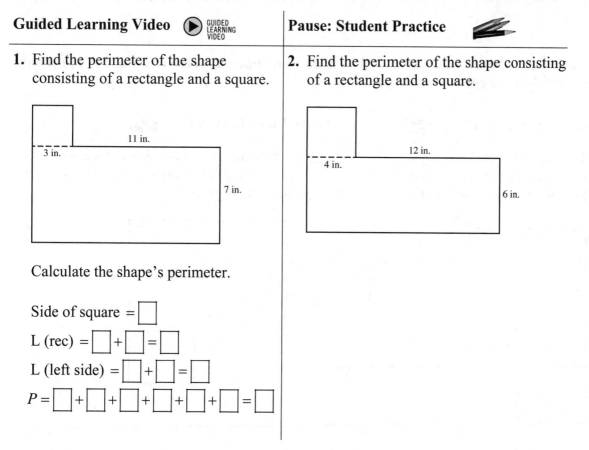

Calculate the shape's perimeter.

Side of square = ☐

L (rec) = ☐ + ☐ = ☐

L (left side) = ☐ + ☐ = ☐

P = ☐ + ☐ + ☐ + ☐ + ☐ + ☐ = ☐

3. Find the perimeter of the shape consisting of a rectangle and a square.

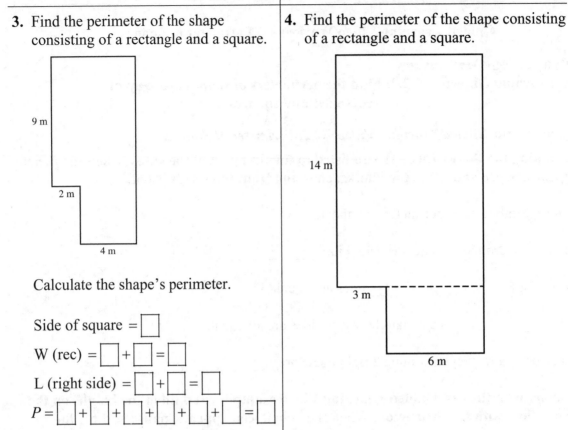

9 m

2 m

4 m

Calculate the shape's perimeter.

Side of square = ☐

W (rec) = ☐ + ☐ = ☐

L (right side) = ☐ + ☐ = ☐

P = ☐ + ☐ + ☐ + ☐ + ☐ + ☐ = ☐

Helpful Hint: Pause the video and write the helpful hint in your own words.

4. Find the perimeter of the shape consisting of a rectangle and a square.

14 m

3 m

6 m

For each Active Video Lesson, when the pencil icon appears, pause and work the problem. Then press play to check your work.

Active Video Lesson 1 | **Active Video Lesson 2**

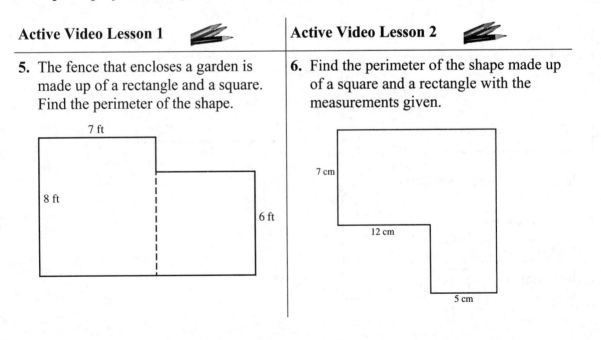

5. The fence that encloses a garden is made up of a rectangle and a square. Find the perimeter of the shape.

7 ft

8 ft

6 ft

6. Find the perimeter of the shape made up of a square and a rectangle with the measurements given.

7 cm

12 cm

5 cm

Name: _____   Date: _____
Instructor: _____   Section: _____

Guided Learning Video Worksheet: Area of a Trapezoid

Text: *Basic College Mathematics*
Student Learning Objective 7.3.2: Find the perimeter and area of a trapezoid,

Follow along with *Guided Learning Video 7.3.2, Area of a Trapezoid.*

Understanding the Big Picture: As you listen to the first part of the video, when the pencil icon appears, pause and fill in the blanks, choosing from the words listed.

equal • parallel • height • perpendicular • trapezoid • bases

1. A(n) _____ is a four-sided figure with two parallel sides called _____ and two adjoining sides that don't have to be _____.

2. In a(n) _____, the lengths of the bases don't have to be _____.

3. The _____ of a trapezoid is the distance between its two _____ sides.

Follow along with the two Guided Learning Video examples, and fill in the blanks on the left as you learn with the instructor. When the pencil icon appears, pause and try the Student Practice on your own.

Guided Learning Video	Pause: Student Practice
1. A trapezoidal-shaped area near a college building is to be planted with grass. The height of the trapezoid is 6 yards. The bases are 8 yards and 4 yards. Find the area of the trapezoid.	2. A garden is in the shape of a trapezoid with bases 40 feet and 24 feet. The height is 15 feet. What is the garden's area?

4 yds
6 yds
8 yds

Use the formula for the area of a trapezoid to calculate the answer.

$A =$

(continued)

$b = 24$ ft
$h = 15$ ft
$B = 40$ ft

$A = \dfrac{}{}$

$A = \dfrac{}{} = \boxed{}$

3. A piece of inlaid woodwork made by a master carpenter is made up of a trapezoid made of cherry and a rectangle made of black walnut. How many square centimeters of wood are needed to make this design?

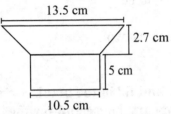

13.5 cm

2.7 cm

5 cm

10.5 cm

Calculate the area of the trapezoid and the rectangle, and then add to calculate the final answer.

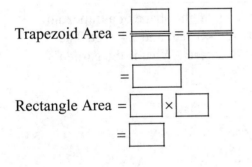

Trapezoid Area $= \dfrac{\boxed{}}{\boxed{}} = \dfrac{\boxed{}}{\boxed{}}$

$= \boxed{}$

Rectangle Area $= \boxed{} \times \boxed{}$

$= \boxed{}$

4. A parcel of land being surveyed is made up of a trapezoid and a rectangle. Determine the area of the parcel of land.

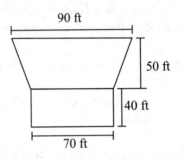

90 ft

50 ft

40 ft

70 ft

Helpful Hint: Pause the video and write the helpful hint in your own words.

Name: _____   Date: _____

Instructor: _____   Section: _____

For each Active Video Lesson, when the pencil icon appears, pause and work the problem. Then press play to check your work.

Active Video Lesson 1	**Active Video Lesson 2**

5. Find the area of a trapezoid with a height of 9 meters and bases of 7 meters and 25 meters.

6. The swimming zone at a lake is in the shape of a trapezoid. The shoreline is 75 feet long. The rope that makes up the other base is 59 feet long, and the height is 50 feet long. What is the area of the swimming zone?

Guided Learning Video Worksheet: Square Roots

Text: *Basic College Mathematics*
Student Learning Objective 7.5.1: Evaluate the square root of a perfect square.

Follow along with *Guided Learning Video 7.5.1, Square Roots.*

Understanding the Big Picture: As you listen to the first part of the video, when the pencil icon appears, pause and fill in the blanks, choosing from the words listed.

negative • positive • identical • taking • square root

1. If a number is the product of two _____ factors, then either factor is its
   _____ .

2. Finding the square root of a number is also referred to as _____ the square root of the number.

3. The square root of a positive number is the _____ number squared to get the original number.

Follow along with the two Guided Learning Video examples, and fill in the blanks on the left as you learn with the instructor. When the pencil icon appears, pause and try the Student Practice on your own.

Guided Learning Video	Pause: Student Practice
1. **a.** Find the square root of 49 . What number times itself equals 49 ? $\boxed{} = \boxed{}$ **b.** Find the square root of 144 . What number times itself equals 144 ? $\boxed{} = \boxed{}$	2. **a.** Find the square root of 100 . **b.** Find the square root of 225 .

3. Evaluate the expression:

   $\sqrt{81} + \sqrt{16}$

   Find the square roots, then add.

   $\sqrt{81} + \sqrt{16}$

   $= \boxed{} + \boxed{} = \boxed{}$

4. Evaluate the expression:

   $\sqrt{64} + \sqrt{196}$

Helpful Hint: Pause the video and write the helpful hint in your own words.

For each Active Video Lesson, when the pencil icon appears, pause and work the problem. Then press play to check your work.

Active Video Lesson 1   **Active Video Lesson 2**

5. Evaluate the expression:

   $\sqrt{169} - \sqrt{25}$

6. Find the square roots:

   a. $\sqrt{36}$

   b. $\sqrt{121}$

Name: _____   Date: _____

Instructor: _____   Section: _____

Guided Learning Video Worksheet: Pythagorean Theorem

Text: *Basic College Mathematics*
Student Learning Objective 7.6.1: Find the hypotenuse of a right triangle given the length of each leg.

Follow along with *Guided Learning Video 7.6.1, Pythagorean Theorem*.

Understanding the Big Picture: As you listen to the first part of the video, when the pencil icon appears, pause and fill in the blanks, choosing from the words listed.

opposite • sum • hypotenuse • ninety • square root

1. The _____ is the longest side of a right triangle.

2. The square of the _____ is the _____ of the squares of each leg of the right triangle.

3. The hypotenuse of the right triangle is always the side _____ the _____ degree angle.

Follow along with the two Guided Learning Video examples, and fill in the blanks on the left as you learn with the instructor. When the pencil icon appears, pause and try the Student Practice on your own.

Guided Learning Video ▶ GUIDED LEARNING VIDEO	**Pause: Student Practice**
1. Find the hypotenuse of the right triangle with legs of 16 centimeters and 12 centimeters.	2. Find the length of the hypotenuse of a right triangle with legs of 15 inches and 20 inches.

16 cm   Hypotenuse   12 cm

15 in.   Hypotenuse   20 in.

Use the Pythagorean Theorem to calculate the length of the hypotenuse.

☐ + ☐

= ☐ + ☐

= ☐

= ☐

3. Find the hypotenuse of the right triangle with legs of 6 inches and 9 inches. Round the answer to the nearest thousandth.

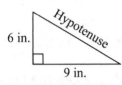

6 in. Hypotenuse 9 in.

Use the Pythagorean Theorem to calculate the length of the hypotenuse.

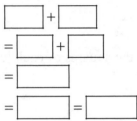

4. Find the hypotenuse of a right triangle with legs of 7 centimeters and 12 centimeters. Round the answer to the nearest hundredth.

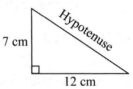

7 cm Hypotenuse 12 cm

Helpful Hint: Pause the video and write the helpful hint in your own words.

For each Active Video Lesson, when the pencil icon appears, pause and work the problem. Then press play to check your work.

Active Video Lesson 1 | **Active Video Lesson 2**

5. Find the hypotenuse. Round to the nearest tenth.

10 ft Hypotenuse 15 ft

6. Find the hypotenuse. Round to the nearest hundredth.

8 m Hypotenuse 10 m

Guided Learning Video Worksheet: Solve Applied Problems Using the Pythagorean Theorem

Text: *Basic College Mathematics*
Student Learning Objective 7.6.3: Solve applied problems using the Pythagorean Theorem.

Follow along with *Guided Learning Video 7.6.3, Solve Applied Problems Using the Pythagorean Theorem.*

Understanding the Big Picture: As you listen to the first part of the video, when the pencil icon appears, pause and fill in the blanks, choosing from the words listed.

right • picture • perpendicular • diagram • sides

1. When solving applied problems involving geometric figures, organize the information in a _____ or draw a _____.

2. The Pythagorean Theorem describes the relationship between the _____ of a _____ triangle.

3. The height and base of a right triangle are _____ to each other.

Follow along with the two Guided Learning Video examples, and fill in the blanks on the left as you learn with the instructor. When the pencil icon appears, pause and try the Student Practice on your own.

Guided Learning Video ▶ GUIDED LEARNING VIDEO	**Pause: Student Practice**
1. The town of Clarion is 7 miles east of the town of Franklin. Salem Lake is 15 miles north of Franklin. What is the straight line distance from Clarion to Salem Lake? Round the distance to the nearest tenth of a mile. Draw a picture that relates to the problem. Then, use the Pythagorean Theorem to solve the problem. $c^2 = \boxed{} + \boxed{}$ $c^2 = \boxed{} + \boxed{}$ $c^2 = \boxed{} + \boxed{}$ $c = \boxed{} = \boxed{} \approx \boxed{}$	2. A pilot flies 28 miles west from Truneville and then 7 miles south to Carelton. What is the straight line distance from Truneville to Carelton? Round the distance to the nearest tenth of a mile.

3. A 12-foot ladder is placed against a wall. The top of the ladder is 10 feet from the floor. What is the distance from the base of the ladder to the wall? Round to the nearest tenth of a foot if necessary.

   Draw a picture that relates to the problem. Then, use the Pythagorean Theorem to solve the problem.

   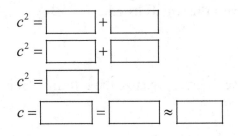

4. A 20-foot ladder is placed against the side of a house at a point 18 feet above level ground. What is the distance from the base of the house to the base of the ladder? Round the distance to the nearest tenth of a foot.

Helpful Hint: Pause the video and write the helpful hint in your own words.

For each Active Video Lesson, when the pencil icon appears, pause and work the problem. Then press play to check your work.

Active Video Lesson 1   **Active Video Lesson 2**

5. A kite is out on 30 yards of string. The kite is directly above a rock. The rock is 27 yards from the boy flying the kite. How far above the rock is the kite? Round to the nearest tenth of a yard.

6. Evie is tiling a kitchen floor. She needs a small piece in the shape of a right triangle to fill in a space underneath a cabinet. This piece of tile will have one leg equal to 7 cm and the other leg equal to 3 cm. What is the length of the longest side of the tile? Round your answer to the nearest tenth of a centimeter.

Name: _____   Date: _____
Instructor: _____   Section: _____

Guided Learning Video Worksheet: Circumference of a Circle

Text: *Basic College Mathematics*
Student Learning Objective 7.7.1: Find the area and circumference of a circle.

Follow along with *Guided Learning Video 7.7.1, Circumference of a Circle.*

Understanding the Big Picture: As you listen to the first part of the video, when the pencil icon appears, pause and fill in the blanks, choosing from the words listed.

radius • center • circumference • equal • diameter

1. A circle is a two-dimensional flat figure for which all points are a(n) _____ distance from a given point called the _____.

2. The _____ is a line segment from one point on the circle to another point on the circle through the circle's _____.

3. The _____ of a circle is a line segment from the circle's center to a point on the circle.

4. The _____ of a circle is the distance around the circle.

Follow along with the two Guided Learning Video examples, and fill in the blanks on the left as you learn with the instructor. When the pencil icon appears, pause and try the Student Practice on your own.

Guided Learning Video	Pause: Student Practice
1. Find the circumference of a circle with a diameter of 8 inches. Use $\pi \approx 3.14$. Use the formula for circumference that involves the diameter to solve the problem. $C = \boxed{} \times \boxed{}$ $C \approx \boxed{} \times \boxed{}$ $C \approx \boxed{}$	2. Find the circumference of a circle with a diameter of 12 centimeters. Use $\pi \approx 3.14$.

3. A wheelbarrow tires has a radius of 5.5 inches. How many feet does the wheelbarrow travel if the wheel makes 5 revolutions? Use $\pi \approx 3.14$, and round the final answer to the nearest tenth of a foot.

Use the formula for circumference that involves the radius to solve the problem.

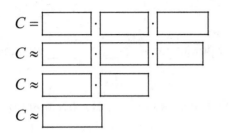

4. A bicycle tire's radius is 16 inches. How many feet does the bicycle travel if the wheel makes 3 revolutions? Use $\pi \approx 3.14$, and round the final answer to the nearest tenth of a foot.

Helpful Hint: Pause the video and write the helpful hint in your own words.

For each Active Video Lesson, when the pencil icon appears, pause and work the problem. Then press play to check your work.

Active Video Lesson 1

5. The tires on a car have a diameter of 27 inches. Each tire makes 750 revolutions. How far, in inches, did the car travel?

Active Video Lesson 2

6. Compute the circumference of a circle with a radius of 2.6 inches. Round your answer to the nearest tenth of an inch.

Guided Learning Video Worksheet: Area of a Circle

Text: *Basic College Mathematics*
Student Learning Objective 7.7.1: Find the area and circumference of a circle.

Follow along with *Guided Learning Video 7.7.1, Area of a Circle.*

Understanding the Big Picture: As you listen to the first part of the video, when the pencil icon appears, pause and fill in the blanks, choosing from the words listed.

area • radius • before • circumference • dividing • diameter

1. The number pi is obtained by _____ a circle's _____ by its
   _____.

2. The _____ of a circle is the product of pi times its _____ squared.

3. When using the formula for the area of a circle, square the radius _____ multiplying by pi.

Follow along with the two Guided Learning Video examples, and fill in the blanks on the left as you learn with the instructor. When the pencil icon appears, pause and try the Student Practice on your own.

Guided Learning Video	Pause: Student Practice
1. Find the area of a circle whose radius is 2 meters. Use 3.14 for π. Use the formula for the area of a circle to solve the problem. $A = \boxed{}\boxed{}^{\boxed{}}$ $A = (\boxed{})(\boxed{})^{\boxed{}}$ $A = (\boxed{})(\boxed{})$ $A = \boxed{}$	2. Find the area of a circle whose radius is 6 centimeters. Use 3.14 for π.

3. Daniella wants to buy a circular braided rug that is 10 feet in diameter. Find the cost of the rug at $30 per square yard. Round the area in square yards to the nearest hundredth before calculating the cost.

Use the formula for the area of a circle to solve the problem.

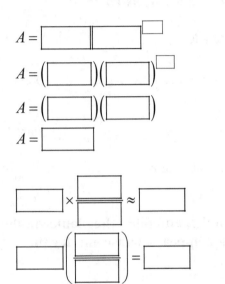

4. A circular shaped pool cover costs $12 per square yard. If the pool is 20 feet in diameter, what is the cost of the pool cover? Round the area in square yards to the nearest tenth before calculating the cost.

Helpful Hint: Pause the video and write the helpful hint in your own words.

For each Active Video Lesson, when the pencil icon appears, pause and work the problem. Then press play to check your work.

Active Video Lesson 1 | **Active Video Lesson 2**

5. The city of Danvers needs to pave a circular turnaround at the end of a dead-end street. The diameter of the area to be paved is 30 feet. Find the cost of paving the area at $20 per square yard.

6. Find the area of a circle whose radius is 8 inches. Use 3.14 for π.

Name: _____   Date: _____
Instructor: _____   Section: _____

Guided Learning Video Worksheet: Volume of a Cylinder

Text: *Basic College Mathematics*
Student Learning Objective 7.8.2: Find the volume of a cylinder.

Follow along with *Guided Learning Video 7.8.2, Volume of a Cylinder.*

Understanding the Big Picture: As you listen to the first part of the video, when the pencil icon appears, pause and fill in the blanks, choosing from the words listed.

area • height • cubic • circumference • cylinder • base

1. A(n) _____ is a form similar to the shape of a can or tube.

2. The volume of a cylinder is measured in _____ units.

3. The volume of a cylinder is the _____ of its circular _____ multiplied times its _____.

Follow along with the two Guided Learning Video examples, and fill in the blanks on the left as you learn with the instructor. When the pencil icon appears, pause and try the Student Practice on your own.

Guided Learning Video ▶ GUIDED LEARNING VIDEO	**Pause: Student Practice** ✏️
1. Find the volume of a cylinder with a radius of 2 inches and a height of 5 inches. Use 3.14 for π. Use the formula for the volume of a cylinder to solve the problem. $V = \boxed{}\left(\boxed{}\right)^{\boxed{}}\left(\boxed{}\right)$ $V = \left(\boxed{}\right)\left(\boxed{}\right)\left(\boxed{}\right)$ $V = \boxed{}$	2. Find the volume of a cylinder with a radius of 4 inches and a height of 16 inches. Use 3.14 for π.

3. Find the volume of a tin can with a radius of 9 centimeters and a height of 21 centimeters. Use 3.14 for π.

Use the formula for the volume of a cylinder to solve the problem.

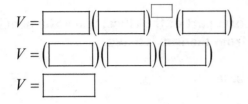

4. Find the volume of a tube with a radius of 1.5 inches and a height of 21 inches. Use 3.14 for π.

Helpful Hint: Pause the video and write the helpful hint in your own words.

For each Active Video Lesson, when the pencil icon appears, pause and work the problem. Then press play to check your work.

Active Video Lesson 1 | **Active Video Lesson 2**

5. Find the volume of a round above-ground swimming pool with a radius of 15 feet and a height of 4 feet. Use 3.14 for π.

6. Find the volume of a silo with a radius of 11 meters and a height of 30 meters. Use 3.14 for π.

Guided Learning Video Worksheet: Volume of a Pyramid

Text: *Basic College Mathematics*
Student Learning Objective 7.8.5: Find the volume of a pyramid.

Follow along with *Guided Learning Video 7.8.5, Volume of a Pyramid.*

Understanding the Big Picture: As you listen to the first part of the video, when the pencil icon appears, pause and fill in the blanks, choosing from the words listed.

square • height • cubic • three • area • base

1. A pyramid's volume is calculated by multiplying the area of its _____ times its _____ and then dividing the product by _____.

2. A pyramid's volume is measured in _____ units.

3. In the formula for the volume of a pyramid, the capital letter B represents the _____ of the _____.

Follow along with the two Guided Learning Video examples, and fill in the blanks on the left as you learn with the instructor. When the pencil icon appears, pause and try the Student Practice on your own.

Guided Learning Video ▶ GUIDED LEARNING VIDEO	Pause: Student Practice
1. Find the volume of a pyramid with height equaling 9 meters, the length of its base equaling 5 meters, and the width of its base equaling 4 meters. Calculate the area of the base, then use the formula for the volume of a pyramid to solve the problem. $B = \boxed{} \times \boxed{}$ $B = (\boxed{})(\boxed{})$ $B = \boxed{}$ $V = \dfrac{(\boxed{})(\boxed{})}{\boxed{}} = \dfrac{\boxed{}}{\boxed{}} = \boxed{}$	2. Find the volume of a pyramid with height equaling 12 meters, the length of its base equaling 8 meters, and the width of its base equaling 3 meters.

3. Find the volume of a pyramid when its height equals 8 inches, the length of its base equals 3 inches, and the width of its base equals 2 inches.

   Calculate the area of the base, then use the formula for the volume of a pyramid to solve the problem.

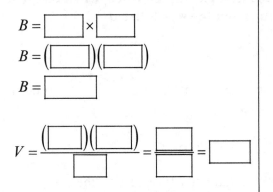

4. Find the volume of a pyramid when its height equals 13 inches, the length of its base equals 15 inches, and the width of its base equals 8 inches.

Helpful Hint: Pause the video and write the helpful hint in your own words.

For each Active Video Lesson, when the pencil icon appears, pause and work the problem. Then press play to check your work.

Active Video Lesson 1	**Active Video Lesson 2**
5. Find the volume of a pyramid when its height equals 10 inches, the length of its base equals 11 inches, and the width of its base equals 6 inches.	6. Find the volume of a pyramid when its height equals 15 feet, the length of its base equals 7 feet, and the width of its base equals 4 feet.

Name: _____   Date: _____

Instructor: _____   Section: _____

Guided Learning Video Worksheet: Double Bar Graph

Text: *Basic College Mathematics*
Student Learning Objective 8.2.2: Read and interpret a double-bar graph.

Follow along with *Guided Learning Video 8.2.2, Double Bar Graph.*

Understanding the Big Picture: As you listen to the first part of the video, when the pencil icon appears, pause and fill in the blanks, choosing from the words listed.

time • data • changes • double • comparisons

1. Making _____ is one of the uses of _____ bar graphs.

2. Bar graphs are helpful for seeing _____ over a period of _____.

3. The use of bar graphs is helpful when the same type of _____ is repeatedly studied.

Follow along with the two Guided Learning Video examples, and fill in the blanks on the left as you learn with the instructor. When the pencil icon appears, pause and try the Student Practice on your own.

Guided Learning Video ▶ GUIDED LEARNING VIDEO	**Pause: Student Practice** 🖊️
1. How many cars were sold in the third quarter of 2010 ?	2. How many cars were sold in the second quarter of 2009 ?

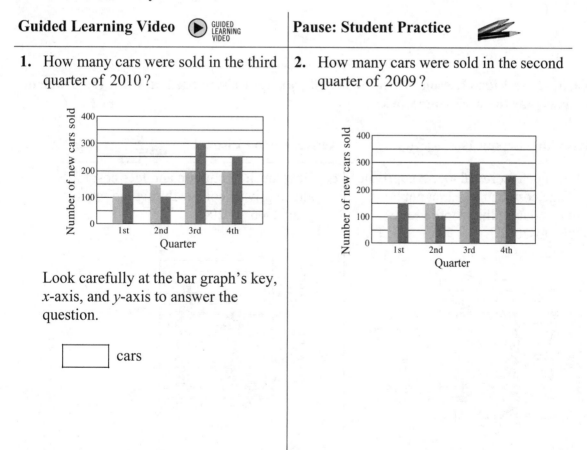

Look carefully at the bar graph's key, *x*-axis, and *y*-axis to answer the question.

☐ cars

3. How many more cars were sold in the fourth quarter of 2010 than in the second quarter of 2010?

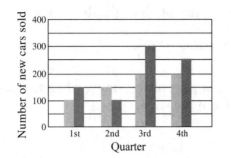

Look carefully at the bar graph's key, *x*-axis, and *y*-axis to answer the question.

☐ − ☐ = ☐ cars

4. How many more cars were sold in the third quarter of 2010 Than in the first quarter of 2010?

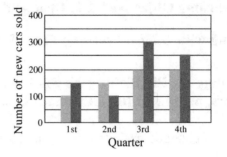

Helpful Hint: Pause the video and write the helpful hint in your own words.

For each Active Video Lesson, when the pencil icon appears, pause and work the problem. Then press play to check your work.

Active Video Lesson 1 | **Active Video Lesson 2**

5. How much more did a year at private college cost than a year at public college during the school year of 2011 to 2012?

6. How much did tuition and fees cost to attend private college during the school year 2006 to 2007?

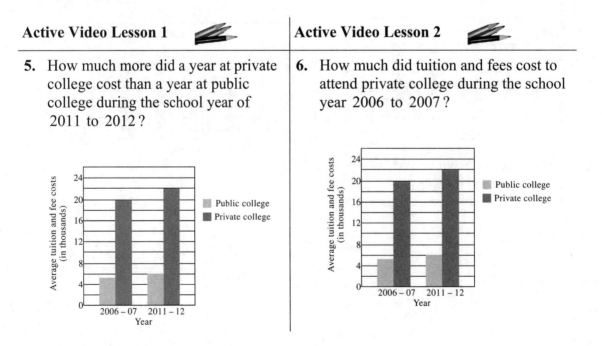

Guided Learning Video Worksheet: Comparison Line Graphs

Text: *Basic College Mathematics*
Student Learning Objective 8.2.4: Read and interpret a comparison line graph.

Follow along with *Guided Learning Video 8.2.4, Comparison Line Graphs.*

Understanding the Big Picture: As you listen to the first part of the video, when the pencil icon appears, pause and fill in the blanks, choosing from the words listed.

read • compared • colored • yellow • sets

1. Two or more _____ of data can be _____ in a line graph.

2. In the video's introduction, the _____ line represents shoppers over fifty-five years of age.

3. Different _____ lines in a comparison line graph help to distinguish between groups of data and make the graph easier to _____.

Follow along with the two Guided Learning Video examples, and fill in the blanks on the left as you learn with the instructor. When the pencil icon appears, pause and try the Student Practice on your own.

Guided Learning Video (▶) GUIDED LEARNING VIDEO	Pause: Student Practice ✏️
1. **a.** Use the graph to determine which month the Bay Shore Restaurant had the fewest customers.	2. **a.** Use the graph to determine which month the Lilly Cafe had the fewest customers.
b. Approximately how many customers came into the Lilly Café during the month of March?	**b.** Approximately how many customers came into the Bay Shore Restaurant during the month of May?

Look carefully at the lines of the graph, the *x*-axis, and the y-axis to answer the questions.

(continued)

a. [   ]

b. [   ] customers

3. **a.** During which months did the Lilly Café have more customers than the Bay Shore Restaurant?

   b. In June, how many more customers did the Bay Shore Restaurant have than the Lilly Cafe?

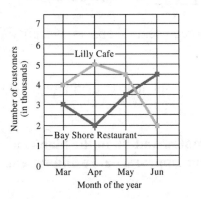

Look carefully at the lines of the graph, the *x*-axis, and the *y*-axis to answer the questions.

a. [   ]

b. [   ] − [   ] = [   ] customers

4. **a.** During which months was the difference between the number of customers at each restaurant equal to 1000?

   b. In April, how many more customers did the Lilly Cafe have than the Bay Shore Restaurant?

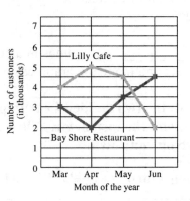

Helpful Hint: Pause the video and write the helpful hint in your own words.

Name: _____   Date: _____

Instructor: _____   Section: _____

For each Active Video Lesson, when the pencil icon appears, pause and work the problem. Then press play to check your work.

Active Video Lesson 1	**Active Video Lesson 2**

5. a. During which years were there more condominiums built than homes?

b. How many more homes were built in 1995 compared to the number of condominiums?

6. a. In which year were the greatest number of homes built?

b. Approximately how many condominiums were built in the year 2000?

New Homes and Condominiums Constructed in Essex County

New Homes and Condominiums Constructed in Essex County

Name: _____   Date: _____

Instructor: _____   Section: _____

Guided Learning Video Worksheet: Histograms

Text: *Basic College Mathematics*
Student Learning Objective 8.3.1: Understand and interpret a histogram.

Follow along with *Guided Learning Video 8.3.1, Histograms.*

Understanding the Big Picture: As you listen to the first part of the video, when the pencil icon appears, pause and fill in the blanks, choosing from the words listed.

equal • frequency • histogram • class • touch • width • interval

1. A _____ is a special type of bar graph where the bars _____.

2. In a histogram, the _____ of each bar is _____.

3. Each bar in a histogram represents a _____, and each bar's width represents an _____.

4. The height of the bar represents the _____ for each _____.

Follow along with the two Guided Learning Video examples, and fill in the blanks on the left as you learn with the instructor. When the pencil icon appears, pause and try the Student Practice on your own.

Guided Learning Video 🔘 GUIDED LEARNING VIDEO	Pause: Student Practice ✏️
1. How many students scored an "A" on the test if the professor considers a test score of 90 to 99 an "A"?	2. How many students scored an "B" on the test if a test score of 80 to 89 is considered a "B"?

Results of Test

Results of Test

Look carefully at the bars of the histogram, the *x*-axis, and the *y*-axis to answer the question.

☐ students

Guided Learning Video (▶) GUIDED LEARNING VIDEO	**Pause: Student Practice**

3. How many students scored 70 or greater on the test?

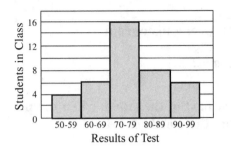

Look carefully at the bars of the histogram, the *x*-axis, and the *y*-axis to answer the question.

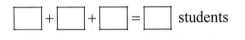

☐ + ☐ + ☐ = ☐ students

4. How many students scored less than 80 on the test?

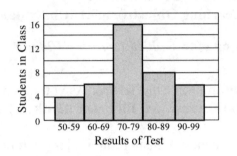

Helpful Hint: Pause the video and write the helpful hint in your own words.

For each Active Video Lesson, when the pencil icon appears, pause and work the problem. Then press play to check your work.

Active Video Lesson 1	**Active Video Lesson 2**

5. Use the histogram to answer the following question. How many light bulbs lasted for fewer than 800 hours?

6. Using the same histogram, how many light bulbs lasted between 1000 and 1199 hours?

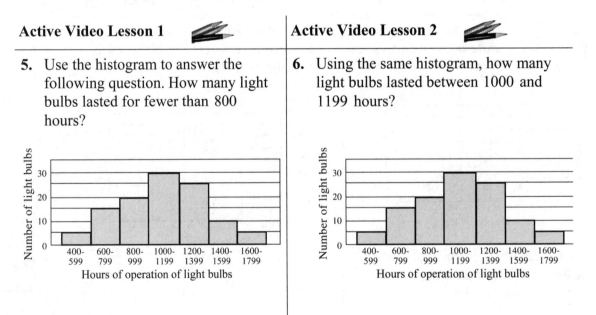

Guided Learning Video Worksheet: Mean

Text: *Basic College Mathematics*
Student Learning Objective 8.4.1: Find the mean of a set of numbers.

Follow along with *Guided Learning Video 8.4.1, Mean.*

Understanding the Big Picture: As you listen to the first part of the video, when the pencil icon appears, pause and fill in the blanks, choosing from the words listed.

average • sum • median • number • mean • dividing

1. A mean is calculated by _____ the _____ of the values by the _____ of values.

2. A middle value of a group of numbers can be described by the group's _____ or _____.

3. A mean is often called an _____.

Follow along with the two Guided Learning Video examples, and fill in the blanks on the left as you learn with the instructor. When the pencil icon appears, pause and try the Student Practice on your own.

Guided Learning Video	Pause: Student Practice
1. Denzel scored 92, 87, 89, 97, and 95 on the tests in his math class. Find the average or mean of Denzel's test scores. Perform the necessary operations for calculating the mean. $\square + \square + \square + \square + \square = \square$ $\dfrac{\square}{\square} = \square$	2. A student's quiz scores in a statistics class are 93, 79, 86, 84, 88, and 92. What is the student's mean quiz score?

3. The miles per gallon Wally's car achieved each week over an eight week period were 24, 23, 25, 25, 23, 26, 26, and 24. What was the mean miles per gallon for the eight weeks Wally kept a record? Round the answer to the nearest mile per gallon.

Perform the necessary operations for calculating the mean.

$$\Box + \Box + \Box + \Box + \Box + \Box + \Box + \Box = \Box$$

$$\frac{\Box}{\Box} = \Box \approx \Box$$

4. Each day, for a five day period, a local bakery recorded the number of customers purchasing coffee. The numbers they recorded were 186, 210, 263, 198, and 233. What was the mean number of customers purchasing coffee over this five day period?

Helpful Hint: Pause the video and write the helpful hint in your own words.

For each Active Video Lesson, when the pencil icon appears, pause and work the problem. Then press play to check your work.

Active Video Lesson 1 ✏ | **Active Video Lesson 2** ✏

5. The daily high temperatures in Denver during one week were 78°, 85°, 81°, 88°, 80°, 75°, and 79°. Find the mean of these temperatures. Round to the nearest whole degree.

6. A group of friends joined an intramural basketball league. The ages of the members of the team were 35, 27, 23, 34, and 26 years old. What was the average age of the members of this basketball team?

Guided Learning Video Worksheet: Median

Text: *Basic College Mathematics*
Student Learning Objective 8.4.2: Find the median of a set of numbers.

Follow along with *Guided Learning Video 8.4.2, Median.*

Understanding the Big Picture: As you listen to the first part of the video, when the pencil icon appears, pause and fill in the blanks, choosing from the words listed.

order • even • middle • same • median • average

1. A(n) _____ is a middle value in a set of numbers with the _____ number of values above and below it when the values in the set are arranged from smallest to largest.

2. The first step in determining a median is to _____ the numbers.

3. If a list contains a(n) _____ number of values, the median is found by calculating the _____ of the two _____ numbers.

Follow along with the two Guided Learning Video examples, and fill in the blanks on the left as you learn with the instructor. When the pencil icon appears, pause and try the Student Practice on your own.

Guided Learning Video ▶ GUIDED LEARNING VIDEO	**Pause: Student Practice** ✏️
1. The daily high temperatures in Chicago during one week in December were $53°$, $31°$, $48°$, $32°$, $47°$, $34°$ and $35°$. Find the median of these temperatures. Arrange the temperatures and identify the middle temperature to determine the median. □, □, □, □, □, □, □ median = □	2. A set of test scores in a calculus course are 83, 67, 79, 77, 91, 98, 84, 86, 88, 73, 70, 62, and 88. Find the median test score.

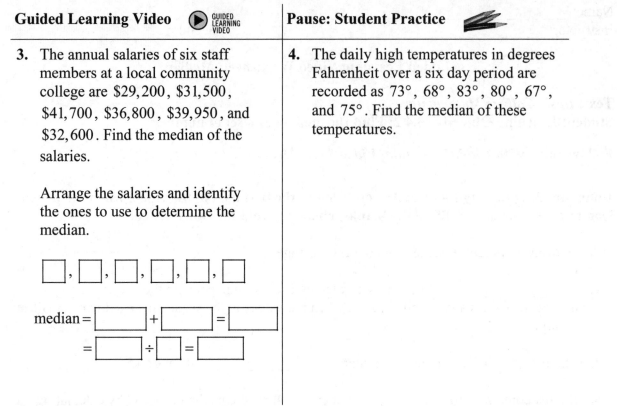
3. The annual salaries of six staff members at a local community college are $29,200, $31,500, $41,700, $36,800, $39,950, and $32,600. Find the median of the salaries.

   Arrange the salaries and identify the ones to use to determine the median.

   ☐ , ☐ , ☐ , ☐ , ☐ , ☐

   median = ☐ + ☐ = ☐

   = ☐ ÷ ☐ = ☐

4. The daily high temperatures in degrees Fahrenheit over a six day period are recorded as 73°, 68°, 83°, 80°, 67°, and 75°. Find the median of these temperatures.

Helpful Hint: Pause the video and write the helpful hint in your own words.

For each Active Video Lesson, when the pencil icon appears, pause and work the problem. Then press play to check your work.

Active Video Lesson 1 | **Active Video Lesson 2**

5. The following is a set of prices for various types of headphones:

   $6.99, $32.85, $19.99, $129.99, $7.99, $8.98, $50.51 and $11.99,

   Find the median price for this set of headphones.

6. As of 2014, the countries with the highest life expectancy are

   Macau 84.5 years
   San Marino 83.2 years
   Monaco 89.6 years
   Andorra 82.7 years
   Japan 84.5 years
   Singapore 84.4 years
   Hong Kong 82.8 years

   Find the median for this set of data.

Guided Learning Video Worksheet: Use Inequality Symbols with Integers

Text: *Basic College Mathematics*
Student Learning Objective 9.1.1: Add two signed numbers with the same sign.

Follow along with *Guided Learning Video 9.1.1, Use Inequality Symbols with Integers*.

Understanding the Big Picture: As you listen to the first part of the video, when the pencil icon appears, pause and fill in the blanks, choosing from the words listed.

positive • integers • right • left • negative • minus

1. Numbers less than zero are called _____ numbers and are written with a _____ sign in front of the number.

2. On a number line, negative numbers are positioned to the _____ of zero, and positive numbers are positioned to the _____ of zero.

3. Positive and negative whole numbers and zero are called _____.

Follow along with the two Guided Learning Video examples, and fill in the blanks on the left as you learn with the instructor. When the pencil icon appears, pause and try the Student Practice on your own.

Guided Learning Video 🔘 GUIDED LEARNING VIDEO	Pause: Student Practice ✏️
1. Graph the numbers −4, −1, 0, and 4 on a number line. Use the number line to plot each number in its proper location. ←++++++++++++++→	2. Graph the numbers −3, −2, 1, and 5 on a number line. ←++++++++++++++→

3. Replace the question mark with the inequality symbol for "greater than" or "less than."

-3 ? 2
5 ? -1
-4 ? -2
-3 ? -5

Use a number line to help identify which inequality symbol to use.

←─┼─┼─┼─┼─┼─┼─┼─┼─┼─┼─┼─┼─→

4. Replace the question mark with the inequality symbol for "greater than" or "less than."

-4 ? -5
0 ? -2
1 ? 3
-2 ? -3

←─┼─┼─┼─┼─┼─┼─┼─┼─┼─┼─┼─┼─→

Helpful Hint: Pause the video and write the helpful hint in your own words.

For each Active Video Lesson, when the pencil icon appears, pause and work the problem. Then press play to check your work.

Active Video Lesson 1 | **Active Video Lesson 2**

5. Replace the question mark with the inequality symbol of "less than" or "greater than."

-1 ? -3
-6 ? -5

←─┼─┼─┼─┼─┼─┼─┼─┼─┼─┼─┼─┼─→

6. Graph -6, -2, 1, and 3 on a number line.

←─┼─┼─┼─┼─┼─┼─┼─┼─┼─┼─┼─┼─→

Guided Learning Video Worksheet: Absolute Value

Text: *Basic College Mathematics*
Student Learning Objective 9.1.1: Add two signed numbers with the same sign.

Follow along with *Guided Learning Video 9.1.1, Absolute Value.*

Understanding the Big Picture: As you listen to the first part of the video, when the pencil icon appears, pause and fill in the blanks, choosing from the words listed.

zero • absolute • vertical • positive • negative • direction • distance

1. Distance is always a _____ number on a number line regardless of the _____ moved.

2. The _____ value of a number is the _____ between the number and _____ on a number line.

3. The symbol for absolute value is two _____ lines on either side of the value.

4. The absolute value of any nonzero number is always _____.

Follow along with the two Guided Learning Video examples, and fill in the blanks on the left as you learn with the instructor. When the pencil icon appears, pause and try the Student Practice on your own.

Guided Learning Video	Pause: Student Practice
1. Simplify. $\left\|-21\right\|$	2. Simplify. $\left\|-14\right\|$
Use the concept of absolute value to determine the answer. $\left\|-21\right\| = \boxed{}$	

Guided Learning Video	Pause: Student Practice
3. Simplify. $-\|-16\|$ Use the concept of absolute value to determine the answer. $-\|-16\| = \square$	**4.** Simplify. $-\|-33\|$ Use the concept of absolute value to determine the answer. $-\|-33\| = \square$

Helpful Hint: Pause the video and write the helpful hint in your own words.

For each Active Video Lesson, when the pencil icon appears, pause and work the problem. Then press play to check your work.

Active Video Lesson 1	Active Video Lesson 2
5. Find the absolute value. $\|-53\|$	**6.** Find the absolute value. $\|0\|$

Guided Learning Video Worksheet: Adding with Same Sign

Text: *Basic College Mathematics*
Student Learning Objective 9.1.1: Add two signed numbers with the same sign.

Follow along with *Guided Learning Video 9.1.1, Adding with Same Sign.*

Understanding the Big Picture: As you listen to the first part of the video, when the pencil icon appears, pause and fill in the blanks, choosing from the words listed.

negative • adding • positive • common • absolute values • same

1. To add two numbers with the same sign, first add the _____ of the numbers.

2. When _____ two numbers with the _____ sign, use the _____ sign in the answer.

3. If all the numbers being added are negative, the answer will be _____.

Follow along with the two Guided Learning Video examples, and fill in the blanks on the left as you learn with the instructor. When the pencil icon appears, pause and try the Student Practice on your own.

Guided Learning Video ▶ GUIDED LEARNING VIDEO	**Pause: Student Practice**
1. Add. $-76+(-32)$ Apply the addition rule for adding two numbers with the same sign. $\boxed{}+\boxed{}$ $=\boxed{}+\boxed{}=\boxed{}$	**2.** Add. $-37+(-64)$

3. Find the total value of surplus or deficit for the two years, 2010 and 2014.

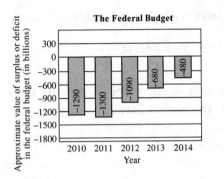

4. Find the total value of the deficit for the two years, 2013 and 2014.

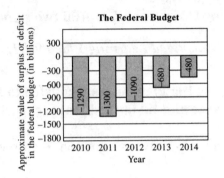

Use the graph to identify the numbers to be added, then apply the addition rule for adding numbers with the same sign.

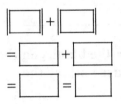

Helpful Hint: Pause the video and write the helpful hint in your own words.

For each Active Video Lesson, when the pencil icon appears, pause and work the problem. Then press play to check your work.

Active Video Lesson 1 | **Active Video Lesson 2**

5. Find the total value of surplus or deficit for the two years, 2011 and 2013.

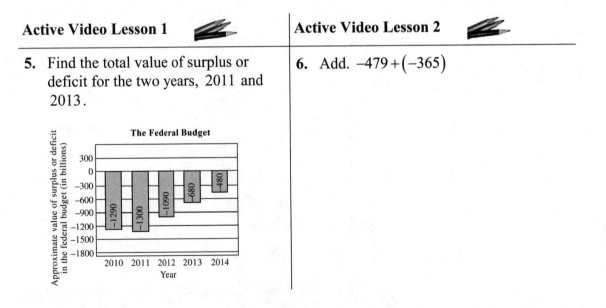

6. Add. $-479 + (-365)$

Guided Learning Video Worksheet: Adding Integers with Different Signs

Text: *Basic College Mathematics*
Student Learning Objective 9.1.2: Add two signed numbers with different signs.

Follow along with *Guided Learning Video 9.1.2, Adding Integers with Different Signs*.

Understanding the Big Picture: As you listen to the first part of the video, when the pencil icon appears, pause and fill in the blanks, choosing from the words listed.

difference • positive • smaller • negative • larger • different

1. A temperature below zero can be represented by a _____ number.

2. When adding two numbers with _____ signs, use the sign of the number with the _____ absolute value in the answer.

3. To add two numbers with different signs, find the _____ between the _____ absolute value and the _____ absolute value.

Follow along with the two Guided Learning Video examples, and fill in the blanks on the left as you learn with the instructor. When the pencil icon appears, pause and try the Student Practice on your own.

Guided Learning Video	Pause: Student Practice
1. Suppose we spend $52 and we earn $38. Calculate the sum.	2. Suppose we spend $73 and we earn $25. Calculate the sum.

Apply the addition rule for adding two numbers with different signs.

$$\boxed{} + \boxed{} = \boxed{}$$

$$\boxed{} + \boxed{} = \boxed{}$$

3. If you were to gain $7 and lose $15, how much money do you have?

Apply the addition rule for adding two numbers with different signs.

$$\boxed{} + \boxed{} = \boxed{}$$
$$\boxed{} + \boxed{} = \boxed{}$$

4. If there is a gain of $150 and a loss of $184, calculate the final balance.

Helpful Hint: Pause the video and write the helpful hint in your own words.

For each Active Video Lesson, when the pencil icon appears, pause and work the problem. Then press play to check your work.

Active Video Lesson 1 ✎ | **Active Video Lesson 2** ✎

5. Add. $8 + (-14)$

6. Add. $-21 + 16$

Name: _____   Date: _____

Instructor: _____   Section: _____

Guided Learning Video Worksheet: Adding Two Signed Numbers With Different Signs

Text: *Basic College Mathematics*
Student Learning Objective 9.1.2: Add two signed numbers with different signs.

Follow along with *Guided Learning Video 9.1.2, Adding Two Signed Numbers With Different Signs.*

Understanding the Big Picture: As you listen to the first part of the video, when the pencil icon appears, pause and fill in the blanks, choosing from the words listed.

difference • positive • smaller • negative • larger • different

1. A temperature below zero can be represented by a _____ number.

2. When adding two numbers with _____ signs, use the sign of the number with the _____ absolute value in the answer.

3. To add two numbers with different signs, find the _____ between the _____ absolute value and the _____ absolute value.

Follow along with the two Guided Learning Video examples, and fill in the blanks on the left as you learn with the instructor. When the pencil icon appears, pause and try the Student Practice on your own.

Guided Learning Video	**Pause: Student Practice** 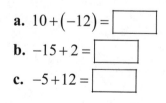

1. Add each of the following.
  a. $10+(-12)$
  b. $-15+2$
  c. $-5+12$
 	
Apply the addition rule for adding two numbers with different signs.
 	
a. $10+(-12)=\boxed{}$
 	
b. $-15+2=\boxed{}$
 	
c. $-5+12=\boxed{}$
	2. Add each of the following.
	  a. $6+(-13)$
	  b. $-23+4$
	  c. $-8+17$

Guided Learning Video ▶ GUIDED LEARNING VIDEO	Pause: Student Practice
3. Add. $-10.7+12.8$	**4.** Add. $-13.9+11.5$

Apply the addition rule for adding two numbers with different signs.

$$\boxed{} + \boxed{} = \boxed{}$$

Helpful Hint: Pause the video and write the helpful hint in your own words.

For each Active Video Lesson, when the pencil icon appears, pause and work the problem. Then press play to check your work.

Active Video Lesson 1	Active Video Lesson 2
5. Add. $6+(-13)$	**6.** Add. $-20.8+15.2$

Name: _____   Date: _____

Instructor: _____   Section: _____

Guided Learning Video Worksheet: Subtracting Integers

Text: *Basic College Mathematics*
Student Learning Objective 9.2.1: Subtract one signed number from another.

Follow along with *Guided Learning Video 9.2.1, Subtracting Integers.*

Understanding the Big Picture: As you listen to the first part of the video, when the pencil icon appears, pause and fill in the blanks, choosing from the words listed.

zero • opposite • same • negative • add • equal

1. The opposite of a positive number is a _____ number with the _____ absolute value.

2. If a number is the opposite of another number, the numbers are an _____ distance from _____ on the number line.

3. To subtract signed numbers, _____ the _____ of the second number to the first number.

Follow along with the two Guided Learning Video examples, and fill in the blanks on the left as you learn with the instructor. When the pencil icon appears, pause and try the Student Practice on your own.

Guided Learning Video	Pause: Student Practice
1. Subtract. $\$24 - \$32$ Apply the rule for subtracting signed numbers. $\$24 - \$32$ $= \boxed{}\ \boxed{}\ \boxed{} = \boxed{}$	2. Subtract. $\$65 - \$80$

Guided Learning Video (▶) GUIDED LEARNING VIDEO	**Pause: Student Practice** ✎
3. Subtract. $-21-5$	4. Subtract. $-37-14$

Apply the rule for subtracting signed numbers.

$-21-5$

$= \boxed{}\boxed{}\boxed{} = \boxed{}$

Helpful Hint: Pause the video and write the helpful hint in your own words.

For each Active Video Lesson, when the pencil icon appears, pause and work the problem. Then press play to check your work.

Active Video Lesson 1 ✎	**Active Video Lesson 2** ✎
5. Subtract. $-19-(-4)$	6. Subtract. $28-(-15)$

Name: _____ Date: _____

Instructor: _____ Section: _____

Guided Learning Video Worksheet: Subtracting Signed Numbers

Text: *Basic College Mathematics*
Student Learning Objective 9.2.1: Subtract one signed number from another.

Follow along with *Guided Learning Video 9.2.1, Subtracting Signed Numbers.*

Understanding the Big Picture: As you listen to the first part of the video, when the pencil icon appears, pause and fill in the blanks, choosing from the words listed.

zero • opposite • same • negative • add • equal

1. The opposite of a positive number is a _____ number with the _____ absolute value.

2. If a number is the opposite of another number, the numbers are an _____ distance from _____ on the number line.

3. To subtract signed numbers, _____ the _____ of the second number to the first number.

Follow along with the two Guided Learning Video examples, and fill in the blanks on the left as you learn with the instructor. When the pencil icon appears, pause and try the Student Practice on your own.

Guided Learning Video ▶ GUIDED LEARNING VIDEO	Pause: Student Practice
1. **a.** The opposite of 7 is ☐.	2. **a.** The opposite of 20 is ☐.
b. The opposite of −9 is ☐.	**b.** The opposite of −17 is ☐.

Guided Learning Video (▶) GUIDED LEARNING VIDEO	**Pause: Student Practice**
3. Subtract. $-13-(-5)$	4. Subtract. $-23-(-12)$

Apply the rule for subtracting signed numbers.

$-13-(-5)$

$=$ [   ] [ ][ ] [   ] $=$ [   ]

Helpful Hint: Pause the video and write the helpful hint in your own words.

For each Active Video Lesson, when the pencil icon appears, pause and work the problem. Then press play to check your work.

Active Video Lesson 1	**Active Video Lesson 2**
5. Subtract. $-9.4-(-5.1)$	6. Subtract. $57-92$

Guided Learning Video Worksheet: Perform Several Integer Operations

Text: *Basic College Mathematics*
Student Learning Objective 9.2.2: Solve problems involving both addition and subtraction of signed numbers.

Follow along with *Guided Learning Video 9.2.2, Perform Several Integer Operations.*

Understanding the Big Picture: As you listen to the first part of the video, when the pencil icon appears, pause and fill in the blanks, choosing from the words listed.

associative • opposite • order • addition • commutative • rewrite

1. When performing several integer operations in the same expression, first _____ all subtraction as _____ of the _____.

2. Subtraction of integers is not _____ or _____.

3. The addition of integers can be completed in any _____ because it is both _____ and _____.

Follow along with the two Guided Learning Video examples, and fill in the blanks on the left as you learn with the instructor. When the pencil icon appears, pause and try the Student Practice on your own.

Guided Learning Video	**Pause: Student Practice**
1. Perform the necessary operations to evaluate $3 - 8 - 6 - 1$. Rewrite all subtraction as addition of the opposite, and then add the integers. $3 - 8 - 6 - 1$ $= \boxed{} + \boxed{} + \boxed{} + \boxed{}$ $= \boxed{} + \boxed{} = \boxed{}$	2. Perform the necessary operations to evaluate $11 - 6 - 13 - 15$.

3. Perform the necessary operations to evaluate $-5-(-7)+(-14)$.

Rewrite all subtraction as addition of the opposite, and then add the integers.

$-5-(-7)+(-14)$

$= \boxed{} + \boxed{} + \boxed{}$

$= \boxed{} + \boxed{} = \boxed{}$

4. Perform the necessary operations to evaluate $-12-(-8)+(-22)$.

Helpful Hint: Pause the video and write the helpful hint in your own words.

For each Active Video Lesson, when the pencil icon appears, pause and work the problem. Then press play to check your work.

Active Video Lesson 1 | **Active Video Lesson 2**

5. Perform the necessary operations to evaluate $-8-(-3)+(-6)$.

6. Perform the necessary operations to evaluate $7-9-4-2$.

Guided Learning Video Worksheet: Multiply Integers

Text: *Basic College Mathematics*
Student Learning Objective 9.3.1: Multiply and divide two signed numbers.

Follow along with *Guided Learning Video 9.3.1, Multiply Integers.*

Understanding the Big Picture: As you listen to the first part of the video, when the pencil icon appears, pause and fill in the blanks, choosing from the words listed.

dot • odd • negative • parentheses • even

1. A positive number times a negative number equals a(n) _____ number.

2. Multiplication can be indicated by placing a raised _____ between numbers or by placing _____ on either side of each number.

3. When multiplying nonzero numbers, the product is positive if there are a(n) _____ number of negative signs and negative if there are a(n) _____ number of negative signs.

Follow along with the two Guided Learning Video examples, and fill in the blanks on the left as you learn with the instructor. When the pencil icon appears, pause and try the Student Practice on your own.

Guided Learning Video	Pause: Student Practice
1. Multiply. $9(-6)$ Determine the sign of the product, then multiply. $9(-6) = \boxed{}$	2. Multiply. $5(-4)$

Guided Learning Video ▶ GUIDED LEARNING VIDEO	**Pause: Student Practice**
3. Multiply. $(-7)(-6)$ Determine the sign of the product, then multiply. $(-7)(-6) = \boxed{}$	**4.** Multiply. $(-3)(-11)$

Helpful Hint: Pause the video and write the helpful hint in your own words.

For each Active Video Lesson, when the pencil icon appears, pause and work the problem. Then press play to check your work.

Active Video Lesson 1	**Active Video Lesson 2**
5. Multiply. $-8 \cdot 7$	**6.** Multiply. $-4 \times (-9)$

Guided Learning Video Worksheet: Divide Integers

Text: *Basic College Mathematics*
Student Learning Objective 9.3.1: Multiply and divide two signed numbers.

Follow along with *Guided Learning Video 9.3.1, Divide Integers.*

Understanding the Big Picture: As you listen to the first part of the video, when the pencil icon appears, pause and fill in the blanks, choosing from the words listed.

same • multiplication • odd • negative • division • even

1. Division statements can be rewritten as _____ statements.

2. The signed number rules for multiplication and _____ are the _____.

3. When dividing nonzero numbers, the answer will be positive if there are a(n) _____ number of negative signs and negative if there are a(n) _____ number of negative signs.

Follow along with the two Guided Learning Video examples, and fill in the blanks on the left as you learn with the instructor. When the pencil icon appears, pause and try the Student Practice on your own.

Guided Learning Video ▶ GUIDED LEARNING VIDEO	**Pause: Student Practice** ✎
1. Divide. $56 \div (-7)$ Follow the rules for dividing integers to calculate the answer. $56 \div (-7) = \boxed{}$	**2.** Divide. $-32 \div 8$

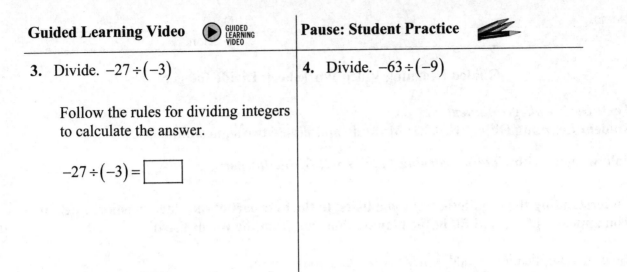
3. Divide. $-27 \div (-3)$

Follow the rules for dividing integers to calculate the answer.

$-27 \div (-3) = \boxed{}$

4. Divide. $-63 \div (-9)$

Helpful Hint: Pause the video and write the helpful hint in your own words.

For each Active Video Lesson, when the pencil icon appears, pause and work the problem. Then press play to check your work.

Active Video Lesson 1

5. Divide. $-48 \div 6$

Active Video Lesson 2

6. Divide. $-54 \div (-9)$

Name: _____   Date: _____

Instructor: _____   Section: _____

Guided Learning Video Worksheet: Multiply Two or More Integers

Text: *Basic College Mathematics*
Student Learning Objective 9.3.2: Multiply three or more signed integers.

Follow along with *Guided Learning Video 9.3.2, Multiply Two or More Integers.*

Understanding the Big Picture: As you listen to the first part of the video, when the pencil icon appears, pause and fill in the blanks, choosing from the words listed.

associative • product • order • negative • total • commutative

1. The multiplication process can be simplified by counting the _____ number of negative signs before multiplying to determine the sign of the _____.

2. In a multiplication expression with three negative signs, the product will be _____.

3. Multiplication is both _____ and _____, so integers can be multiplied in any _____.

Follow along with the two Guided Learning Video examples, and fill in the blanks on the left as you learn with the instructor. When the pencil icon appears, pause and try the Student Practice on your own.

Guided Learning Video	Pause: Student Practice
1. Multiply. $(-2)(-3)(-7)$ Determine the sign of the product, and then multiply. $(-2)(-3)(-7) = \boxed{}$	2. Multiply. $(-5)(-4)(-6)$

Guided Learning Video ▶ GUIDED LEARNING VIDEO	**Pause: Student Practice** ✏
3. Multiply. $(-2)(-1)(7)(-4)$	**4.** Divide. $(-3)(-4)(-2)(-1)$

Determine the sign of the product, and then multiply.

$$(-2)(-1)(7)(-4)$$

$$= \boxed{} = \boxed{}$$

Helpful Hint: Pause the video and write the helpful hint in your own words.

For each Active Video Lesson, when the pencil icon appears, pause and work the problem. Then press play to check your work.

Active Video Lesson 1 ✏	**Active Video Lesson 2** ✏
5. Multiply. $(-3)(-2)(2)(-1)(-3)$	**6.** Multiply. $(4)(-2)(5)$

Guided Learning Video Worksheet: Order of Operations with Signed Numbers

Text: *Basic College Mathematics*
Student Learning Objective 9.4.1: Calculate with signed numbers using more than one operation.

Follow along with *Guided Learning Video 9.4.1, Order of Operations with Signed Numbers.*

Understanding the Big Picture: As you listen to the first part of the video, when the pencil icon appears, pause and fill in the blanks, choosing from the words listed.

exponents • right • parentheses • left • priorities

1. The order of operations is a list of _____ for working with the numbers in computation problems.

2. When there is more than one operation in a problem, first perform any operations inside
_____.

3. Evaluating _____ is the second priority in the order of operations.

4. The operations of multiplication and division are performed in the order they appear from
_____ to _____.

Follow along with the two Guided Learning Video examples, and fill in the blanks on the left as you learn with the instructor. When the pencil icon appears, pause and try the Student Practice on your own.

Guided Learning Video ⏵ GUIDED LEARNING VIDEO	Pause: Student Practice
1. Simplify. $-6(-3)-4(3-7)^2$	**2.** Simplify. $-5(-2)-6(5-8)^2$

Use the order of operations to simplify the expression.

$-6(-3)-4(3-7)^2$

= [ ]

= [ ]

= [ ]

= [ ] = [ ]

3. Simplify. $\left(\dfrac{1}{3}\right)^3 + 2\left(\dfrac{5}{3} - \dfrac{1}{6}\right) \div \left(-\dfrac{9}{4}\right)$

Use the order of operations to simplify the expression.

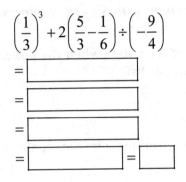

$$\left(\dfrac{1}{3}\right)^3 + 2\left(\dfrac{5}{3} - \dfrac{1}{6}\right) \div \left(-\dfrac{9}{4}\right)$$

$= \boxed{}$

$= \boxed{}$

$= \boxed{}$

$= \boxed{} = \boxed{}$

4. Simplify. $\left(\dfrac{1}{2}\right)^2 - 4\left(\dfrac{1}{4} - \dfrac{3}{2}\right) \div \left(-\dfrac{3}{2}\right)$

Helpful Hint: Pause the video and write the helpful hint in your own words.

For each Active Video Lesson, when the pencil icon appears, pause and work the problem. Then press play to check your work.

Active Video Lesson 1 | **Active Video Lesson 2**

5. Multiply. $3(4-6)^3 + 12 \div (-4) + 2$

6. Multiply. $\left(\dfrac{1}{5}\right)^2 + 4\left(\dfrac{1}{5} - \dfrac{3}{10}\right) \div \left(-\dfrac{2}{3}\right)$

Name: _____   Date: _____
Instructor: _____   Section: _____

Guided Learning Video Worksheet: Order of Operations: Signed Decimals

Text: *Basic College Mathematics*
Student Learning Objective 9.4.1: Calculate with signed numbers using more than one operation.

Follow along with *Guided Learning Video 9.4.1, Order of Operations: Signed Decimals.*

Understanding the Big Picture: As you listen to the first part of the video, when the pencil icon appears, pause and fill in the blanks, choosing from the words listed.

powers • right • parentheses • equal • priorities • left

1. The order of operations is a list of _____ for working with the numbers in computation problems.

2. Performing operations with _____ is the second step in the order of operations.

3. Multiplication and division have _____ priority in the order of operations.

4. The operations of multiplication and division are performed in the order they appear from _____ to _____.

Follow along with the two Guided Learning Video examples, and fill in the blanks on the left as you learn with the instructor. When the pencil icon appears, pause and try the Student Practice on your own.

Guided Learning Video ▶ GUIDED LEARNING VIDEO	**Pause: Student Practice**
1. Simplify. $(0.8)^2 - 2.3(-4.5)$	2. Simplify. $(0.4)^2 - 5.1(-3.2)$

Use the order of operations to simplify the expression.

$(0.8)^2 - 2.3(-4.5)$

= _____

= _____

= _____

= _____ = _____

3. Simplify. $$1.7 + 0.54 \div 0.6 - (-0.2)^3 - (1.1 + 0.3)$$ Use the order of operations to simplify the expression. $$1.7 + 0.54 \div 0.6 - (-0.2)^3 - (1.1 + 0.3)$$ $= \boxed{}$ $= \boxed{}$ $= \boxed{}$ $= \boxed{} = \boxed{}$	**4.** Simplify. $$3.9 + 0.81 \div 0.9 - (5.3 + 0.6) - (0.3)^3$$

Helpful Hint: Pause the video and write the helpful hint in your own words.

For each Active Video Lesson, when the pencil icon appears, pause and work the problem. Then press play to check your work.

Active Video Lesson 1 ✏️	**Active Video Lesson 2** ✏️
5. Simplify. $$-1.3 + 0.8 \times 0.07 - (0.4)^2 + (3.5 - 0.9)$$	**6.** Simplify. $0.3 + 2.5 - (0.7 - 0.9)^3$

Name: _____   Date: _____
Instructor: _____   Section: _____

Guided Learning Video Worksheet: Change Scientific Notation to Standard Notation

Text: *Basic College Mathematics*
Student Learning Objective 9.5.2: Change numbers in scientific notation to standard notation.

Follow along with *Guided Learning Video 9.5.2, Change Scientific Notation to Standard Notation.*

Understanding the Big Picture: As you listen to the first part of the video, when the pencil icon appears, pause and fill in the blanks, choosing from the words listed.

ten • one • positive • negative • scientific notation

1. To perform a calculation involving very large or very small numbers, _____ is often used to express the numbers.

2. If a number written in scientific notation has a(n) _____ power of ten, the number is greater than or equal to _____.

3. If a number written in scientific notation has a(n) _____ power of ten, the number is less than _____.

Follow along with the two Guided Learning Video examples, and fill in the blanks on the left as you learn with the instructor. When the pencil icon appears, pause and try the Student Practice on your own.

Guided Learning Video	Pause: Student Practice
1. Write in standard notation. 8.3×10^3 Move the decimal point to write the number in standard notation. $8.3 \times 10^3 = \boxed{}$	2. Write in standard notation. 5.7×10^4

Guided Learning Video ▶ GUIDED LEARNING VIDEO	Pause: Student Practice 🖎
3. Write in standard notation. 6.1×10^{-3} Move the decimal point to write the number in standard notation. $6.1 \times 10^{-3} = $ [　　　　]	**4.** Write in standard notation. 2.98×10^{-4}

Helpful Hint: Pause the video and write the helpful hint in your own words.

For each Active Video Lesson, when the pencil icon appears, pause and work the problem. Then press play to check your work.

Active Video Lesson 1 🖎	Active Video Lesson 2 🖎
5. Write in standard notation. 9.36×10^{-5}	**6.** Write in standard notation. 7.451×10^{4}

Guided Learning Video Worksheet: Combining Like Terms

Text: *Basic College Mathematics*
Student Learning Objective 10.1.2: Combine like terms containing a variable.

Follow along with *Guided Learning Video 10.1.2, Combining Like Terms.*

Understanding the Big Picture: As you listen to the first part of the video, when the pencil icon appears, pause and fill in the blanks, choosing from the words listed.

subtraction • one • coefficients • term • like • addition

1. A _____ is a number, a variable, or a product of a number and one or more variables.

2. The terms of an expressions are separated by _____ and _____ symbols.

3. Terms with identical variables with identical exponents are called _____ terms.

4. To combine _____ terms, combine the numerical _____ using the rules for adding signed numbers.

5. The numerical coefficient of a variable term is understood to be equal to _____ when it is not written.

Follow along with the two Guided Learning Video examples, and fill in the blanks on the left as you learn with the instructor. When the pencil icon appears, pause and try the Student Practice on your own.

Guided Learning Video	Pause: Student Practice
1. Simplify by combining like terms. $-2a + 4b - 5a$ Use the rules for combining like terms to simplify the expression. $-2a + 4b - 5a$ $= \boxed{} + \boxed{} + \left(\boxed{}\right)$ $= \boxed{} + \left(\boxed{}\right) + \boxed{}$ $= \boxed{} + \boxed{}$	2. Simplify by combining like terms. $-5x + 8y - 6x$

3. Simplify by combining like terms.
$9x + y + (-6y) + 7xy$

Use the rules for combining like terms to simplify the expression.

$9x + y + (-6y) + 7xy$

$= \boxed{} + \left(\boxed{}\right) + \boxed{}$

4. Simplify by combining like terms.
$5a + b + (-3b) + 2ab$

Helpful Hint: Pause the video and write the helpful hint in your own words.

For each Active Video Lesson, when the pencil icon appears, pause and work the problem. Then press play to check your work.

5. Simplify by combining like terms.
$-8x + 2xy - 4x$

6. Simplify by combining like terms.
$4x + 9y - 7x - 3y$

Name: _____   Date: _____

Instructor: _____   Section: _____

Guided Learning Video Worksheet: The Distributive Property

Text: *Basic College Mathematics*
Student Learning Objective 10.2.1: Remove parentheses using the distributive property.

Follow along with *Guided Learning Video 10.2.1, The Distributive Property.*

Understanding the Big Picture: As you listen to the first part of the video, when the pencil icon appears, pause and fill in the blanks, choosing from the words listed.

multiplied • parentheses • subtraction • distributive • addition

1. The _____ property is used to remove _____.

2. The distributive property can be used over _____ and _____.

3. Distributing a number to all terms inside a set of parentheses means that all terms inside the parentheses are _____ by the number.

Follow along with the two Guided Learning Video examples, and fill in the blanks on the left as you learn with the instructor. When the pencil icon appears, pause and try the Student Practice on your own.

Guided Learning Video	**Pause: Student Practice**
1. Simplify. $4(x+3)$	2. Simplify. $-5(x-6y)$

Use the distributive property to simplify the expression.

$4(x+3)$

$= \boxed{}(\boxed{}) + \boxed{}(\boxed{})$

$= \boxed{} + \boxed{}$

Guided Learning Video	Pause: Student Practice

3. Simplify. $-3(5x-2y-8z)$

Use the distributive property to simplify the expression.

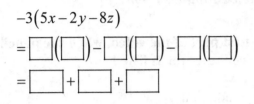

$-3(5x-2y-8z)$

$= \boxed{}(\boxed{}) - \boxed{}(\boxed{}) - \boxed{}(\boxed{})$

$= \boxed{} + \boxed{} + \boxed{}$

4. Simplify. $-4(6x-3y+2z)$

Helpful Hint: Pause the video and write the helpful hint in your own words.

For each Active Video Lesson, when the pencil icon appears, pause and work the problem. Then press play to check your work.

Active Video Lesson 1	Active Video Lesson 2

5. Simplify. $5(9x-y)$

6. Simplify. $-7(8x+6)$

Guided Learning Video Worksheet: The Distributive Property with Combining like Terms

Text: *Basic College Mathematics*
Student Learning Objective 10.2.2: Simplify expressions by removing parentheses and combining like terms.

Follow along with *Guided Learning Video 10.2.2, The Distributive Property with Combining Like Terms.*

Understanding the Big Picture: As you listen to the first part of the video, when the pencil icon appears, pause and fill in the blanks, choosing from the words listed.

opposite • one • multiplied • distributive • equal

1. A negative sign in front of a set of parentheses is _____ to a numerical coefficient of negative _____.

2. If a set of parentheses is preceded by a negative sign, all terms inside the parentheses are changed to their signed _____.

3. When a set of parentheses is preceded by a positive sign, all terms inside the parentheses are _____ by positive _____.

Follow along with the two Guided Learning Video examples, and fill in the blanks on the left as you learn with the instructor. When the pencil icon appears, pause and try the Student Practice on your own.

Guided Learning Video	Pause: Student Practice
1. Simplify. $4(2x - y) + 5(x + 3y)$ Use the distributive property and combine like terms. $4(2x - y) + 5(x + 3y)$ $= \rule{3cm}{0.4pt}$ $= \rule{2cm}{0.4pt}$	2. Simplify. $3(x + 4y) - 2(5x + y)$

Guided Learning Video ▶ GUIDED LEARNING VIDEO	**Pause: Student Practice**
3. Simplify. $9(7s+4t)-8(6s-3)$	**4.** Simplify. $-6(3x-y+2)+4(4x+3y)$

Use the distributive property and combine like terms.

$9(7s+4t)-8(6s-3)$

$=$ ⬚

$=$ ⬚

Helpful Hint: Pause the video and write the helpful hint in your own words.

For each Active Video Lesson, when the pencil icon appears, pause and work the problem. Then press play to check your work.

Active Video Lesson 1	**Active Video Lesson 2**
5. Simplify. $6(5x-7y+3z)-2(9x-4z)$	**6.** Simplify. $2(-3a+2b)-5(a-2b)$

Guided Learning Video Worksheet: Addition Principle of Equality

Text: *Basic College Mathematics*
Student Learning Objective 10.3.1: Solve equations using the addition property.

Follow along with *Guided Learning Video 10.3.1, Addition Principle of Equality.*

Understanding the Big Picture: As you listen to the first part of the video, when the pencil icon appears, pause and fill in the blanks, choosing from the words listed.

solution • addition • inverse • same • equal • expressions

1. An equation is a statement indicating two _____ are _____.

2. A number's signed opposite is its additive _____.

3. The numerical _____ of an equation is the number(s) that makes the equation a true statement.

4. The _____ principle of equality states that if the _____ number is added to both sides of an equation, the results on both sides are _____ in value.

Follow along with the two Guided Learning Video examples, and fill in the blanks on the left as you learn with the instructor. When the pencil icon appears, pause and try the Student Practice on your own.

Guided Learning Video	Pause: Student Practice
1. Solve. $x - 9 = 4$ Use the addition principle of equality and the additive inverse to solve the equation. $x - 9 = 4$ $\boxed{} = \boxed{}$ $\boxed{} = \boxed{}$ $\boxed{} = \boxed{}$	2. Solve. $x - 12 = 3$

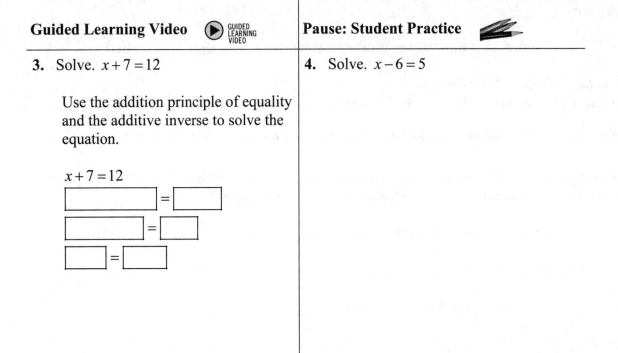
3. Solve. $x + 7 = 12$

Use the addition principle of equality and the additive inverse to solve the equation.

$x + 7 = 12$

$$\boxed{} = \boxed{}$$

$$\boxed{} = \boxed{}$$

$$\boxed{} = \boxed{}$$

4. Solve. $x - 6 = 5$

Helpful Hint: Pause the video and write the helpful hint in your own words.

For each Active Video Lesson, when the pencil icon appears, pause and work the problem. Then press play to check your work.

Active Video Lesson 1	**Active Video Lesson 2**
5. Solve. $15 = x + 8$	**6.** Solve. $6 = x - 11$

Name: _____ Date: _____
Instructor: _____ Section: _____

Guided Learning Video Worksheet: The Division Principle of Equality

Text: *Basic College Mathematics*
Student Learning Objective 10.4.1: Solve equations using the division property.

Follow along with *Guided Learning Video 10.4.1, The Division Principle of Equality.*

Understanding the Big Picture: As you listen to the first part of the video, when the pencil icon appears, pause and fill in the blanks, choosing from the words listed.

nonzero • divide • equal • coefficient • undefined • division

1. The _____ principle of equality states if both sides of an equation are divided by the same nonzero number, the results on both sides are _____ in value.

2. Division by zero is _____; therefore, when applying the division principle, the divisor must be a(n) _____ number.

3. To solve an equation of the form $ax = b$, _____ both sides by the _____ of x.

Follow along with the two Guided Learning Video examples, and fill in the blanks on the left as you learn with the instructor. When the pencil icon appears, pause and try the Student Practice on your own.

Guided Learning Video ⏵ GUIDED LEARNING VIDEO	Pause: Student Practice
1. Solve. $7x = 63$ Use the division principle of equality to solve the equation. $7x = 63$ $\dfrac{\Box}{\Box} = \dfrac{\Box}{\Box}$ $\Box = \Box$	**2.** Solve. $5x = 65$

Guided Learning Video GUIDED LEARNING VIDEO

3. Solve. $-8x = 56$

Use the division principle of equality to solve the equation.

$-8x = 56$

$\frac{\square}{\square} = \frac{\square}{\square}$

$\square = \square$

Pause: Student Practice

4. Solve. $-4x = 24$

Helpful Hint: Pause the video and write the helpful hint in your own words.

For each Active Video Lesson, when the pencil icon appears, pause and work the problem. Then press play to check your work.

Active Video Lesson 1

5. Solve. $-6x = -72$

Active Video Lesson 2

6. Solve. $-3x = 51$

Name: _____   Date: _____

Instructor: _____   Section: _____

Guided Learning Video Worksheet: Multiplication Principle of Equality

Text: *Basic College Mathematics*
Student Learning Objective 10.4.2: Solve equations using the multiplication property.

Follow along with *Guided Learning Video 10.4.2, Multiplication Principle of Equality.*

Understanding the Big Picture: As you listen to the first part of the video, when the pencil icon appears, pause and fill in the blanks, choosing from the words listed.

numerical • reciprocal • equal • coefficient • multiplication • division

1. The _____ principle of equality states if both sides of an equation are multiplied by the same nonzero number, the results on both sides are _____ in value.

2. When solving and equation of the form $ax = b$, if the coefficient of the variable is a fraction, multiply both sides by the _____ of the _____.

3. In an equation of the form $ax = b$, where a and b are real numbers, a is called the _____ coefficient of the variable.

Follow along with the two Guided Learning Video examples, and fill in the blanks on the left as you learn with the instructor. When the pencil icon appears, pause and try the Student Practice on your own.

Guided Learning Video ▶ GUIDED LEARNING VIDEO	Pause: Student Practice
1. Solve. $\frac{1}{9}x = -6$ Use the multiplication principle of equality to solve the equation. $\frac{1}{9}x = -6$ $\square \dfrac{\boxed{}}{\boxed{}} = \square\left(\boxed{}\right)$ $\boxed{} = \boxed{}$ $\boxed{} = \boxed{}$	2. Solve. $\frac{1}{5}x = -8$

3. Solve. $\dfrac{x}{-8} = 7$

    Use the multiplication principle of equality to solve the equation.

$$\dfrac{x}{-8} = 7$$

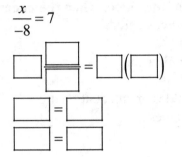

4. Solve. $\dfrac{x}{-6} = 3$

Helpful Hint: Pause the video and write the helpful hint in your own words.

For each Active Video Lesson, when the pencil icon appears, pause and work the problem. Then press play to check your work.

Active Video Lesson 1

5. Solve. $-\dfrac{1}{7}x = 11$

Active Video Lesson 2

6. Solve. $\dfrac{x}{12} = 8$

Guided Learning Video Worksheet: Solving Equations $ax + b = c$

Text: *Basic College Mathematics*
Student Learning Objective 10.5.1: Use two properties to solve an equation.

Follow along with *Guided Learning Video 10.5.1, Solving Equations* $ax + b = c$.

Understanding the Big Picture: As you listen to the first part of the video, when the pencil icon appears, pause and fill in the blanks, choosing from the words listed.

value • reciprocal • addition • true • variable • division

1. When solving equations, sometimes both the _____ principle of equality and the _____ principle of equality will be used to get the _____ alone on one side of the equation.

2. To solve equations, determine the _____ of the variable that will make the equation _____.

3. Solving an equation is oftentimes made easier if the _____ principle of equality is used before the _____ principle of equality.

Follow along with the two Guided Learning Video examples, and fill in the blanks on the left as you learn with the instructor. When the pencil icon appears, pause and try the Student Practice on your own.

Guided Learning Video ▶ GUIDED LEARNING VIDEO	Pause: Student Practice
1. Solve. $3x - 7 = 8$	2. Solve. $8x - 3 = 37$

Use the addition and division principles of equality to solve the equation.

$3x - 7 = 8$

$\boxed{} = \boxed{}$

$\boxed{} = \boxed{}$

$\dfrac{\boxed{}}{\boxed{}} = \dfrac{\boxed{}}{\boxed{}}$

$\boxed{} = \boxed{}$

3. Solve. $-27 = 6x + 9$

Use the addition and division principles of equality to solve the equation.

$-27 = 6x + 9$

$$\boxed{} = \boxed{}$$

$$\boxed{} = \boxed{}$$

$$\frac{\boxed{}}{\boxed{}} = \frac{\boxed{}}{\boxed{}}$$

$$\boxed{} = \boxed{}$$

4. Solve. $-54 = 5x + 6$

Helpful Hint: Pause the video and write the helpful hint in your own words.

For each Active Video Lesson, when the pencil icon appears, pause and work the problem. Then press play to check your work.

Active Video Lesson 1 | **Active Video Lesson 2**

5. Solve. $4y + 5 = 29$

6. Solve. $-8x - 12 = 20$

Guided Learning Video Worksheet: Solving Equations of the Form $ax + b = cx + d$

Text: *Basic College Mathematics*
Student Learning Objective 10.5.2: Solve equations where the variable is on both sides of the equals sign.

Follow along with *Guided Learning Video 10.5.2, Solving Equations of the Form* $ax + b = cx + d$.

Understanding the Big Picture: As you listen to the first part of the video, when the pencil icon appears, pause and fill in the blanks, choosing from the words listed.

opposite • variable • simplified • addition • terms • numerical

1. The process of solving an equation is easier if each side of the equation is _____ first.

2. To simplify each side of an equation, combine any like _____ and simplify _____ work.

3. The _____ principle of equality should be used to get all _____ terms on one side of the equation and all numerical terms on the _____ side.

Follow along with the two Guided Learning Video examples, and fill in the blanks on the left as you learn with the instructor. When the pencil icon appears, pause and try the Student Practice on your own.

Guided Learning Video ▶ GUIDED LEARNING VIDEO	**Pause: Student Practice**
1. Solve. $6x + 5 = 4x - 15$	2. Solve. $8x + 3 = 4x - 21$

Use the procedure for solving equations to determine the solution.

$6x + 5 = 4x - 15$

☐   ☐

☐ = ☐

$\dfrac{☐}{☐} = \dfrac{☐}{☐}$

☐ = ☐

| **Guided Learning Video** 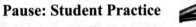 | **Pause: Student Practice** |

3. Solve. $2y - 9 + 4y + 5 = 11y - 7$

   Use the procedure for solving
   equations to determine the solution.

   $2y - 9 + 4y + 5 = 11y - 7$

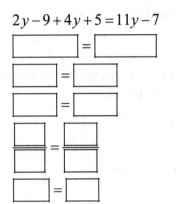

4. Solve. $6x - 5 + 3x + 2 = 4x + 12$

Helpful Hint: Pause the video and write the helpful hint in your own words.

For each Active Video Lesson, when the pencil icon appears, pause and work the problem. Then press play to check your work.

Active Video Lesson 1	**Active Video Lesson 2**
5. Solve. $7z - 6 - 4z + 11 = 2z - 8$	6. Solve. $10x - 7 = 13x + 5$

Name: _____   Date: _____
Instructor: _____   Section: _____

Guided Learning Video Worksheet: Solve Equations with Parentheses

Text: *Basic College Mathematics*
Student Learning Objective 10.5.3: Solve equations with parentheses.

Follow along with *Guided Learning Video 10.5.3, Solve Equations with Parentheses.*

Understanding the Big Picture: As you listen to the first part of the video, when the pencil icon appears, pause and fill in the blanks, choosing from the words listed.

opposite • multiplication • variable • like • addition • combine • division

1. To simplify each side of an equation, _____ any _____ terms and simplify any numerical work.

2. At the final step in the procedure to solve equations, the _____ or _____ principle of equality is used to solve for the variable.

3. The _____ principle of equality is used to isolate all numerical values on one side of the equation and all _____ terms on the _____ side.

Follow along with the two Guided Learning Video examples, and fill in the blanks on the left as you learn with the instructor. When the pencil icon appears, pause and try the Student Practice on your own.

Guided Learning Video ▶ GUIDED LEARNING VIDEO	Pause: Student Practice
1. Solve. $7(2x-1)-11x=5$ Use the procedure for solving equations to determine the solution. $7(2x-1)-11x=5$ ☐ = ☐ ☐ = ☐ ☐ = ☐ ☐ = ☐ ☐ = ☐	2. Solve. $3(4x-1)-8x=17$

3. Solve. $5(2-y)=8-(3+6y)$

Use the procedure for solving equations to determine the solution.

$5(2-y)=8-(3+6y)$

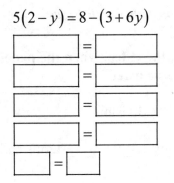

4. Solve. $2(3-y)=9-(5+3y)$

Helpful Hint: Pause the video and write the helpful hint in your own words.

For each Active Video Lesson, when the pencil icon appears, pause and work the problem. Then press play to check your work.

Active Video Lesson 1 | **Active Video Lesson 2**

5. Solve. $4(x-1)=-6(x+2)+48$

6. Solve. $13y-20=3(5y+2)-8$

Guided Learning Video Worksheet: Translate English Statements into Equations

Text: *Basic College Mathematics*
Student Learning Objective 10.6.1: Translate English into mathematical equations using two given variables.

Follow along with *Guided Learning Video 10.6.1, Translate English Statements into Equations.*

Understanding the Big Picture: As you listen to the first part of the video, when the pencil icon appears, pause and fill in the blanks, choosing from the words listed.

multiplication • subtraction • division • translated • addition • minus • equivalents

1. English words can be _____ into their math _____ and then used to solve math problems.

2. English phrases like "increased by," "more than," and "sum of" can all be represented by a(n) _____ symbol.

3. Phrases like "less than," "decreased by," and "difference between" indicate the operation of _____ and are represented by a(n) _____ sign.

4. The word "of" indicates _____ and the word "ratio" indicates _____.

Follow along with the two Guided Learning Video examples, and fill in the blanks on the left as you learn with the instructor. When the pencil icon appears, pause and try the Student Practice on your own.

Guided Learning Video ▶ GUIDED LEARNING VIDEO	Pause: Student Practice
1. Translate the English statement into an equation.	2. Translate the English statement into an equation.
"The sum of 15 and what number is 35?"	"The sum of 8 and what number is 23?"
Identify the math operation and use the variable x to represent the number.	
☐	

3. Translate the English statement into an equation.

"The quotient of 56 and what number is 7?"

Identify the math operation and use the variable x to represent the number.

4. Translate the English statement into an equation.

"The difference between 45 and what number is 17?"

Helpful Hint: Pause the video and write the helpful hint in your own words.

For each Active Video Lesson, when the pencil icon appears, pause and work the problem. Then press play to check your work.

Active Video Lesson 1 🖎 | **Active Video Lesson 2** 🖎

5. Translate the English statement into an equation.

"What number decreased by 4 is the same as 9?"

6. Translate the English statement into an equation.

"8 times what number is equal to 72?"

Name: _____   Date: _____

Instructor: _____   Section: _____

Guided Learning Video Worksheet: Write an Algebraic Expression to Compare Quantities using One Variable

Text: *Basic College Mathematics*

Student Learning Objective 10.6.2: Write algebraic expressions for several quantities using one given variable.

Follow along with *Guided Learning Video 10.6.2, Write an Algebraic Expression to Compare Quantities Using One Variable.*

Understanding the Big Picture: As you listen to the first part of the video, when the pencil icon appears, pause and fill in the blanks, choosing from the words listed.

basis • same • subtracted • variable • before • commutative

1. The number appearing _____ the phrase "less than" would be the number being _____.

2. The operation of addition is _____; therefore, it's easier to write the values and mathematical symbols in the _____ order as they appear in the English statement.

3. When two or more quantities are described in terms of another different quantity, let the _____ represent the quantity that is the _____ of comparison.

Follow along with the two Guided Learning Video examples, and fill in the blanks on the left as you learn with the instructor. When the pencil icon appears, pause and try the Student Practice on your own.

Guided Learning Video (▶) GUIDED LEARNING VIDEO	**Pause: Student Practice**
1. Use a variable and algebraic expressions to describe the quantities in the English expression. "Olivia is 4 years older than Jake." Identify the quantities to be expressed algebraically, and then write each in terms of the same variable. [___] = [___] [___] = [___]	2. Use a variable and algebraic expressions to describe the quantities in the English expression. "Sally's height is 5 inches less than Neal's height."

3. Use a variable and algebraic expressions to describe the quantities in the English expression. Use the variable x.

   "The measure of the first angle is 15° less than the measure of the second angle. The measure of the third angle is 12° more than twice the measure of the second angle."

   Identify the quantities to be expressed algebraically, and then write each in terms of the single variable x.

⬚	=	⬚
⬚	=	⬚
⬚	=	⬚

4. Use a variable and algebraic expressions to describe the quantities in the English expression. Use the variable x.

   "The measure of the second side of a triangle is 3 inches less than the measure of the first side. The measure of the third side is 6 inches less than triple the first side."

Helpful Hint: Pause the video and write the helpful hint in your own words.

For each Active Video Lesson, when the pencil icon appears, pause and work the problem. Then press play to check your work.

Active Video Lesson 1 | **Active Video Lesson 2**

5. Write algebraic expressions for the length and width of the rectangle.

   "The length of the rectangle is 5 inches shorter than double the width."

6. "Savannah's salary is $750 more than Ryan's salary."

   Write algebraic expressions for Savannah's salary and Ryan's salary.

Guided Learning Video Worksheet: Solve Problems Involving Comparisons

Text: *Basic College Mathematics*
Student Learning Objective 10.7.1: Solve problems involving comparisons.

Follow along with *Guided Learning Video 10.7.1, Solve Problems Involving Comparisons.*

Understanding the Big Picture: As you listen to the first part of the video, when the pencil icon appears, pause and fill in the blanks, choosing from the words listed.

equation • quantity • formulas • variable • pictures • values

1. In the procedure to solve applied problems, writing down _____ and drawing _____ are helpful strategies.

2. When solving applied problems, _____ expressions must be written for each _____ described in the problem.

3. A(n) _____ must be written that uses all of the _____ expressions written for the quantities.

4. After solving the _____, determine the _____ asked for in the problem.

Follow along with the two Guided Learning Video examples, and fill in the blanks on the left as you learn with the instructor. When the pencil icon appears, pause and try the Student Practice on your own.

Guided Learning Video	Pause: Student Practice
1. A 10-foot piece of wood is cut into two pieces. The longer piece is 2 feet longer than the shorter piece. What is the length of each piece? Use a math blueprint to help organize the information and solve the problem. Short piece = [ ] Long piece = [ ] **(continued)**	2. A service technician for a cable television provider spends a total of 102 minutes at two different addresses. The service call at the second address takes 6 fewer minutes than the service call at the first address. How many minutes were spent at each address?

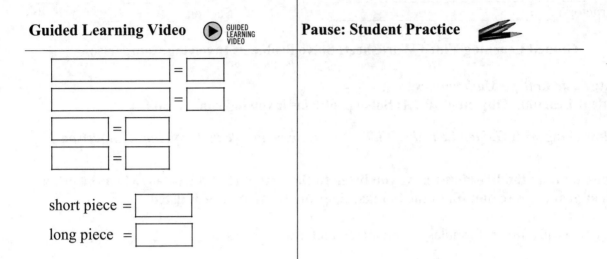

short piece = ☐

long piece = ☐

3. Three friends work together at a local store. Laura earns $3 per hour more than Jessica, and Wendy earns $2 per hour less than Jessica. Together the girls earn $34 per hour. How much does each girl earn per hour?

4. The total student enrollment in three sections of an algebra course is 78 students. Section J has 4 fewer students than section K, and section L has 13 more students than section K. What is the student enrollment for each section?

Jessica = ☐

Laura = ☐

Wendy = ☐

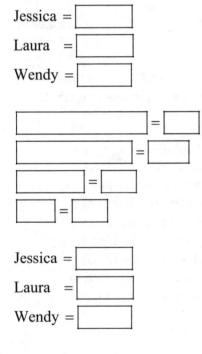

Jessica = ☐

Laura = ☐

Wendy = ☐

Helpful Hint: Pause the video and write the helpful hint in your own words.

Name: _____ Date: _____

Instructor: _____ Section: _____

**For each Active Video Lesson, when the pencil icon appears, pause and work the problem.
Then press play to check your work.**

Active Video Lesson 1	Active Video Lesson 2
4. There were 659 automobile accidents in the tri-county region last year. In Dover County, there were 53 more accidents than in Clark County. Everett County had 45 fewer accidents than Clark County. How many automobile accidents occurred in each county last year?	**5.** The Center City Animal Hospital treated a total of 17,580 dogs and cats last year. The hospital treated 1958 more dogs than cats. How many of each animal were treated last year?

Answers
Basic College Mathematics: Guided Learning Video Worksheets

Student Learning Objective 1.2.4 Answers
Understanding the Big Picture: 1. total 2. carrying 3. vertically
Guided Learning Video: 1. 71 3. 1237
Student Practice: 2. 73 4. 1314
Active Video Lesson: 5. 1721 6. 142

Student Learning Objective 1.3.3 Answers
Understanding the Big Picture: 1. minuend 2. subtrahend 3. borrowing, larger
Guided Learning Video: 1. 235 3. 158
Student Practice: 2. 257 4. 229
Active Video Lesson: 5. 2452 6. 4992

Student Learning Objective 1.3.5 Answers
Understanding the Big Picture: 1. subtraction 2. difference, minus 3. comparison
Guided Learning Video: 1. 36 homes 3. 16,031,282
Student Practice: 2. 3 homes 4. 6,780,865
Active Video Lesson: 5. 11,438,021 6. 39 homes

Student Learning Objective 1.4.3 Answers
Understanding the Big Picture: 1. power 2. zeros, right 3. end
Guided Learning Video: 1. 202,600 homes 3. 115,000
Student Practice: 2. 4,035,000 4. 192,000
Active Video Lesson: 5. 603,000 6. 8,079,000

Student Learning Objective 1.4.4 Answers
Understanding the Big Picture: 1. second 2. partial products 3. add, partial products
Guided Learning Video: 1. 2184 square miles 3. 57,408
Student Practice: 2. 1728 square miles 4. 42,696
Active Video Lesson: 5. 35,816 6. 6586

Student Learning Objective 1.4.6 Answers
Understanding the Big Picture: 1. multiplication, addition 2. operation 3. average
Guided Learning Video: 1. 260 miles 3. 819,520 cars
Student Practice: 2. 476 miles 4. 8592 cars
Active Video Lesson: 5. $4200 6. 2403 yards

Student Learning Objective 1.5.3 Answers
Understanding the Big Picture: 1. equal, division 2. divisor, dividend, quotient 3. long division
Guided Learning Video: 1. 41 3. 34 R17
Student Practice: 2. 42 4. 39 R7
Active Video Lesson: 5. 120 R3 6. 35

Student Learning Objective 1.6.1 Answers
Understanding the Big Picture: 1. shorthand 2. base 3. exponent 4. superscript
Guided Learning Video: 1. 64 3. 5^4
Student Practice: 2. 512 4. 6^4
Active Video Lesson: 5. 32 6. 7^3

Student Learning Objective 1.6.2 Answers
Understanding the Big Picture: 1. priorities 2. exponents 3. adding, subtracting
Guided Learning Video: 1. 79 3a. 6 b. 56
Student Practice: 2. 11 4a. 7 b. 192
Active Video Lesson: 5. 25 6. 23

Student Learning Objective 1.7.1 Answers
Understanding the Big Picture: 1. approximate 2. place 3. number line
Guided Learning Video: 1a. 400 b. 100 3. 25,800
Student Practice: 2a. 600 b. 400 4. 67,900
Active Video Lesson: 5. 63,290 6. 627,000

Student Learning Objective 1.7.2 Answers
Understanding the Big Picture: 1. estimate 2. approximation 3. rounded
Guided Learning Video: 1a. ≈ 2600 people 3. $\approx 110,000$
Student Practice: 2. ≈ 3000 people 4. 150,000
Active Video Lesson: 5. $\approx 300,000$ square feet 6. ≈ 16 gallons/room

Student Learning Objective 1.8.2 Answers
Understanding the Big Picture: 1. organize, plan, blueprint 2. understand, calculate 3. check
Guided Learning Video: 1. current salary of $31,200$ is greater than $28,000$ 3. \$137
Student Practice: 2. second job offer's salary $46,800$ is greater than $45,760$ 4. \$1987
Active Video Lesson: 5. \$2401 6. 15 miles per gallon

Student Learning Objective 2.1.3 Answers
Understanding the Big Picture: 1. fractions 2. denominator 3. numerator
Guided Learning Video: 1. $\dfrac{9}{20}$ 3. $\dfrac{688}{847}$
Student Practice: 2. $\dfrac{7}{17}$ 4. $\dfrac{41}{135}$
Active Video Lesson: 5. $\dfrac{3}{16}$ 6. $\dfrac{9}{43}$

Student Learning Objective 2.2.1 Answers
Understanding the Big Picture: 1. factoring 2. prime 3. factors
Guided Learning Video: 1. $2 \cdot 3 \cdot 7$ 3. $2^2 \cdot 3 \cdot 5$
Student Practice: 2. $2 \cdot 3^2 \cdot 5$ 4. $2^2 \cdot 3 \cdot 7$
Active Video Lesson: 5. $2^3 \cdot 7$ 6. $2 \cdot 3 \cdot 5 \cdot 7$

Student Learning Objective 2.2.2 Answers

Understanding the Big Picture: 1. equivalent 2. reduced, factor 3. divide

Guided Learning Video: 1a. $\dfrac{3}{7}$ b. $\dfrac{4}{5}$ 3. $\dfrac{2}{5}$

Student Practice: 2a. $\dfrac{3}{5}$ b. $\dfrac{5}{9}$ 4. $\dfrac{7}{10}$

Active Video Lesson: 5. $\dfrac{2}{5}$ 6. $\dfrac{8}{13}$

Student Learning Objective 2.2.3 Answers

Understanding the Big Picture: 1. equivalent 2. one 3. identity

Guided Learning Video: 1. $\dfrac{15}{21}$ 3. $\dfrac{16x}{36x}$

Student Practice: 2. $\dfrac{12}{32}$ 4. $\dfrac{28x}{60x}$

Active Video Lesson: 5. $\dfrac{18x}{48x}$ 6. $\dfrac{63}{77}$

Student Learning Objective 2.3.1 Answers

Understanding the Big Picture: 1. proper 2. improper 3. zero, mixed

Guided Learning Video: 1. $\dfrac{59}{6}$ 3. $\dfrac{119}{8}$

Student Practice: 2. $\dfrac{23}{5}$ 4. $\dfrac{155}{9}$

Active Video Lesson: 5. $\dfrac{137}{7}$ 6. $\dfrac{58}{7}$

Student Learning Objective 2.3.2 Answers

Understanding the Big Picture: 1. divide, numerator, denominator 2. quotient 3. remainder

Guided Learning Video: 1. $3\dfrac{3}{4}$ 3. $7\dfrac{5}{8}$

Student Practice: 2. $3\dfrac{5}{6}$ 4. $8\dfrac{1}{9}$

Active Video Lesson: 5. $6\dfrac{4}{9}$ 6. $8\dfrac{2}{3}$

Student Learning Objective 2.4.1 Answers

Understanding the Big Picture: 1. multiplication 2. numerators 3. denominators

Guided Learning Video: 1. $\dfrac{8}{63}$ 3. $\dfrac{3}{10}$

Student Practice: 2. $\dfrac{10}{99}$ 4. $\dfrac{3}{20}$

Active Video Lesson: 5. $\dfrac{2}{7}$ 6. $\dfrac{3}{8}$

Student Learning Objective 2.4.3 Answers

Understanding the Big Picture: 1. improper 2. divide 3. mixed number

Guided Learning Video: 1. $14\frac{2}{3}$ 3. $27\frac{1}{3}$

Student Practice: 2. $16\frac{4}{5}$ 4. $31\frac{1}{3}$

Active Video Lesson: 5. $11\frac{2}{3}$ square inches 6. $8\frac{1}{2}$

Student Learning Objective 2.5.1 Answers

Understanding the Big Picture: 1. reciprocal 2. inverted 3. second

Guided Learning Video: 1. $\frac{63}{80}$ 3. $\frac{5}{6}$

Student Practice: 2. $\frac{28}{55}$ 4. $\frac{10}{13}$

Active Video Lesson: 5. $\frac{3}{4}$ 6. $\frac{9}{10}$

Student Learning Objective 2.5.3 Answers

Understanding the Big Picture: 1. improper 2. multiplication

Guided Learning Video: 1. $1\frac{1}{3}$ 3. $\frac{2}{3}$

Student Practice: 2. $\frac{15}{22}$ 4. $\frac{2}{5}$

Active Video Lesson: 5. $1\frac{1}{2}$ 6. $\frac{22}{25}$

Student Learning Objective 2.6.2 Answers

Understanding the Big Picture: 1. compare 2. prime 3. greatest, one
Guided Learning Video: 1. 36 3. 54
Student Practice: 2. 75 4. 72
Active Video Lesson: 5. 30 6. 72

Student Learning Objective 2.6.3 Answers

Understanding the Big Picture: 1. equivalent 2. value 3. building

Guided Learning Video: 1. $\frac{20}{36}$ 3. $\frac{32}{56}$

Student Practice: 2. $\frac{24}{40}$ 4. $\frac{63}{72}$

Active Video Lesson: 5. $\frac{21}{45}$ 6. $\frac{55}{60}$

Student Learning Objective 2.7.2 Answers

Understanding the Big Picture: 1. common 2. numerators 3. simplify

Guided Learning Video: 1. $\dfrac{10}{21}$ 3. $\dfrac{47}{90}$

Student Practice: 2. $\dfrac{3}{10}$ 4. $\dfrac{67}{140}$

Active Video Lesson: 5. $\dfrac{41}{75}$ 6. $\dfrac{5}{84}$

Student Learning Objective 2.8.2 Answers

Understanding the Big Picture: 1. least 2. fractions, whole numbers

Guided Learning Video: 1. $4\dfrac{14}{15}$ 3. $1\dfrac{5}{6}$

Student Practice: 2. $2\dfrac{9}{20}$ 4. $5\dfrac{11}{15}$

Active Video Lesson: 5. $1\dfrac{13}{20}$ 6. $1\dfrac{3}{4}$

Student Learning Objective 2.8.3 Answers

Understanding the Big Picture: 1. operations 2. numerator, denominator 3. add, subtract, multiplying, dividing

Guided Learning Video: 1. $\dfrac{14}{27}$ 3. $\dfrac{5}{12}$

Student Practice: 2. $\dfrac{9}{16}$ 4. $\dfrac{19}{36}$

Active Video Lesson: 5. $\dfrac{11}{36}$ 6. $\dfrac{1}{2}$

Student Learning Objective 2.9.1 Answers

Understanding the Big Picture: 1. fractions 2. reading, picture 3. facts

Guided Learning Video: 1. $16\dfrac{1}{5}$ tons 3. $6\dfrac{1}{2}$ feet

Student Practice: 2. $22\dfrac{1}{12}$ hours 4. $1\dfrac{3}{4}$ feet

Active Video Lesson: 5. 36 square feet 6. 8 packages

Student Learning Objective 3.1.2 Answers

Understanding the Big Picture: 1. decimal 2. numerator, equal, zeros 3. delete

Guided Learning Video: 1. 6.59 3. 0.078

Student Practice: 2. 4.06 4. 0.045

Active Video Lesson: 5. 8.003 6. 0.07

Student Learning Objective 3.1.3 Answers

Understanding the Big Picture: 1. whole, point 2. and, point 3. last 4. numerator

Guided Learning Video: 1. $6\dfrac{317}{1000}$ 3. $\dfrac{3}{4}$

Student Practice: 2. $4\dfrac{629}{1000}$ 4. $\dfrac{8}{25}$

Active Video Lesson: 5. $\dfrac{27}{200}$ 6. $4\dfrac{29}{100}$

Student Learning Objective 3.2.1 Answers
Understanding the Big Picture: 1. increase 2. less 3. inequality
Guided Learning Video: 1. $5.78 < 5.79$ 3. $0.666 > 0.66$
Student Practice: 2. $9.35 > 9.33$ 4. $0.88 < 0.888$
Active Video Lesson: 5. $0.57 < 0.571$ 6. $2.032 > 2.031$

Student Learning Objective 3.2.3 Answers
Understanding the Big Picture: 1. place 2. less 3. increase, one
Guided Learning Video: 1. 14.26 3. 0.703
Student Practice: 2. 63.54 4. 0.095
Active Video Lesson: 5. $9.87 6. 3.2

Student Learning Objective 3.3.1 Answers
Understanding the Big Picture: 1. fractions 2. lining 3. same, sum
Guided Learning Video: 1. 11.8 3. 34.509
Student Practice: 2. 13.3 4. 50.406
Active Video Lesson: 5. 44.838 6. $954.38

Student Learning Objective 3.3.2 Answers
Understanding the Big Picture: 1. borrow 2. zeros, right, same 3. difference
Guided Learning Video: 1a. 11.6 b. 18.59 3. 8.231
Student Practice: 2a. 21.2 b. 51.97 4. 16.408
Active Video Lesson: 5. 14.955 6. 48.06

Student Learning Objective 3.4.1 Answers
Understanding the Big Picture: 1. whole 2. total 3. equal 4. zeros, left
Guided Learning Video: 1. 0.056 3. 247.987
Student Practice: 2. 0.024 4. 150.65
Active Video Lesson: 5. 0.036 6. 72.768

Student Learning Objective 3.4.2 Answers
Understanding the Big Picture: 1. multiplication 2. zeros, places, right 3. less
Guided Learning Video: 1. 8,174,000 3. 91,600
Student Practice: 2. 667,300 4. 4950
Active Video Lesson: 5. 1,700 6. 62,800 m

Student Learning Objective 3.5.2 Answers
Understanding the Big Picture: 1. divisor, right 2. dividend, divisor 3. quotient, dividend
Guided Learning Video: 1. 1.79 3. 3.15
Student Practice: 2. 1.53 4. 2.65
Active Video Lesson: 5. 2.35 6. 4.5

Student Learning Objective 3.6.1 Answers
Understanding the Big Picture: 1. numerator, denominator 2. terminating 3. repeat, repeating
Guided Learning Video: 1. 0.875 3. $0.\overline{4}$
Student Practice: 2. 0.375 4. $0.8\overline{3}$
Active Video Lesson: 5. $0.41\overline{6}$ 6. 0.5375

Student Learning Objective 3.6.2 Answers
Understanding the Big Picture: 1. priorities 2. simplify, exponents 3. multiplication
Guided Learning Video: 1. 19.86 3. 6.1952
Student Practice: 2. 55.54 4. 28.9
Active Video Lesson: 5. 0.111 6. 3.16

Student Learning Objective 3.7.2 Answers
Understanding the Big Picture: 1. numerical 2. solution 3. process
Guided Learning Video: 1. $743.22 3. 10.8825 seconds
Student Practice: 2. $1093.75 4. 51 hours
Active Video Lesson: 5. 17 containers 6. $278,384.40

Student Learning Objective 4.1.1 Answers
Understanding the Big Picture: 1. ratio, comparison 2. fraction, simplified 3. colon
Guided Learning Video: 1. $\dfrac{7}{8}$ 3. $\dfrac{3}{7}$

Student Practice: 2. $\dfrac{2}{3}$ 4. $\dfrac{3}{16}$

Active Video Lesson: 5. $\dfrac{7}{29}$ 6. $\dfrac{2}{3}$

Student Learning Objective 4.1.2 Answers
Understanding the Big Picture: 1. different, rate 2. unit, one 3. division
Guided Learning Video: 1. 11 calories per gram of fat 3. $2 profit per pound of coffee
Student Practice: 2. 8 grams of fat per tablespoon 4. $7 per gallon of ice cream
Active Video Lesson: 5. 69 mph 6. 8 gigabytes per month

Student Learning Objective 4.2.1 Answers
Understanding the Big Picture: 1. proportional 2. ratios, equal 3. numerators, denominators
Guided Learning Video: 1. $\dfrac{6}{7} = \dfrac{18}{21}$ 3. $\dfrac{5 hours}{350 miles} = \dfrac{8 hours}{560 miles}$

Student Practice: 2. $\dfrac{9}{11} = \dfrac{45}{55}$ 4. $\dfrac{330 miles}{15 gallons} = \dfrac{1320 miles}{60 gallons}$

Active Video Lesson: 5. $\dfrac{3 credit\ hours}{\$669} = \dfrac{5 credit\ hours}{\$1115}$ 6. $\dfrac{5 rotations}{2 min} = \dfrac{30 rotations}{12 min}$

Student Learning Objective 4.2.2 Answers
Understanding the Big Picture: 1. cross 2. proportion 3. numerators, denominators
Guided Learning Video: 1. $60 = 60$, yes 3. $285,000 \neq 300,000$, no
Student Practice: 2. $684 = 684$, yes 4. $140 \neq 150$, no
Active Video Lesson: 5. $\$30 = \$30$, yes 6. $336 \neq 378$, no

Student Learning Objective 4.3.2 Answers
Understanding the Big Picture: 1. cross, denominators 2. equation 3. unknown 4. positions
Guided Learning Video: 1. $n = 6$ 3. $n = 10\,mg$
Student Practice: 2. $n = 12$ 4. $n = 4\,grams$
Active Video Lesson: 5. $n = 0.7$ 6. $n = \$45$

Student Learning Objective 4.4.1 Answers
Understanding the Big Picture: 1. facts 2. proportion 3. sample, requested
Guided Learning Video: 1. $n = 120$ compact cars 3. $n = 4.8$ minutes
Student Practice: 2. $n = 136$ female students 4. $n = 30$ minutes
Active Video Lesson: 5. $n = 19$ miles 6. $n = 160$ deer

Student Learning Objective 5.1.3 Answers
Understanding the Big Picture: 1. percent 2. ratios, denominators 3. parts
Guided Learning Video: 1. 72.6% 3. 580%
Student Practice: 2. 53.9% 4. 310%
Active Video Lesson: 5. 290% 6. 16.5%

Student Learning Objective 5.2.2 Answers
Understanding the Big Picture: 1. decimal 2. decimal 3. two, multiplying, hundred
Guided Learning Video: 1. 55% 3. 22.5%
Student Practice: 2. 28% 4. 37.5%
Active Video Lesson: 5. 64% 6. 56.25%

Student Learning Objective 5.2.3 Answers
Understanding the Big Picture: 1. decimal 2. decimal 3. two, multiplying, hundred
Guided Learning Video: 1. 55% 3. 22.5%
Student Practice: 2. 65% 4. 12.5%
Active Video Lesson: 5. 64% 6. 56.25%

Student Learning Objective 5.3A.2 Answers
Understanding the Big Picture: 1. amount, percent, base 2. multiplication 3. percent, decimal
Guided Learning Video: 1. 83.2 3. 2450 freshman
Student Practice: 2. 328.5 4. 5750 ballots
Active Video Lesson: 5. 510 students 6. 12,700 people

Student Learning Objective 5.3A.3 Answers
Understanding the Big Picture: 1. amount, percent, base 2. divide, amount, base 3. two, right
Guided Learning Video: 1. 25% 3. 8%
Student Practice: 2. 60% 4. 6%
Active Video Lesson: 5. 80% 6. 40%

Student Learning Objective 5.3B.2 Answers
Understanding the Big Picture: 1. base, percent 2. amount, base, percent 3. two
Guided Learning Video: 1. 27 3. 140
Student Practice: 2. 42 4. 228
Active Video Lesson: 5. 800 6. 153

Student Learning Objective 5.4.1 Answers
Understanding the Big Picture: 1. equations, proportions 2. amount, base 3. two
Guided Learning Video: 1. $2700 3. $13.63
Student Practice: 2. $7500 4. $87.72
Active Video Lesson: 5. $4170 6. $174.11

Student Learning Objective 5.4.3 Answers
Understanding the Big Picture: 1. amount, rate 2. percent 3. subtracted
Guided Learning Video: 1. $58.50 3. $121.50, $148.50
Student Practice: 2. $206 4. $292.50
Active Video Lesson: 5. $305.50 6. $1005

Student Learning Objective 5.5.1 Answers
Understanding the Big Picture: 1. commission 2. amount, sales 3. dividing, rate
Guided Learning Video: 1. $19,120 3. $350
Student Practice: 2. $128,500 4. $19,750
Active Video Lesson: 5. $145,000 6. $5874

Student Learning Objective 5.5.3 Answers
Understanding the Big Picture: 1. interest 2. principal, borrowed 3. rate, amount 4. yearly
Guided Learning Video: 1. $1200 3. $40.50, $1840.50
Student Practice: 2. $2250 4. $93.75, $2593.75
Active Video Lesson: 5. $165 6. $455

Student Learning Objective 6.1.2 Answers
Understanding the Big Picture: 1. unit, one 2. numerator, denominator 3. numerator
Guided Learning Video: 1. 155 minutes 3. 1.25 quarts
Student Practice: 2. 10,668 feet 4. 10.5 cups

Active Video Lesson: 5. $3\frac{1}{2}$ packages 6. 72 inches

Student Learning Objective 6.2.2 Answers
Understanding the Big Picture: 1. meter 2. shorter 3. kilometers
Guided Learning Video: 1. 80 mm 3. 1.3 km
Student Practice: 2. 320 mm 4. 0.000045 km
Active Video Lesson: 5. 0.152 km 6. 356 cm

Student Learning Objective 6.3.1 Answers
Understanding the Big Picture: 1. liter 2. quart 3. centimeters 4. smaller
Guided Learning Video: 1. 28,000 milliliters 3. 4 kiloliters
Student Practice: 2. 70,000 deciliters 4. 0.135 liters
Active Video Lesson: 5. 3.82 liters 6. 120 deciliters

Student Learning Objective 6.3.2 Answers
Understanding the Big Picture: 1. mass 2. gravity, weight 3. gram
Guided Learning Video: 1. 6250 milligrams 3. 7.9 centigrams
Student Practice: 2. 3200 grams 4. 0.1750 kilograms
Active Video Lesson: 5. 0.983 kilograms 6. 14,800 decigrams

Student Learning Objective 6.4.1 Answers
Understanding the Big Picture: 1. approximations 2. multiply, one 3. numerator
Guided Learning Video: 1. ≈ 3.784 liters 3. ≈ 6.6 pounds
Student Practice: 2. ≈ 8.48 quarts 4. ≈ 14.12 ounces
Active Video Lesson: 5. ≈ 53.3 miles 6. 25.4 centimeters

Student Learning Objective 6.4.2 Answers
Understanding the Big Picture: 1. Celsius 2. Fahrenheit 3. hundred, zero
Guided Learning Video: 1. $\approx 424°F$ 3. $\approx 27.8°C$
Student Practice: 2. $\approx 302°F$ 4. $\approx 17.8°C$
Active Video Lesson: 5. $\approx 71.6°F$ 6. $\approx 4.4°C$

Student Learning Objective 6.5.1A Answers
Understanding the Big Picture: 1. relationship 2. numerator 3. multiplied
Guided Learning Video: 1. 24 feet 3. $1.50
Student Practice: 2. 45 feet 4. $9
Active Video Lesson: 5. $32.50 6. 9 gallons

Student Learning Objective 6.5.1B Answers
Understanding the Big Picture: 1. equal, same 2. greater 3. milliliter
Guided Learning Video: 1. 5000 packets 3. $0.35 per centiliter
Student Practice: 2. 670 bottles 4. $0.23 per milliliter
Active Video Lesson: 5. $0.09 per milligram 6. 8.3 meters

Student Learning Objective 7.1.1 Answers
Understanding the Big Picture: 1. ray 2. angle, vertex 3. complementary 4. supplementary
Guided Learning Video: 1. 38° 3. 53°
Student Practice: 2. 81° 4. 73°
Active Video Lesson: 5. 41°, 131° 6. 27°, 117°

Student Learning Objective 7.2.2 Answers
Understanding the Big Picture: 1. perimeter 2. opposite 3. square 4. perpendicular
Guided Learning Video: 1. 48 in. 3. 38 m
Student Practice: 2. 52 in. 4. 58 m
Active Video Lesson: 5. 42 ft 6. 58 cm

Student Learning Objective 7.3.2 Answers
Understanding the Big Picture: 1. trapezoid, bases, perpendicular 2. trapezoid, equal 3. height, parallel
Guided Learning Video: 1. 36 square yards 3. 84.9 square centimeters
Student Practice: 2. 480 square feet 4. 6800 square feet
Active Video Lesson: 5. 144 square meters 6. 3350 square feet

Student Learning Objective 7.5.1 Answers
Understanding the Big Picture: 1. identical, square root 2. taking 3. positive
Guided Learning Video: 1a. 7 b. 12 3. 13
Student Practice: 2a. 10 b. 15 4. 22
Active Video Lesson: 5. 8 6a. 6 b. 11

Student Learning Objective 7.6.1 Answers
Understanding the Big Picture: 1. hypotenuse 2. hypotenuse, sum 3. opposite, ninety
Guided Learning Video: 1. 20 cm 3. ≈10.817 in.
Student Practice: 2. 25 in. 4. ≈13.89 cm
Active Video Lesson: 5. ≈18.0 ft 6. ≈12.81 m

Student Learning Objective 7.6.3 Answers
Understanding the Big Picture: 1. diagram, picture 2. sides, right 3. perpendicular
Guided Learning Video: 1. ≈16.6 mi 3. ≈6.6 ft
Student Practice: 2. ≈28.9 mi 4. ≈8.7 ft
Active Video Lesson: 5. ≈13.1 yds 6. ≈7.6 cm

Student Learning Objective 7.7.1A Answers
Understanding the Big Picture: 1. equal, center 2. diameter, center 3. radius 4. circumference
Guided Learning Video: 1. ≈25.12 in. 3. ≈14.4 ft
Student Practice: 2. ≈37.68 cm 4. ≈25.1 ft
Active Video Lesson: 5. ≈63,585 in. 6. ≈16.3 in.

Student Learning Objective 7.7.1B Answers
Understanding the Big Picture: 1. dividing, circumference, diameter 2. area, radius 3. before
Guided Learning Video: 1. ≈12.56 sq meters. 3. $261.60
Student Practice: 2. ≈113.04 sq cm 4. $418.80
Active Video Lesson: 5. $1570 6. ≈200.96 sq in.

Student Learning Objective 7.8.2 Answers
Understanding the Big Picture: 1. cylinder 2. cubic 3. area, base, height
Guided Learning Video: 1. ≈62.8 cubic inches 3. ≈5341.14 cubic centimeters
Student Practice: 2. ≈803.84 cubic inches 4. ≈148.365 cubic inches
Active Video Lesson: 5. ≈2826 cubic feet 6. ≈11,398.2 cubic meters

Student Learning Objective 7.8.5 Answers
Understanding the Big Picture: 1. base, height, three 2. cubic 3. area, base
Guided Learning Video: 1. 60 cubic meters 3. 16 cubic inches
Student Practice: 2. 96 cubic meters 4. 520 cubic inches
Active Video Lesson: 5. 220 cubic inches 6. 140 cubic feet

Student Learning Objective 8.2.2 Answers
Understanding the Big Picture: 1. comparisons, double 2. changes, time 3. data
Guided Learning Video: 1. 300 cars 3. 150 cars
Student Practice: 2. 150 cars 4. 150 cars
Active Video Lesson: 5. $16,000 6. $20,000

Student Learning Objective 8.2.4 Answers
Understanding the Big Picture: 1. sets, compared 2. yellow 3. colored, read
Guided Learning Video: 1a. April b. 4000 customers 3a. March, April, May b. 2500 customers
Student Practice: 2a. June b. 3500 customers 4a. March, May b. 3000 customers
Active Video Lesson: 5a. 2005 and 2010 b. 400 homes 6a. 1990 b. 600 condominiums

Student Learning Objective 8.3.1 Answers
Understanding the Big Picture: 1. histogram, touch 2. width, equal 3. class, interval
4. frequency, class
Guided Learning Video: 1. 6 students 3. 30 students
Student Practice: 2. 8 students 4. 26 students
Active Video Lesson: 5. 20 bulbs 6. 30 bulbs

Student Learning Objective 8.4.1 Answers
Understanding the Big Picture: 1. dividing, sum, number 2. mean, median 3. average
Guided Learning Video: 1. 92 3. ≈ 25 mpg
Student Practice: 2. 87 4. 218 customers
Active Video Lesson: 5. $\approx 81°$ 6. 29 years of age

Student Learning Objective 8.4.2 Answers
Understanding the Big Picture: 1. median, same 2. order 3. even, average, middle
Guided Learning Video: 1. 35° 3. $34,700
Student Practice: 2. 83 4. 74°
Active Video Lesson: 5. $15.99 6. 84.4 years

Student Learning Objective 9.1.1A Answers
Understanding the Big Picture: 1. negative, minus 2. left, right 3. Integers

Guided Learning Video: 1. 3. $-3 < 2$, $5 > -1$, $-4 < -2$, $-3 > -5$

Student Practice: 2. 4. $-4 > -5$, $0 > -2$, $1 < 3$, $-2 > -3$

Active Video Lesson: 5. $-1 > -3$, $-6 < -5$ 6.

Student Learning Objective 9.1.1B Answers
Understanding the Big Picture: 1. positive, direction 2. absolute, distance, zero 3. vertical
4. positive
Guided Learning Video: 1. 21 3. -16
Student Practice: 2. 14 4. -33
Active Video Lesson: 5. 53 6. 0

Student Learning Objective 9.1.1C Answers
Understanding the Big Picture: 1. absolute values 2. adding, same, common 3. negative
Guided Learning Video: 1. -108 3. $-\$1,770,000,000,000$
Student Practice: 2. -101 4. $-\$1,160,000,000,000$
Active Video Lesson: 5. $-\$1,980,000,000,000$ 6. -844

Student Learning Objective 9.1.2A Answers
Understanding the Big Picture: 1. negative 2. different, larger 3. difference, larger, smaller
Guided Learning Video: 1. $-\$14$ 3. $-\$8$
Student Practice: 2. $-\$48$ 4. $-\$34$
Active Video Lesson: 5. -6 6. -5

Student Learning Objective 9.1.2B Answers
Understanding the Big Picture: 1. negative 2. different, larger 3. difference, larger, smaller
Learning Video: 1a. −2 b. −13 c. 7 3. 2.1
Student Practice: 2a. −7 b. −19 c. 9 4. −2.4
Active Video Lesson: 5. −7 6. −5.6

Student Learning Objective 9.2.1A Answers
Understanding the Big Picture: 1. negative, same 2. equal, zero 3. add, opposite
Guided Learning Video: 1. −$8 3. −26
Student Practice: 2. −$15 4. −51
Active Video Lesson: 5. −15 6. 43

Student Learning Objective 9.2.1B Answers
Understanding the Big Picture: 1. negative, same 2. equal, zero 3. add, opposite
Guided Learning Video: 1a. −7 b. 9 3. −8
Student Practice: 2a. −20 b. 17 4. −11
Active Video Lesson: 5. −4.3 6. 35

Student Learning Objective 9.2.2 Answers
Understanding the Big Picture: 1. rewrite, addition, opposite 2. commutative, associative
3. order, commutative, associative
Guided Learning Video: 1. −12 3. −12
Student Practice: 2. −23 4. −26
Active Video Lesson: 5. −11 6. −8

Student Learning Objective 9.3.1A Answers
Understanding the Big Picture: 1. negative 2. dot, parentheses 3. even, odd
Guided Learning Video: 1. −54 3. 42
Student Practice: 2. −20 4. 33
Active Video Lesson: 5. −56 6. 36

Student Learning Objective 9.3.1B Answers
Understanding the Big Picture: 1. multiplication 2. division, same 3. even, odd
Guided Learning Video: 1. −8 3. 9
Student Practice: 2. −4 4. 7
Active Video Lesson: 5. −8 6. 6

Student Learning Objective 9.3.2 Answers
Understanding the Big Picture: 1. total, product 2. negative 3. commutative, associative, order
Guided Learning Video: 1. −42 3. −56
Student Practice: 2. −120 4. 24
Active Video Lesson: 5. 36 6. −40

Student Learning Objective 9.4.1A Answers
Understanding the Big Picture: 1. priorities 2. parentheses 3. exponents 4. left, right
Guided Learning Video: 1. −46 3. $-\dfrac{35}{27}$

Student Practice: 2. −44 4. $-\dfrac{37}{12}$

Active Video Lesson: 5. −25 6. $-\dfrac{14}{25}$

Student Learning Objective 9.4.1B Answers
Understanding the Big Picture: 1. priorities 2. powers 3. equal 4. left, right
Guided Learning Video: 1. 10.99 3. 1.208
Student Practice: 2. 16.48 4. −1.127
Active Video Lesson: 5. 1.196 6. 2.808

Student Learning Objective 9.5.2 Answers
Understanding the Big Picture: 1. scientific notation 2. positive, ten 3. negative, one
Guided Learning Video: 1. 8300 3. 0.0061
Student Practice: 2. 57,000 4. 0.000298
Active Video Lesson: 5. 0.0000936 6. 74,510

Student Learning Objective 10.1.2 Answers
Understanding the Big Picture: 1. term 2. addition, subtraction 3. like 4. like, coefficients
5. one
Guided Learning Video: 1. $-7a+4b$ 3. $9x+(-5y)+7xy$

Student Practice: 2. $-11x+8y$ 4. $5a+(-2b)+2ab$

Active Video Lesson: 5. $-12x+2xy$ 6. $-3x+6y$

Student Learning Objective 10.2.1 Answers
Understanding the Big Picture: 1. distributive, parentheses 2. addition, subtraction 3. multiplied
Guided Learning Video: 1. $4x+12$ 3. $-15x+6y+24z$
Student Practice: 2. $-5x+30y$ 4. $-24x+12y-8z$
Active Video Lesson: 5. $45x-5y$ 6. $-56x-42$

Student Learning Objective 10.2.2 Answers
Understanding the Big Picture: 1. equal, one 2. opposite 3. multiplied, one
Guided Learning Video: 1. $13x+11y$ 3. $15s+36t+24$
Student Practice: 2. $-7x+10y$ 4. $-2x+18y-12$
Active Video Lesson: 5. $12x-42y+26z$ 6. $-11a+14b$

Student Learning Objective 10.3.1 Answers
Understanding the Big Picture: 1. expressions, equal 2. inverse 3. solution 4. addition, same, equal
Guided Learning Video: 1. $x=13$ 3. $x=5$
Student Practice: 2. $x=15$ 4. $x=11$
Active Video Lesson: 5. $x=7$ 6. $x=17$

Student Learning Objective 10.4.1 Answers
Understanding the Big Picture: 1. division, equal 2. undefined, nonzero 3. divide, coefficient
Guided Learning Video: 1. $x=9$ 3. $x=-7$

Student Practice: 2. $x = 13$ 4. $x = -6$
Active Video Lesson: 5. $x = 12$ 6. $x = -17$

Student Learning Objective 10.4.2 Answers

Understanding the Big Picture: 1. multiplication, equal 2. reciprocal, coefficient 3. numerical
Guided Learning Video: 1. $x = -54$ 3. $x = -56$
Student Practice: 2. $x = -40$ 4. $x = -18$
Active Video Lesson: 5. $x = -77$ 6. $x = 96$

Student Learning Objective 10.5.1 Answers

Understanding the Big Picture: 1. addition, division, variable 2. value, true 3. addition, division
Guided Learning Video: 1. $x = 5$ 3. $x = -6$
Student Practice: 2. $x = 5$ 4. $x = -12$
Active Video Lesson: 5. $y = 6$ 6. $x = -4$

Student Learning Objective 10.5.2 Answers

Understanding the Big Picture: 1. simplified 2. terms, numerical 3. addition, variable, opposite

Guided Learning Video: 1. $x = -10$ 3. $y = \dfrac{3}{5}$

Student Practice: 2. $x = -6$ 4. $x = 3$
Active Video Lesson: 5. $z = -13$ 6. $x = -4$

Student Learning Objective 10.5.3 Answers

Understanding the Big Picture: 1. combine, like 2. multiplication, division 3. addition, variable, opposite
Guided Learning Video: 1. $x = 4$ 3. $y = -5$
Student Practice: 2. $x = 5$ 4. $y = -2$
Active Video Lesson: 5. $x = 4$ 6. $y = -9$

Student Learning Objective 10.6.1 Answers

Understanding the Big Picture: 1. translated, equivalents 2. addition 3. subtraction, minus
4. multiplication, division
Guided Learning Video: 1. $15 + x = 35$ 3. $56 \div x = 7$
Student Practice: 2. $8 + x = 23$ 4. $45 - x = 17$
Active Video Lesson: 5. $x - 4 = 9$ 6. $8x = 72$

Student Learning Objective 10.6.2 Answers

Understanding the Big Picture: 1. before, subtracted 2. commutative, same 3. variable, basis
Guided Learning Video: 1. $j =$ Jake's age; $j + 4 =$ Olivia's age 3. $x =$ degrees in the second angle; $x - 15 =$ degrees in the first angle; $2x + 12 =$ degrees in the third angle
Student Practice: 2. $n =$ Neal's height; $n - 5 =$ Sally's height 4. $x =$ measure of the first side; $x - 3 =$ measure of the second side; $3x - 6 =$ measure of the third side
Active Video Lesson: 5. $w =$ width; $2w - 5 =$ length 6. $r =$ Ryan's salary; $r + 750 =$ Savannah's salary

Student Learning Objective 10.7.1 Answers

Understanding the Big Picture: 1. formulas, pictures 2. variable, quantity 3. equation, variable 4. equation, values

Guided Learning Video: 1. short piece = 4 feet; long piece = 6 feet 3. Jessica = \$11/hr; Laura = \$14/hr; Wendy = \$9/hr

Student Practice: 2. first address = 54 min; second address = 48 min 4. section J = 19 students; section K = 23 students; section L = 36 students

Active Video Lesson: 5. Clark County = 217 accidents; Dover County = 270 accidents; Everett County = 172 accidents 6. 7811 cats, 9769 dogs